U0493463

新材料新能源学术专著译丛

# 低温燃料电池材料

## Materials for Low-Temperature Fuel Cells

[澳]布拉德利·兰德维格(Bradley Ladewig)
[澳]蒋三平(San Ping Jiang) 主编
[美]严玉山(Yushan Yan)
廖世军 宋慧宇 杜丽 译

国防工业出版社
·北京·

**著作权合同登记　图字：军－2016－108 号**

**图书在版编目（CIP）数据**

低温燃料电池材料／（澳）布拉德利·兰德维格（Bradley Ladewig），（澳）蒋三平，（美）严玉山主编；廖世军，宋慧宇，杜丽译. —北京：国防工业出版社，2019.12

（新材料新能源学术专著译丛）

书名原文：Materials for Low－Temperature Fuel Cells

ISBN 978－7－118－11591－8

Ⅰ. ①低… Ⅱ. ①布… ②蒋… ③严… ④廖… ⑤宋… ⑥杜… Ⅲ. ①低温燃料电池－材料－研究 Ⅳ. ①TM911.46

中国版本图书馆 CIP 数据核字(2019)第 227185 号

Materials for Low-Temperature Fuel Cells by Bradley Ladewig, San Ping Jiang, Yushan Yan and Max Lu.

ISBN:978－3－527－33042－3.



※

国防工業出版社 出版发行

（北京市海淀区紫竹院南路 23 号　邮政编码 100048）

三河市腾飞印务有限公司印刷

新华书店经售

*

**开本** 710×1000　1/16　插页彩 2　**印张** 14½　**字数** 260 千字

2019 年 12 月第 1 版第 1 次印刷　**印数** 1—2000 册　**定价** 106.00 元

**（本书如有印装错误，我社负责调换）**

国防书店：(010)88540777　　发行邮购：(010)88540776

发行传真：(010)88540755　　发行业务：(010)88540717

# 译者序

随着世界对能源需求的持续增长，研究开发新型高效的可再生能源和电力已经成为亟待解决的难题。燃料电池是一种将化学能直接转换为电能的电化学装置，燃料和空气分别送进燃料电池，电就被奇妙地生产出来。作为迄今最有效的能量转换技术，燃料电池是新能源和可再生能源链中必不可少的一部分。根据工作温度的不同，燃料电池分为高温燃料电池和低温燃料电池，低温燃料电池是指操作温度在200℃以下的燃料电池，通常包含碱性燃料电池、质子交换膜燃料电池等，具有清洁、高效、可移动、操作条件温和等诸多优点。

目前，燃料电池已成为一种重要的替代电源，其最主要，也最适合的用途是军事应用。高效、多面性、使用时间长等特点极适合于军事工作对电力的需要。燃料电池可以以多种形态为绝大多数军事装置——从战场上的移动手提装备到海陆运输——提供动力。微型燃料电池比普通的固体电池具有更大的优越性，其使用期长意味着在战场只需少量的备品供应。此外，对于燃料电池而言，添加燃料也是轻而易举的事情。同样，燃料电池的高效性能极大地减少车辆所需的燃料用量，从而使车辆行驶得更远，或在遥远的地区活动更长的时间。这样，战地所需的支持车辆、人员和装备的数量便可以显著减少。

本书根据Wiley-VCH出版的*Materials for Low-Temperature Fuel Cells*一书翻译而来，全面、系统地介绍了低温燃料电池中关键材料的发展现状及存在的关键技术问题。本书所有的章节均由国际知名专家撰写，主编为澳大利亚莫纳什大学的Bradley Ladewig副教授、科廷大学蒋三平教授（澳籍）和特拉华大学严玉山教授（美籍）。Bradley Ladewig副教授现任职于澳大利亚莫纳什大学化学工程系，主要从事膜材料的开发及清洁能源的应用技术，研究内容包括直接甲醇燃料电池膜材料、质子交换膜燃料电池系统中的热电联测及模型化，以及脱盐膜的开发等。近期工作还包括直接碳燃料电池、气体吸附金属有机框架材料和膜组分以及基于纸和线基质的低成本微流体传感器。蒋三平教授为国际著名燃料电池专家，博士毕业于英国伦敦城市大学，现任澳大利亚科廷大学化学工程系教授、燃料与能源技术研究院副院长，是阳光海岸大学的兼职教授。蒋三平教授在工业届和学术届拥有很高的知名度，主要从事固体氧化物燃料电池、质子交换膜和直接甲醇燃料电池、直接乙醇燃料电池及电解等研究工作。严玉山教授是特拉华大学化学与生物分子工程系的杰出教授，曾担任加州大学河滨分校首席科学家、大学学者和系主任，是联信公司的高级工程师，开创了纳米水、全周期能源、液态沸石材料和氢能等研究的先河，研究领域为分子筛薄膜和电化学装置，包括燃料电池、电解池、太阳能制氢和氧化还原液流电池。

本书共10章，着重介绍各类低温燃料电池中的关键材料。第1章为绪论；第2章介绍了碱性阴离子交换膜燃料电池；第3章重点讲述了质子交换膜催化剂的载体材料；第4章重点说明了低温直接醇类燃料电池阳极催化剂；第5章介绍了直接甲醇燃料电池的膜材料；第6章主要讲述了氢氧根离子交换膜及离聚物；第7章详细说明了微生物燃料电池的关键材料；第8章介绍了生物电化学系统；第9章介绍了微流体燃料电池关键材料；第10章介绍了直接醇类燃料电池催化剂的进展。纵观全书，其论述内容视角独特，深入浅出，是作者对其多年深入研究和工程实践的精心总结，具有很高的参考价值。

本书内容比较完善，读者只需要有基本的燃料电池研究背景就能读懂。本书的翻译工作持续了一年多，进行了多次校对和调整，绝大部分内容是根据原文直接翻译的，为了使读者读起来更加通畅，有些地方进行了调整。全书由华南理工大学廖世军教授、宋慧宇教授、杜丽副教授共同翻译而成，其中，廖世军教授负责第1~4章内容的翻译工作，宋慧宇教授负责第5~7章内容的翻译工作，杜丽副教授负责第8~10章内容的翻译工作。同时，华南理工大学舒婷老师参与了本书部分章节的翻译工作，重庆大学魏子栋教授、中山大学童叶翔教授在本书的翻译、出版过程中给予悉心指导与帮助，在此向他们一并表示衷心的感谢。本书获得装备科技译著出版基金资助，在此表示感谢。

鉴于译者水平有限，书中难免有不妥之处，敬请读者批评指正。

译者

2018年10月

# 原著丛书主编前言

## ——关于 Wiley 的“可持续能源新材料和发展”丛书

伴随着全球性关于清洁化石能源、氢能和可再生能源以及对水的再利用和再循环的技术研发,可持续能源的发展正吸引着从科研团体到工业界越来越多的关注。根据 REN21[①] 的《可再生能源全球状况报告 2012》(第 17 页),全球可再生能源的总投资从 2010 年的 2110 亿美元增加到 2011 年的 2570 亿美元。2011 年投资排在前位的国家为中国、德国、美国、意大利和巴西。在解决能源安全、石油价格上涨以及气候变化的挑战性问题方面,材料创新是立足之本。

在这种情况下,有必要通过权威的信息来源系统地梳理在材料科学和工程中涉及能源和环境的最新科学突破和知识进步。这正是出版关于可持续能源新材料和发展的 Wiley 丛书的目的。这是关于材料科学在能源领域应用的有抱负的出版计划。丛书的每一册都由国际顶级科学家亲笔撰写,并有望成为今后许多年的标准参考书籍。

该丛书涵盖了材料科学的进步和可再生能源的创新,化石能源的清洁利用,温室气体的减排及相关环境技术。丛书各分册如下:

《超级电容器:材料、系统和应用》;

《水处理功能纳米材料和膜》;

《高温燃料电池材料》;

《低温燃料电池材料》;

《先进热电材料:基础与应用》;

《先进锂离子电池:最新趋势和前景》;

《光催化和水纯化:从基础到最新应用》。

在《低温燃料电池材料》一书中,我想要感谢作者们和编辑们,是他们的巨大努力和辛勤工作使得本书得以及时完成和出版。本书每一章都是撰稿人的心血结晶,并毫无疑问会得到读者的认可和重视。

我想要感谢编委会成员们,感谢他们在题目范围、推荐作者及评估方案方面

① REN21:21 世纪可再生能源政策网络。

的建议和帮助。

我还想感谢从 2008 年就一起工作的 Wiley-VCH 出版社的编辑们，他们是 Esther Levy 博士、Gudrun Walter 博士、Bente Flier 博士、Martin Graf-Utzmann 博士，感谢他们在这个项目上提供的专业帮助和大力支持。

希望读到本书的您会觉得本书内容丰富、深入浅出并值得一看，可以作为工作中的参考。我们将努力在这个具有成长性的领域里出版更多的丛书。

G. Q. Max Lu<br>澳大利亚，布里斯班<br>2012. 7. 31

# 原著丛书主编简介

逯高清教授
"可持续能源新材料和发展"丛书主编

逯高清教授的研究专长主要在材料化学和纳米技术领域。他以纳米颗粒和纳米多孔材料应用到清洁能源和环境技术的相关工作而闻名。曾在《自然》《美国化学学会杂志》《德国应用化学》和《先进材料》等杂志发表了500多篇高影响力的论文。他还拥有20项国际专利。逯高清教授是科学信息研究所(ISI)高度引用的材料科学的作者,超过17500次引文(h指数为63)。他获得了国内外众多知名奖项,包括中国科学院国际合作奖(2011年)、奥里卡奖、墨菲金质勋章、勒菲尔奖、埃克森美孚奖、化学奖章、澳大利亚(2004年、2010年和2012年)100强最有影响力的工程师,以及世界前50强最有影响力的中国人(2006年)。他曾两度获得澳大利亚研究理事会联合会奖学金(2003年和2008年)。他是澳大利亚技术科学与工程学院(ATSE)和化学工程师协会(IChemE)的当选研究员。他是12个主要国际期刊的编辑和编辑委员会成员,包括《胶体和界面科学与碳杂志》。

逯高清教授自2009年起担任澳大利亚昆士兰大学主管研究的副校长;在此之前,2008年10月至2009年6月担任校长助理;2012年担任常务副校长。2003年—2009年,他还担任由澳大利亚研究理事会成立的国家纳米材料卓越中心的首任主任。

逯高清教授曾担任过许多政府委员会和咨询小组成员,包括总理科学、工程与创新理事会(2004年、2005年和2009年)和ARC专家学院(2002年—2004年)的成员。他是IChemE澳大利亚董事会的前任主席,也是ATSE前董事。他作为前任董事会成员的单位包括Uniseed有限公司、ARC纳米技术网络和昆士兰中国委员会。目前,他是澳大利亚同步加速器、国家eResearch协作工具和资源以及研究数据存储基础架构的董事会成员。他还获得了国家新兴技术论坛成员的部长级任命。

# 本书主编简介

Bradley Ladewig 副教授是澳大利亚莫纳什大学化学工程系的一名学者，他领导一个研究小组开发薄膜材料和清洁能源应用技术。他是具有广泛经验的化学工程研究人员，包括直接甲醇燃料电池的膜开发，热电联产燃料电池组合的测试和建模以及海水淡化膜开发。最近，他在直接碳燃料电池、金属有机骨架材料作为气体吸附剂和膜组件以及基于纸张和线基材的低成本微流体传感器领域进行了多项合作项目。他也是化学工程师学会的会员。

蒋三平教授是澳大利亚科廷大学化学工程系教授燃料与能源技术研究院、副院长，澳大利亚阳光海岸大学兼职教授，也是西南大学、中南大学、哈尔滨工业大学、广州大学、华中科技大学、武汉理工大学、中国科技大学(USTC)、四川大学和山东大学的客座教授。蒋教授在学术界和行业领域拥有丰富的经验，曾在南洋理工大学、澳大利亚 CSIRO(联邦科学与工业研究组织)制造科学技术部和陶瓷燃料电池有限公司(CFCL)担任职务。他的研究领域包括固体氧化物燃料电池、质子交换膜和直接甲醇燃料电池，直接乙醇燃料电池与电解。他发表了 270 多篇文章，h 指数为 50，已累计引用约 8500 次。

严玉山教授是特拉华大学化学与生物分子工程系杰出工程学教授。他曾担任加州大学河滨分校首席科学家、大学学者、系主任等职务，以及 AlliedSignal 公司的高级工程师。他对诸多技术的启动和形成起到了重要作用，如纳米水、全周期能源、沸石溶液材料和氢能的研究。他的研究重点是沸石薄膜和电化学装置，包括燃料电池、电解槽、太阳能制氢和氧化还原液流电池。他发表了 160 多篇期刊论文，h 指数为 52。他获得了国际沸石协会颁发的唐纳德·布雷克奖，并获得美国科学促进会会员资格。

# 撰稿人名单

Benjamin M. Asquith

澳大利亚莫纳什大学化学工程系,墨尔本 Clayton 校区,维多利亚州,3800

德国德累斯顿莱布尼茨研究所,德累斯顿电子股份有限公司,胡和大街 6 号,01069

Daniel J. L. Brett

英国伦敦大学伦敦分校化学工程系,电化学创新实验室,托林顿大楼,WC1E 7JE

Yanzhen Fan

美国俄勒冈州立大学生物与生态工程系,吉尔摩尔大楼 116 房间,科瓦利斯市,OR 97331

Shuang Gu

美国纽瓦克特拉华大学化学与生物分子工程系,学院街 150 号, DE 19716

Falk Harnisch

澳大利亚布里斯班昆士兰大学先进水管理中心, QLD 4072

德国不伦瑞克 TU 不伦瑞克环境与可持续化学研究所, 38106

Rhodri Jervis

英国伦敦大学学院化学工程系,电化学创新实验室,托林顿大楼,WC1E 7JE

姜鲁华

中国科学院大连化学物理研究所大连清洁能源国家实验室,燃料电池与电池部,大连中山路 457 号,116023

Robert B. Kaspar
美国特拉华大学化学与生物分子工程系，学院街 150 号，DE 19716

Bradley P. Ladewig
澳大利亚莫纳什大学化学工程系，墨尔本 Clayton 校区，维多利亚州，3800

Wenzhen Li
美国密歇根理工大学化学工程系，霍顿汤森德大街 1400 号，MI 49931

Hong Liu
美国俄勒冈州立大学生物与生态工程系，吉尔摩尔大楼 116 房间，科瓦利斯市，OR 97331

Jochen Meier-Haack
德国德累斯顿莱布尼茨研究所，德累斯顿电子股份有限公司，胡和大街 6 号，01069

Nam-Trung Nguyen
新加坡南洋理工大学机械航空航天工程学院，南洋大道 50 号，639798

Korneel Rabaey
澳大利亚昆士兰大学先进水管理中心，布里斯班，QLD 4072
比利时根特大学微生物生态与技术实验室（LabMET），9000

Seyed Ali Mousavi Shaegh
新加坡南洋理工大学机械航空航天工程学院，南洋大道 50 号，639798

孙公权
中国科学院大连化学物理研究所大连清洁能源国家实验室，燃料电池与电池部，大连中山路 457 号，116023

Junhua Wang
美国纽瓦克特拉华大学化学与生物分子工程系，学院街 150 号，19716

Shuangyin Wang
新加坡南洋理工大学化学与生物医学工程学院，南洋大道 50 号，639798

Xin Wang
新加坡南洋理工大学化学与生物医学工程学院,南洋大道 50 号,639798

Yushan Yan
美国纽瓦克特拉华大学化学与生物分子工程系,学院街 150 号,19716

Bingzi Zhang
美国纽瓦克特拉华大学化学与生物分子工程系,学院街 150 号,19716

# 目录

## 第1章　低温燃料电池的关键材料:绪论



## 第2章　碱性阴离子交换膜燃料电池



## 第3章　用于质子交换膜燃料电池的催化剂载体材料

## 第4章 低温直接醇类燃料电池阳极催化剂

## 第5章　直接甲醇燃料电池膜材料



## 第6章　氢氧根离子交换膜及离聚物

## 第7章 微生物燃料电池材料

## 第8章 生物电化学系统



## 第9章 微流体燃料电池材料

## 第10章 直接醇类燃料电池催化剂的研究进展

# 第 1 章

## 低温燃料电池的关键材料:绪论

Bradley P. Ladewig, Benjamin M. Asquith, Jochen Meier-Haack

国际学术界和企业界已经多次承诺,将使用具有较低运行温度的燃料电池作为多功能、高效率的动力能源。低温燃料电池作为能够高效地将化学能转化为电能的装置的潜力已为人们所知多年,然而,尽管全球科学界付出了巨大的努力,这类电池至今仍未能实现大规模的商业化。

对于低温燃料电池,目前全球的重要共识之一是:在一系列关键材料的挑战得到很好的解决之前,这种燃料电池不可能得到很好的发展和应用。本书中,我们展示了低温燃料电池的现状,回顾了燃料电池的发展,指出低温燃料电池材料面临的挑战。根据低温燃料电池的实际构成,我们所采用的燃料电池的工作温度在 200℃以下。大多数情况下,低温燃料电池的工作温度远低于 100℃;然而,考虑高性能聚合物膜的研究进展(尤其是基于聚苯并咪唑的聚合物膜,见第 5 章),在一些系统中,现在已有实现气相进料的可能,并且,在相对提高的温度下,电池的动力学状况将大为改善,可能输出更高的功率密度。

本书不追求成为一本包罗万象的可以解决低温燃料电池每一个材料的百科全书。读者要寻求一个全面的参考资料,应该参考 Wolf Vielstich 编辑的优秀手册[1]。相反,我们力求突出重点,关注燃料电池研究的挑战。我们重点关注新兴领域,尤其侧重于碱性交换(或氢氧根离子交换)膜燃料电池。这些燃料电池完全和过去几十年发表的数千篇专门集中在质子交换或阳离子交换膜的低温燃料电池的文章相背离(最明显的是早期应用在氯仿工业的阳离子高性能交换膜)。在碱性燃料电池交换膜研制过程中,关键材料是一个很大的挑战,不仅仅是发展新的聚合物膜,更需要有选择性地传输氢氧根离子。当然还有一些更微妙的催化剂选择问题,这些都会在本书的一些细节当中提及。

在此必须提及的另外两个领域是微生物燃料电池和微流体燃料电池,从某些角度来说,这两个新领域可以被认为是低温燃料电池在极端尺寸范围内的操作。

微生物燃料电池的起源是对废水处理中的电化学和生物电化学系统的探索。根据其性质,反应器的体积可能是几十立方米(远大于常规低温燃料电池)。

相比之下,微流体燃料电池在尺寸范围的另一端,在过去 10 年里随着微流体领域的广泛发展,微流体燃料电池已经进入燃料电池的研究领域,并引起了关注。驱动微流体燃料电池发展的动力不仅仅在于微流体装置的制造技术的发展及广泛普及,还在于这种微流体燃料电池在功能设备(如传感器、医疗器械等)方面的重要应用前景。

本书涉及的话题范围广泛,希望读者能够认识和理解,这本书的潜在主题就是突出低温燃料电池的关键材料面临的挑战,并专业和简洁地展示燃料电池技术当前的技术状况。

## 参考文献

[1] Vielstich, W. (2009) *Handbook of Fuel Cells*, 6 *Volume Set*, John Wiley & Sons,Inc. , Hoboken, NJ.

# 第 2 章

# 碱性阴离子交换膜燃料电池

Rhodri Jervis, Daniel J. L. Brett

## 2.1 燃料电池

燃料电池被认为是很有前途的绿色能源转换技术,它直接将化学能转换成电能,操作过程中无须部件移动,且没有颗粒或温室气体排出。它比燃烧更高效,与电池相比具有能量密度高、充电次数少等优点。然而,燃料电池在电子能源技术方面仍存在一些缺点,成本是阻碍其商业化和广泛应用的主要障碍。膜电极组件(MEA)是燃料电池的主要组成部分,包括催化剂(通常是 Pt 基催化剂)和离子聚合物膜,这两部分成本决定了燃料电池的总成本。本章主要讲述降低碱性阴离子交换膜(AAEM)燃料电池成本的途径,使其在各能源领域实现商业化。在说明 AAEM 燃料电池的主要区别及它们可能提供的更有利的技术前,我们首先讨论常见的酸性质子交换膜燃料电池的基本原理和其进行电化学反应的热力学和动力学特性。

这种燃料电池的基本研究可以追溯到 1839 年,当时出生于斯旺西(Swansea)的物理学家 William Grove 先生意识到电解水的可逆过程是可能发生的。然而,基于该理念的发展过程很缓慢,直到 19 世纪 60 年代“阿波罗”太空计划才使得燃料电池以碱性液体电解质燃料电池的形式成为现实。液体电解质燃料电池在便携性上有很多缺点,这也使得最近的关注焦点转向固体电解质,尤其是聚合物电解质膜(PEM)燃料电池。聚合物电解质膜主要采用离子交联聚合物,该聚合物单体中包含一种离子官能团,它作为电解质可以使质子在一个非水相体系中传输。近年来膜技术的进步,特别是产业化标准的全氟磺酸膜性能的提高使得 PEM 燃料电池成为燃料电池领域的研究焦点。通常使用的碱性 PEM 燃料电池使用能够传导 $OH^-$ 离子的碱性膜,利用在碱性条件下的优异的阴极动力学特性,最终减少催化剂在燃料电池成本中的比重。

## 2.2 PEM 燃料电池原理

燃料电池的基本原理是通过直接电化学氧化还原反应产生电流。在氢/氧燃料电池中，氧化还原反应由两个电化学半反应构成，即阳极发生的氢氧化反应(HOR)

$$H_2 \rightleftharpoons 2H^+ + 2e^- \tag{2.1}$$

和阴极发生的氧还原反应(ORR)

$$\frac{1}{2}O_2 + 2H^+ + 2e^- \rightleftharpoons H_2O \tag{2.2}$$

总的氧化还原反应为

$$H_2 + \frac{1}{2}O_2 \rightleftharpoons H_2O \tag{2.3}$$

氢气通过流场流入阳极，空气/氧气通过流场流入阴极，再通过气体扩散层(GDL)到达催化剂层，气体、催化剂和电解质构成三相边界。HOR 和 ORR 分别发生在阳极和阴极的三相界面处，中间由聚合物电解质膜隔开。氢氧化生成的质子通过电解质扩散到阴极与氧气发生电化学反应产生电子，电子被绝缘聚合物拦截，通过外部电路产生电流(图 2.1)。

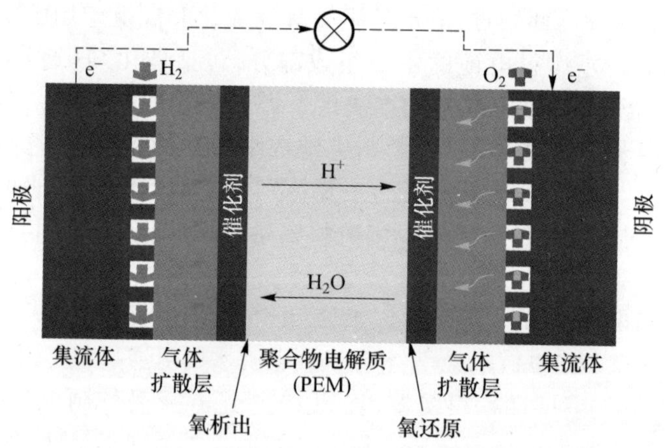


图 2.1 酸性 PEM 燃料电池原理图

### 2.2.1 平衡动力学

燃料电池电极上发生的 HOR 和 ORR 两个反应遵循电荷守恒原则，在阳极产生的电子和质子必须在阴极消耗掉。虽然通过两电极的电流相同，但两个反应产生这一电流所需的活化极化反应不同。电流是在电化学半反应中电子产生和消耗的速率，是电化学反应速率的直接量度。如果 1mol 反应物反应产生

$n$ mol电子，则电流 $i$ 为

$$i = nF\frac{\mathrm{d}N}{\mathrm{d}t} \tag{2.4}$$

式中：$\frac{\mathrm{d}N}{\mathrm{d}t}$为反应速率(mol/s)。

当界面发生反应时，电流通常用单位面积上的电流 $j = i/A$ 表示，且单位面积上的反应速率为 $v$，$j$ 称为电流密度(这也有利于对比不同燃料电池中不同几何面积的反应速率)。重新整理速率方程，得

$$v = \frac{j}{nF} \tag{2.5}$$

反应速率为产物的表面浓度乘以反应速率常数，即

$$v_1 = C_r k_1 \tag{2.6}$$

从统计力学可以得到速率常数是过渡态吉布斯自由能的函数，因为反应物必须以过渡态形式存在才能反应。速率常数为

$$k_1 = \frac{k_B T}{h}\mathrm{e}^{-\Delta G_1/(RT)} \tag{2.7}$$

式中：$k_B$ 为玻耳兹曼常数；$T$ 为温度(K)；$h$ 为普朗克常量；$R$ 为摩尔气体常数；$\Delta G_1$ 为反应活化能，即反应物自由能与过渡态自由能之差，如图2.2所示。

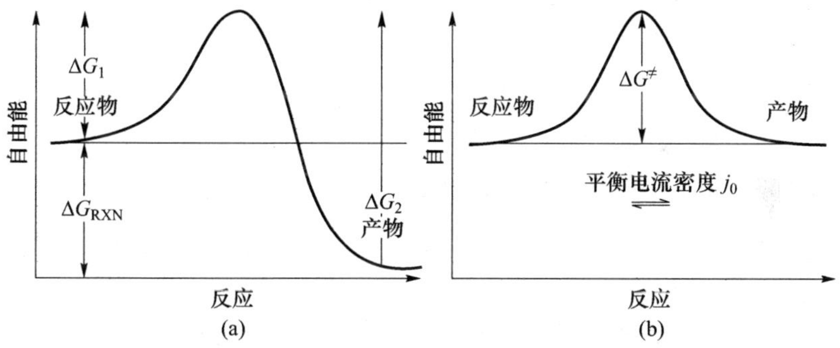


图2.2　反应活化能。(a)产物自由能比反应物自由能低，有利于反应进行；(b)双电层累积平衡，则电化学平衡

$v_1$ 是电化学半反应的正反应速率。逆反应的反应速率 $v_2$ 用 $\Delta G_1$ 代替 $\Delta G_2$，即产物与活化能的自由能之差。整个反应的反应速率 $v$ 为 $v_1 - v_2$，因此净电流为

$$j = nF(v_1 - v_2) \tag{2.8}$$

平衡时，正、逆反应速率相等，总电流密度 $j$ 为0。然而，$j$ 包括正反应电流密度 $j_1 = nFv_1$ 和逆反应电流密度 $j_2 = nFv_2$，当达平衡时两者相等，称作交换电流密度 $j_0$。从图2.2(b)可以看出，平衡时正、逆反应活化能 $\Delta G^{\neq}$ 相等。因此，交换电流密度表示为

$$j_0 = nFC_r f \mathrm{e}^{-\frac{\Delta G^{\neq}}{RT}} \tag{2.9}$$

式(2.7)中的前指数项已包含在 $f$ 中。在这个例子中,由于过渡态是对称的,所以正、逆反应活化能 $\Delta G^{\neq}$ 是相同的。然而,情况并非总是如此,因此一般情况下会引入对称因子 $\beta$。$\beta$ 用来度量过渡态的不对称性,它是衡量促进正反应电极电位变化的分数。$\beta$ 的取值范围为 0~1,促进逆反应的电位分数为 $1-\beta$。如图 2.2 所示,如果过渡态是对称的,那么 $\beta$ 为 0.5,电极电势的增加对正、逆反应的促进作用相同。大多数催化剂均为这种情况,通过降低活化能来同等地促进氧化半反应和还原半反应(归因于低能量的过渡态)。因此,除单电子转移反应以外,$\beta$ 为 0.5 的假设是安全合理的。如果多电子和连续电子转移反应的某一步反应非常缓慢,且为速率控制步骤,那么 $\beta$ 仍为 0.5;然而,情况并非总是如此。正因如此,$\beta$ 通常由实验参数 $\alpha$ 即电荷转移系数来代替,与对称因子 $\beta$ 不同,$\alpha$ 的正、逆反应系数之和不必为 1。因此,常用实验值 $\alpha_n$ 代替 $\beta$。

根据热力学式 $E=-\Delta G/(nF)$ 和电荷转移系数,交换电流密度表达式为

$$j_0 = nFC_r f_1 \mathrm{e}^{\alpha_1 nFE_0/(RT)} = nFC_P f_2 \mathrm{e}^{-\alpha_2 nFE_0/(RT)} \tag{2.10}$$

式中:$\alpha_1$,$\alpha_2$ 分别为正、逆反应的电荷转移系数;$E_0$ 为给定半反应的可逆平衡电位。

将式(2.10)取自然对数,整理,得

$$E_0 = \frac{RT}{nF}\ln\left(\frac{f_1}{f_2}\right) - \frac{RT}{nF}\ln\left(\frac{C_r}{C_P}\right) \tag{2.11}$$

该方程称为能斯特方程,适用于所有热力学平衡的动力学理论。

### 2.2.2 Butler – Volmer 动力学

当动力学达到平衡时,没有净电流的流动;因此,若要产生有效电流(或电荷转移速率),正反应必须要比逆反应更容易产生电子。为此,必须改变由双电层产生的平衡电势,使得产物的过电势低于反应物(图 2.2)。要做到这一点,需通过牺牲一定的电极电势,促使反应物和产物之间产生电势差,从而产生电流。这就是活化过电势($\eta_{act}$)产生的原因,电压降用来克服电荷双电层的均衡效应和诱导动态平衡的净电荷转移。如果我们考虑对一个电化学半反应应用过电势 $\eta$,那么电势为

$$E = \eta + E_0 \tag{2.12}$$

总反应的净电流密度为

$$j = j_1 - j_2 = nFC_r f_1 \mathrm{e}^{\frac{\alpha_1 nF(\eta + E_0)}{RT}} - nFC_P f_2 \mathrm{e}^{\frac{-\alpha_2 nF(\eta + E_0)}{RT}} \tag{2.13}$$

由式(2.10)可以看出,我们对交换电流密度的两种定义都出现在式(2.13)中,所以可以将其简化为

$$j = j_0\left[\mathrm{e}^{\alpha_1 nF\eta/(RT)} - \mathrm{e}^{-\alpha_2 nF\eta/(RT)}\right] \tag{2.14}$$

这就是 Butler-Volmer(BV)方程,并广泛用于模拟燃料电池反应。注意到,上述推理过程只涉及单电极的半反应,BV 方程应分别应用于每个电极。阳极氢氧化反应和阴极氧还原反应的交换电流密度、电荷转移系数和活化过电势不同,但都遵循电荷守恒定律,即整个电池的电流与阴极电流和阳极电流相等:$i_{cell} = i_c = i_a$。因此,每个电极上的 BV 方程式相等,即

$$j_{0,a}\left(e^{\frac{\alpha_{1,a}nF\eta_a}{RT}} - e^{\frac{-\alpha_{2,a}nF\eta_a}{RT}}\right) = j_{0,c}\left(e^{\frac{\alpha_{1,c}nF\eta_c}{RT}} - e^{\frac{-\alpha_{2,c}nF\eta_c}{RT}}\right) \tag{2.15}$$

因此,阴极和阳极的过电势会达到燃料电池所需的电流密度。随着阴极的负向极化和阳极的正向极化使得电池的可逆电势损失越来越多直至消耗殆尽,达到极限电流。

### 2.2.3　交换电流密度

交换电流密度是一个极为重要的参数,主要影响电化学反应的动力学。它反映了电荷转移的速率,因此需要较高的交换电流密度 $j_0$ 使电极动力学变得容易,从而减少活化电势的损失。式(2.10)给出的交换电流密度定义较为简单,它没有考虑反应过程中反应物和产物浓度的变化,也没有明确温度对交换电流密度 $j_0$ 的影响。同时,催化剂的表面积和负载量也会影响交换电流密度,所以 $j_0$ 可定义为

$$j_0 = j_0^0 a\left(\frac{C_r}{C_r^0}\right) e^{\left\{-\frac{E_0}{RT}\left[1-(T/T_0)\right]\right\}} \tag{2.16}$$

式中:$j_0^0$ 为在一定温度 $T_0$ 和一定浓度 $C_r^0$ 下测量的参比交换电流密度;$a$ 为粗糙度系数,是催化剂实际电化学表面积与电极几何表面积之比。

从式(2.16)可以看出,交换电流密度是温度的指数函数,尽管 BV 式(2.14)表明随温度升高,活化损失增大,但实际上温度升高使得电流密度 $j_0$ 大大增加,动力学损失减少,且呈非线性。因此,在高温下运行燃料电池将大大减少活化损失。

尽管交换电流密度不是催化剂的本征属性,但它与催化剂的种类密切相关。高性能燃料电池可以通过以下方式增大催化剂/反应物的交换电流密度 $j_0$:

(1) 增大反应物浓度;

(2) 降低活化能 $E_0$;

(3) 升高温度;

(4) 增大粗糙度系数 $a$。

增加反应物浓度对反应热力学影响较小(归因于能斯特方程的对数形式),但它对反应动力学影响显著(呈线性关系)。因此,燃料电池常在纯氧中运行(空气中氧分压仅为 0.2),所以当电池在空气中运行时,交换电流密度降低。

活化能降低说明催化剂影响了交换电流密度 $j_0$。催化剂可以通过为反应提

供较大的表面和稳定的过渡态来降低活化能垒。催化剂与吸附物间的键强是寻找合适催化剂的主要考虑因素。键的强弱要恰到好处,在反应之初催化剂表面要能够吸附反应物,增加反应概率,但结合键能不能太强,否则键难以断裂形成产物,且吸附了大量的反应物,使得催化剂表面可用的活性位点数量下降。这通常用火山图表示,铂族金属通常具有较佳的键强(图 2.3)。

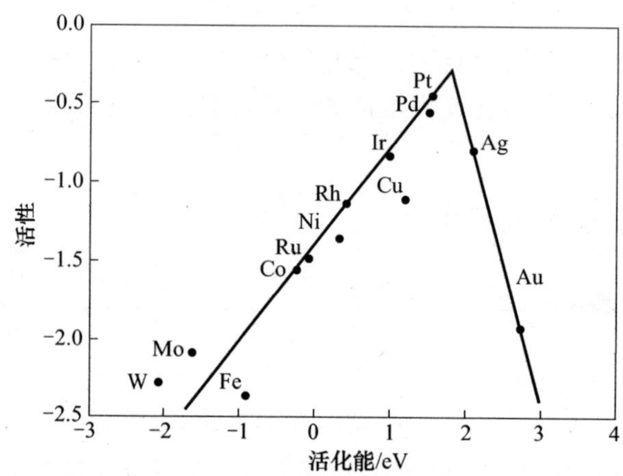


图 2.3　不同过渡金属的 ORR 活性火山图,氧原子与催化剂间结合能的函数[1]

一般来说,活化能也受到反应复杂性的影响,复杂反应机理比简单反应机理的活化损失更大。基于此原因,ORR 比非常简单的 HOR 慢很多,在酸性 PEM 燃料电池中,ORR 是造成活化损失的主要原因(图 2.4)。

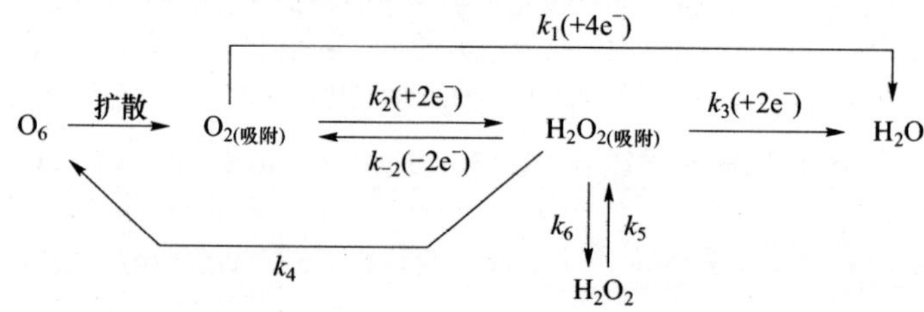


图 2.4　氧还原可能的反应路径。$4e^-$ 路径($k_1$)较为有效且不产生具有破坏性的过氧化氢中间体[2]

根据吸附步骤的本质,酸性溶液中的 HOR 的反应机理为 Tafel－Volmer 机理或者 Heyrovsky－Volmer 机理。若吸附为纯化学过程,那么反应机理为Tafel－Volmer 机理;若吸附为化学与电化学相结合的过程,那么反应机理为 Heyrovsky－Volmer 机理[3]。

$$H_2 + M_{cat} \rightleftharpoons 2M_{cat}H_{chem}\,(\text{Tafel}) \tag{2.17}$$

$$M_{cat}H_{chem} \rightleftharpoons M_{cat} + H^+ + e^-\,(\text{Volmer}) \tag{2.18}$$

$$H_2 + M_{cat} \rightleftharpoons M_{cat}H_{chem} + H^+ + e^- \text{(Heyrovsky)} \tag{2.19}$$

综上所述，温度升高，交换电流密度将增大。其物理原因是随着温度的升高，能够发生反应生成产物的反应物分子比例升高。

增加粗糙度系数 $a$ 相当于增加催化剂的电化学表面积，也就是增加反应表面活性位点的数量。这将显著提高交换电流密度，使反应瞬间完成。碳载铂催化剂（广泛应用于燃料电池的催化剂）电极的粗糙度系数在600～2000之间变化[4]。

### 2.2.4 燃料电池极化曲线

Butler－Volmer方程中描述的活化过电势可被认为是燃料电池的理论热力学电压降，用来确保在有效电流密度下运行。它并不是燃料电池操作过程中发生的唯一电压损失，还有由于欧姆电阻、质量传输限制和开路造成的损失。这些过电位的根源将在书中其他部分更详细地阐述。燃料电池极化产生的各种过电势和电压损失可用极化曲线来描述，即燃料电池中电压随电流密度的变化。电池电压表示为

$$E_{cell} = E_{thermo} - \eta_{act} - \eta_{ohm} - \eta_{conc} - \eta_{OCV} \tag{2.20}$$

式中：$E_{thermo}$ 为在给定温度和压力下热力学预测的电位；$\eta_{act}$，$\eta_{ohm}$，$\eta_{conc}$，$\eta_{OCV}$ 分别为由活化、欧姆电阻、质量传递和内部电流造成的损失。

图2.5所示为一个典型的极化曲线。值得注意的是，尽管每一种损失的影响可大致划分为一定的区域，但损失是操作电流（$\eta_{act}$，$\eta_{ohm}$ 和 $\eta_{conc}$）的函数，且对整个曲线均有影响。

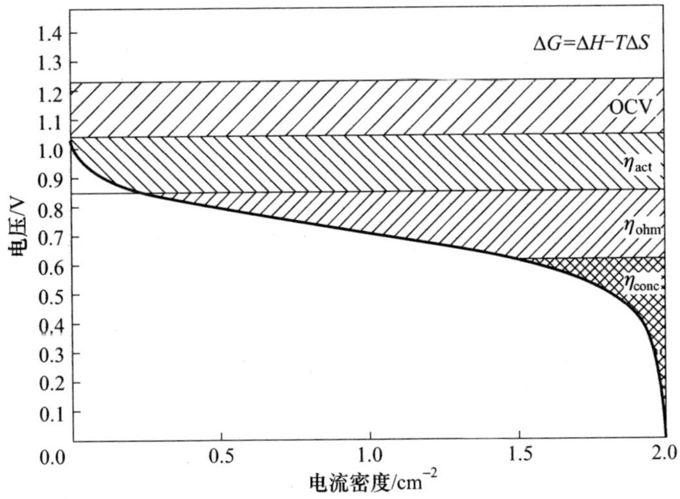


图2.5 燃料电池极化曲线，以前述的燃料电池行为为模型，展示了电压降的区域，数值主要受到 $H_2$ 的生成焓的影响；区域分别为熵、内部电流、动力学损失、电阻损失和质量传输损失

## 2.3 碱性燃料电池

近20年来,酸性PEM燃料电池成为低温燃料电池研究的焦点。酸性PEM电池有很多值得关注的优势,但这种电池也存在一些不足,主要是在酸性介质中其ORR的动力学较为缓慢。在碱性介质中,其ORR的动力学则较为容易[5-10],碱性燃料电池(AFC)能够使用更便宜的、非贵金属的阴极催化剂,如Ag[11,12]、Au(银和金在强酸性媒介中均不稳定)[13]、Ni[14,15],另外,即使使用Pt催化剂,碱性PEM燃料电池也可使得阴极Pt的载量大幅度降低。

### 2.3.1 ORR的反应机理

催化剂表面发生的氧还原反应,其关键步骤是需要4对质子和电子转移的O—O键断裂,而且可能有很多副反应和副产物(图2.4)[6]。尽管对其反应路径已达成共识,主要通过直接四电子还原路径或有中间产物过氧化物参与的2+2串联的四电子路径[16-18],但对ORR的复杂性及其诸多潜在的副产物仍知之甚少。

直接四电子路径(式(2.21))要在第一个电子发生转移前断裂O—O键(被认为是速率控制步骤[12]),且氧的位错能较大,催化剂与氧分子间的强键能有利于补偿氧分子内部化学键断裂所需的能量(并导致严重过电势)。串联四电子路径是有效的反应路径(式(2.22)和式(2.23)):两电子还原为过氧化物,然后过氧化物经两电子还原为水(因为$O_2^-$和$O_2^{2-}$阴离子的位错能低于$O_2$)[11]。

Marković等研究表明,在酸性介质中Pt表面发生的ORR为串联路径,但吸附的阴离子和欠电位沉积的氢会阻止第二步过氧化物两电子还原反应,这意味着反应一定程度上停留在中间产物过氧化物上[19-21]。强烈吸附的阴离子减少了用于氧吸附和O—O键断裂的活性位点[18],且经分析可知转移电子数$n$小于4(部分反应物并不完全经四电子过程还原为水)。这些多步反应和可能分解的过氧化物中间体意味着ORR的动力学缓慢,常常只有Pt具有合理的催化作用。

文献中常提及在碱性介质中ORR动力学较为容易[10,22-25],但是没有详细解释为什么会这样,通常归因于阴离子吸附效应,这来源于电解质的选择性和水溶液电解质中的实验[11,26,27]。尽管Srinivasan等的研究表明,碱性燃料电池中的动力学较酸性PEM燃料电池快[28],但由于没有阴离子移动的聚合物电解质应用于燃料电池中,这些基础性ORR研究似乎不适用于工作燃料电池的情形[27]。Pt催化剂的形态和类型(单晶、多晶或Pt/C)对动力学和ORR的交换电流密度有较大影响[27,29-31],在ORR中块状Pt表面比Pt纳米粒子的ORR活性高[32,33]。然而,尽管一些研究表明在酸性和碱性介质中Pt具有差不多一样的活性[34]，且 Pt 或 Pt/C 仍是碱性介质中 ORR 最好的催化剂[35]，但是，在碱性燃料电池中，有可能使用更为廉价的阴极催化剂[5-9]。

Blizanac 等[11,12]采用旋转环盘电极（RRDE）检测中间产物来研究银单晶电极上 pH 值对 ORR 的影响，发现在碱性电解质中 ORR 几乎完全通过四电子路径来进行，几乎检测不到过氧化物中间体。相反，酸性电解质中需要高过电势来阻止占主导的两电子路径（生成过氧化物）反应，且 ORR 的起始电位也比碱性介质高。Au 表面的 ORR 在碱性溶液中的反应活性明显高于酸性溶液，在 $Au_{100}$表面为四电子路径[36]。在碱性溶液中，直接反应路径为

$$O_2 + 2H_2O + 4e^- \longrightarrow 4OH^- \tag{2.21}$$

连续反应路径为

$$O_2 + H_2O + 2e^- \longrightarrow HO_2^- + OH^- \tag{2.22}$$

$$HO_2^- + H_2O + 2e^- \longrightarrow 3OH^- \tag{2.23}$$

碳载体负载的催化剂（如应用于燃料电池中的那些催化剂）能够表现出与大多数本体金属非常不同的 ORR 行为。Yang 等[37]研究表明在碱性介质中 Pd/C 催化剂对 ORR 有较高的催化活性，且碳载体本身对 $O_2$ 两电子还原为过氧化物具有催化活性（然后中间体迁移到 Pd 粒子表面进行接下来的两电子还原为水的过程）。研究表明，所有的碳材料在碱性介质中对 ORR 均具有一定的活性（在酸性介质中没有），通常是两电子还原为过氧化物，随后氧化的碳表面需在更高的过电势下完成接下来的两电子反应，还原为水[24,38-40]。

因此，碱性燃料电池被看作是能够使用更便宜的阴极催化剂，从而降低燃料电池成本的一条商业化的途径。

## 2.3.2　碱性介质中的氢氧化反应

高于室温时，氢的氧化反应动力学在酸性 PEM 燃料电池中通常较为迅速，对整个活化过电势贡献可以忽略不计，因此通常将活化过电势完全归因于 ORR[41]。人们发现阳极催化剂的负载量可低至 $0.05mg_{Pt}/cm^2$[42]而几乎对于燃料电池的性能没有多少不利的影响，因此，对于酸性 PEM 燃料电池而言，阴极催化剂是研究的主要关注点。然而，对于碱性 PEM 燃料电池，情况则完全不一样，在多晶 Pt 表面发生的 HOR 的交换电流密度比在酸性介质中要低两个数量级[43]，所以在碱性燃料电池中的阳极活化极化不可忽略[28]。事实上，为了在碱性燃料电池中实现能够应用更便宜的阴极催化剂的优势，有关阳极 HOR 的催化的研究是至关重要的，并且这种改进的需求常常被忽视[34,44]。

HOR 的反应机理遵循简单的 Tafel – Volmer 机理或 Heyrovsky – Volmer 机理，但是 $OH^-$为中间产物的反应除外。

$$H_2 + M_{cat} \rightleftharpoons 2M_{cat}H_{chem}\,(\text{Tafel}) \tag{2.24}$$

$$M_{cat}H_{chem} + OH^- \rightleftharpoons M_{cat} + H_2O + e^- \text{(Volmer)} \tag{2.25}$$

$$H_2 + M_{cat} + OH^- \rightleftharpoons M_{cat}H_{chem} + H_2O + e^- \text{(Heyrovsky)} \tag{2.26}$$

碱性介质中单晶 Pt 的 HOR 研究提出 Tafel 或 Heyrovsky 步骤为速率控制步骤[45,46]。电极表面吸附的氢原子有两种形式：一种是欠电位强烈吸附的氢原子($H_{UPD}$)；另一种是反应中间媒介高于过电势弱吸附的氢原子($H_{OPD}$)[47]。碱性介质中 HOR 较慢是因为 OH 物种的吸附覆盖了 $H_2$ 的吸附位点并阻碍其反应[18,45]，甚至改变了 $H_{OPD}$层的能量。

Sheng 等[34]利用旋转圆盘电极(RDE)首次比较研究了 Pt/C 电极在酸性和碱性介质中的 HOR。RDE 研究用于区分基于动力学和扩散对电流的影响，如 Levich - Koutecky 方程所示：

$$\frac{1}{i} = \frac{1}{i_k} + \frac{1}{i_d} = \frac{1}{i_k} + \frac{1}{BC_0\omega^{1/2}} \tag{2.27}$$

式中：$B$ 为常数，表达式为 $0.62nFD^{2/3}\nu^{-1/6}$，其中，$D$ 为扩散系数，$\nu$ 为电解液运动黏度；$C_0$ 为反应物的溶解度；$\omega$ 为圆盘电极的旋转速率($rad^{-1}$)。

这样，改变电极的旋转速率，并将电流的倒数对 $\omega^{1/2}$ 作图，则截距为动力学控制电流的倒数，因此，将总电流展开为扩散控制和动力学控制的电流，可得到关于反应动力学的有用信息。Sheng 等[34]研究表明，对于酸性介质中的 HOR，RDE 实验即使在高转速下(给定高扩散极限电流下)也非常接近扩散电流(意味着本质上无限快的动力学)，这说明酸性溶液中的 HOR 太快，不适合用 RDE 研究，会导致文献中对交换电流密度的低估[3]。另外，碱性介质中的 HOR 明显背离了仅为扩散行为的影响，意味着动力学电流可通过 RDE 实验恰当地加以阐明。如果用于 PEM 阳极的超低负载 Pt[42]用于 AFC 中，则碱性介质中的缓慢动力学是 PEM 电池阳极产生过电势(约 130mV)的主要原因。强调阳极催化剂的研制是为了获得 AFC 潜在的更便宜的阴极催化剂。另外关于在碱性介质中的 HOR 的研究来自 Cabot 等的研究，在他们的 RDE 实验中，采用气体扩散电极(GDE)近似地代替燃料电池的电极研究了碱性介质中在贵金属表面上发生的 HOR[25,44]。他们得出以下结论：在低过电势(接近开路电压)时，在 Tafel - Volmer 机理中 Tafel 反应为速率控制步骤，而在高过电势时，$H_2$ 的扩散为速率控制步骤。这些研究也表明，对于 GDE 电极来说，HOR 的交换电流密度在碱性介质中较低。

总之，碱性介质中 ORR 的动力学使得使用廉价的阴极催化剂成为可能，但 HOR 显著的过电势需要开发阳极催化剂来充分利用碱性燃料电池的这一优势。

### 2.3.3 水电解液碱性燃料电池

在燃料电池早期的开发中，碱性燃料电池优越的 ORR 动力学使得其成为主导技术，如 Bacon[48] 的开创性工作以及碱性燃料电池用于“阿波罗”太空任务。碱性燃料电池使用氢氧化钾水溶液为电解液，质量浓度约为 30%，通常盛装在电池基体中（图 2.6）。就像其他氢/氧燃料电池一样，HOR 发生在阳极，ORR 发生在阴极，尽管在碱性介质中的半反应式与酸性介质中稍有不同，即以 $OH^-$ 为中间媒介而非质子：

$$H_2 + 2OH^- \rightleftharpoons 2H_2O + 2e^- \text{ (HOR)} \tag{2.28}$$

$$\frac{1}{2}O_2 + H_2O + 2e^- \rightleftharpoons 2OH^- \text{ (ORR)} \tag{2.29}$$

总氧化还原反应式相同：

$$H_2 + \frac{1}{2}O_2 \rightleftharpoons H_2O \tag{2.30}$$

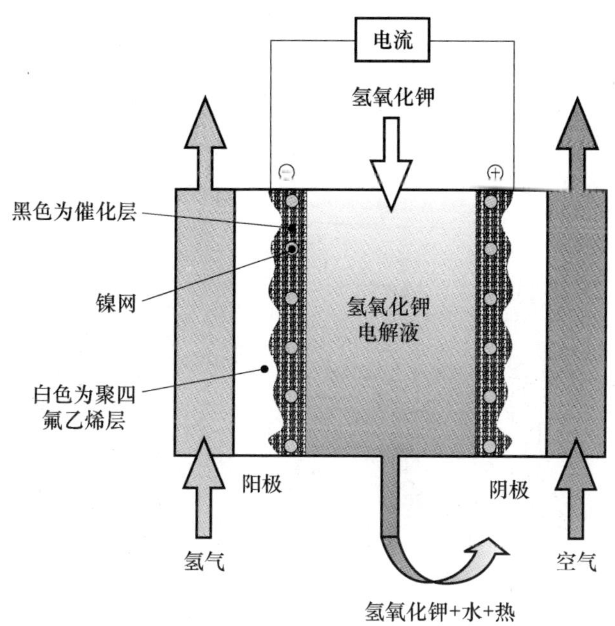


图 2.6　水溶液碱性燃料电池示意图[10]

注意到，在酸性介质中，阳极产生的 2mol 水，其中 1mol 被阴极消耗，从而阳极上还有 1mol 水。这将引起阳极潜在水淹和水管理问题。强腐蚀性的水性电解液很危险，且易与二氧化碳反应生成碳酸盐：

$$CO_2 + 2OH^- \longrightarrow CO_3^{2-} + H_2O \tag{2.31}$$

碳酸根离子可与金属阳离子结合生成沉淀析出，从而堵塞电极，致使电解液传导能力下降，严重影响电池性能，所以系统中使用的是不含二氧化碳的纯氧气和氢气。基于这些缺点和质子交换膜技术的改进，碱性燃料电池被 PEM 燃料电池取代成为研究的热点。尽管碱性燃料电池技术[49] 近来有所改进，但本章将聚焦于碱性固体聚合物电解质的研究进展。关于碱性燃料电池的更多细节，可参考文献[10,23]。

## 2.3.4 AAEM 燃料电池

碱性阴离子交换膜(AAEM)燃料电池的进展揭示了发展可媲美酸性固体聚合物电解质燃料电池的碱性电池的可能性。这将能够利用碱性阴极动力学的优势,同时消除使用水性电解液的缺点。为清晰起见,本章规定:PEM 仅指质子交换膜燃料电池(酸性),AAEM 指阴离子交换膜氢/氧燃料电池,且 AFC 专指水性电解液碱性氢/氧燃料电池。阴离子交换膜也用于碱性直接醇燃料电池中,讨论时将它们写为 ADMFC(碱性直接甲醇燃料电池)或 ADEFC(碱性直接乙醇燃料电池)。

### 2.3.4.1 AAEM 原理

AAEM 燃料电池与 PEM 燃料电池原理基本相同,即使用固体电解质直接将氢气经电化学转化为电力,主要不同在于基本半反应式(式(2.28)和式(2.29))和氢氧化物离子传导膜。AAEM 燃料电池原理图如图 2.7 所示。

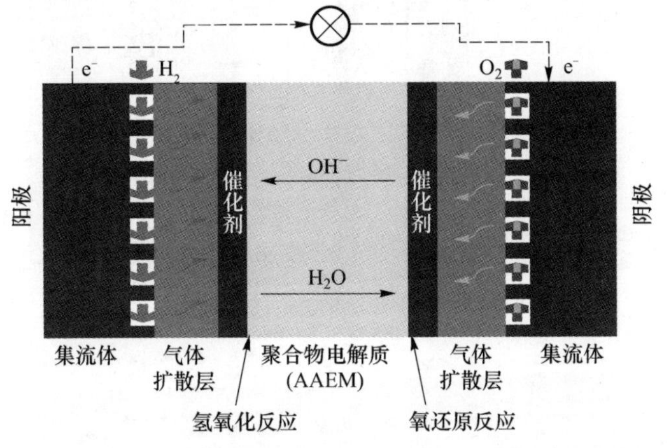


图 2.7 AAEM 碱性燃料电池示意图(对比图 2.1)

AAEM 燃料电池的外围组件与 PEM 燃料电池一样,并起着相同的作用。在 AAEM 中对水的处理较为不同,阳极产生水,阴极消耗水,意味着阳极可能发生水淹,阴极的润湿程度不够。"阳极自增湿"可能会提供更简单的增湿方式,即只有阴极需要增湿(尽管最近一个有趣的研究显示,当只使用干阴极流和加湿阳极流时,即阴极加湿的方式仅是水通过膜扩散到阴极,性能受到限制[50])。

AAEM 燃料电池与 PEM 系统的另一个主要区别在于膜对氢氧根离子的传导。氢离子的扩散系数约是氢氧根离子的 4 倍,AAEM 的离子电阻率通常比全氟磺酸膜高很多,可能需要在膜上添加氢氧化钾溶液以增大电导率[7]。

### 2.3.4.2 碱性膜

在燃料电池中使用阴离子交换膜是一个相对初期的技术，因此没有像全氟磺酸那样居此领域主要地位[51]。Merle[52]在碱性膜进展中综述了大量的碱性膜和相对较少的商业化膜。因为 AAEM 是固态的，不包含移动的阳离子，所以即使二氧化碳与氢氧根形成了固体碳酸盐，也很少会沉积在膜上。近年来，由于较少的碳酸盐问题及缺乏高浓度的水系电解质，AAEM 的研究主要集中在固体碱性燃料电池技术和膜的研发上（图 2.8）。

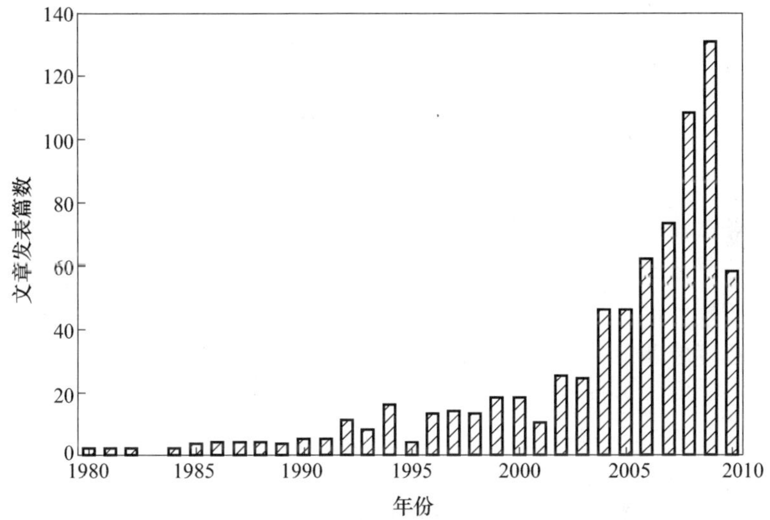


图 2.8 碱性燃料电池近年来发表刊物数量的增加情况[52]

正因为全氟磺酸膜含有带负电的磺酸基从而能传导带正电的质子，所以 AAEM 必须含有带正电的基团来传导带负电的氢氧根。如前所述，AAEM 的传导能力低于 PEM[53]。增强传导能力的一种方式是聚合物中包含更多的阳离子基团，但这样常使得膜的机械强度和化学稳定性下降。在 AAEM 中最常用的官能团是季铵基团 $R_4N^+$[54]，在有氢氧根存在时会经历 E2 消除反应（霍夫曼消除）而失去活性基团（图 2.9）。

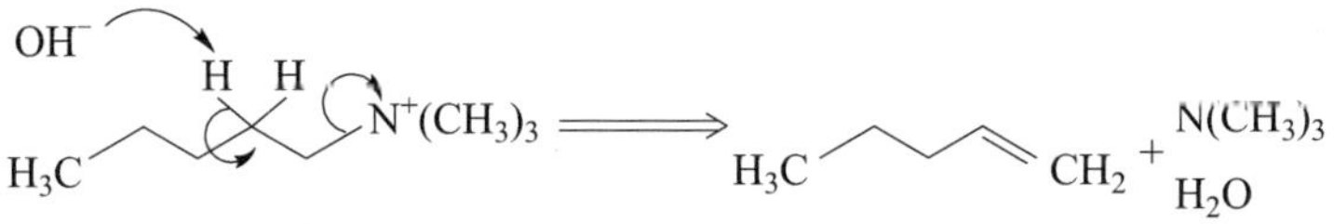

$$OH^- + H_3C\text{—}CH_2\text{—}CH(H)\text{—}CH(H)\text{—}CH_2\text{—}N^+(CH_3)_3 \Longrightarrow H_3C\text{—}CH_2\text{—}CH_2\text{—}CH{=}CH_2 + N(CH_3)_3 + H_2O$$

图 2.9 季铵基团的霍夫曼消除

很多这方面的研究工作表明，霍夫曼消除反应在 60℃ 以下时进行缓慢，但在高温下反应较快，使得大多数 AAEM 在 60℃ 以上性能显著衰减[55]。由于没有获得在高温下操作的动力学优越性（PEM 燃料电池常在 80℃时运行），这对当今的碱性膜电池显然是一个限制。季铵基团的弱碱性使得 AAEM 电池中氢氧根离子的移动比 PEM 电池中质子的移动慢。而且,尽管没有固体碳酸盐的沉积,但仍形成了碳酸根,使得膜中氢氧根离子的浓度降低,这是电导率降低的一个影响因素[52,56]。

文献[52,57]中报道了诸多不同种类的离子交换膜,但应用于燃料电池最具前景的膜是在聚合物骨架中接入季铵离子基团[58-66]或对现有聚合物进行化学修饰[51,67-70]。2007 年,Varcoe 研制出第一个电导率超过固体碱性燃料电池运行所需预期值 10mS/cm 的 AAEM,不过膜需要在高湿度下才能保持良好的传导率[71]。先前的研究经常不得不将膜浸在氢氧化钾溶液中以获得良好的传导性能[7,72]。2011 年,Tanaka 等[73]报道了传导率最高的膜,该膜用于肼燃料电池中,80℃下最大传导率为 114mS/cm。性能良好的 AAEM 应满足以下性能:

(1) 化学、机械和热力学稳定性高;

(2) 氢氧根离子传导能力强,大于 10mS/cm;

(3) 导电性低;

(4) 透气性低;

(5) 厚度较薄;

(6) 不同湿度下性能良好。

PEM 燃料电池中的电极含有一定量的全氟磺酸聚合物溶液,作为黏结剂,同时也可向三相界面处输送一些氢离子[74]。对于 AAEM 燃料电池,需要在催化剂层中使用类似于离子交联聚合物的物质[54],这成为开发碱性燃料电池良好电极主要考虑的因素[64]。

总之,开发 AAEM 膜是最近研究和发展的一个重要领域,但它的电导率和稳定性仍然是阻碍其发展的重要因素。然而,好的 AAEM 膜的性能接近质子交换膜,因此 AAEM 燃料电池的发展还需要开发催化剂,使其在性能和成本上都能与质子交换膜燃料电池相媲美。

#### 2.3.4.3 AAEM 燃料电池举例

作为一门新兴且有待发展的技术,文献中报道的 AAEM 燃料电池例子较少。Surrey 大学的 Varcoe 和 Slade 做了一些关于碱性膜和 AAEM 燃料电池的开创性工作[8,13,51,58,59,61,62,64-66,71,75-78]。他们主要采用辐射接枝季铵基团的膜。使用 Pt 作为阴极和阳极,纯氧作为氧化剂,获得最大功率密度为 55mW/cm$^2$(电压为 0.6V 时,电流密度约为 90mA/cm$^2$)[66]。当使用 Au 或 Ag 作为阳极催化剂时性能下降(图 2.10)[8]。运行耐碳酸盐膜的 $H_2$/空气燃料电池产生了更低的功率密度,约为 38mW/cm$^2$,但膜以碳酸盐形式比氢氧化物形式表现出了更好的性能,或许是由于 ORR 在碱性碳酸盐中比在 KOH 水溶液中更快[75]。2009 年,

Zhou 等也观察到类似的规律,尽管功率密度很低,只有 $4mW/cm^2$[79]。

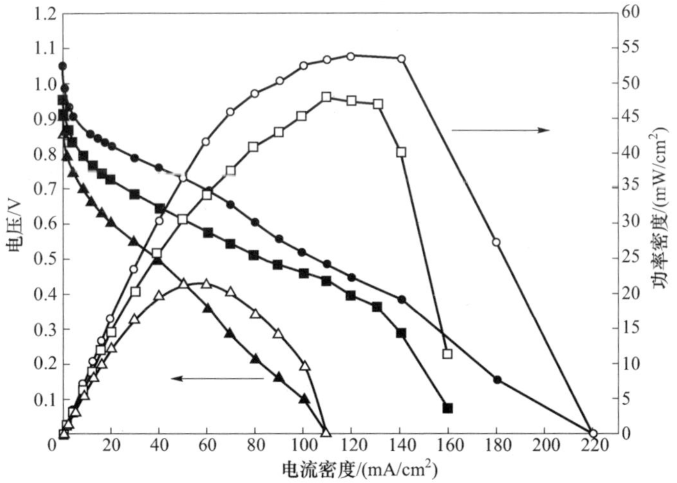

图 2.10　纯氧下 Pt(○)、Ag(□)和 Au(△)阴极的燃料电池极化曲线(实心)和功率密度(空心)[8]

最近他们用 Pt 为阴极,使用最薄的膜(完全水化后为 17μm)在纯氧气下测定产生的峰值功率密度为 $230mW/cm^2$(图 2.11)[13]。通过降低膜厚度来获得较好的性能归因于增加了水从阳极到阴极(消耗水,式(2.29))的交叉流动。在这项研究中,他们也对 Au 和 Ag 阴极进行了测试,与 Pt 相比性能降低。

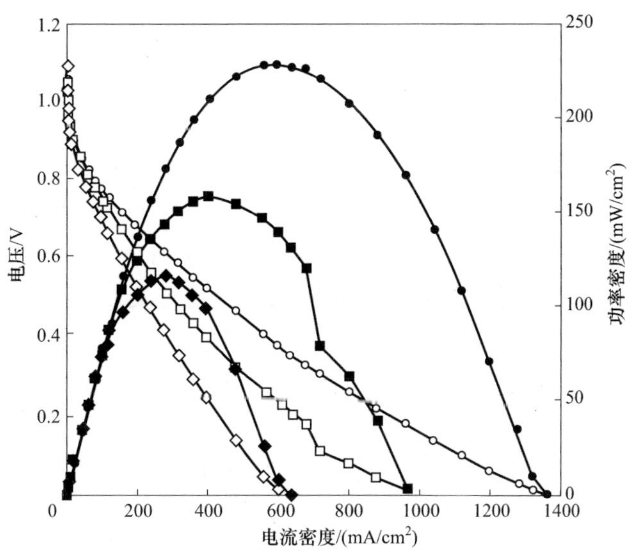


图 2.11　Pt 阴极在纯氧中膜厚度对燃料电池极化曲线(空心)和功率密度(实心)性能的影响。性能依次为 85μm(◇)、46μm(□)和 17μm(○)[13]

他们还开发了一种用于 AAEM 燃料电池中的参比电极，可以解耦整个电池中阴极和阳极的极化。结果表明，阳极极化明显高于阴极极化，与 PEM 燃料电池中的情形恰好相反（图 2.12）。此外，根据碱性介质中缓慢的 HOR 动力学，人们认为即使在低电流密度下，AAEM 燃料电池阳极的水淹仍可导致较高的质量传输极化（注意：在这项研究中使用的膜厚为 80μm）。他们还证明，水作为 ORR 的反应物（与 PEM 不同），阴极电极设计对获得良好的电池性能至关重要。使用不含聚四氟乙烯的阴极，他们获得的功率密度为 $125mW/cm^2$，且表明碳在碱性电池中的催化作用较小[78]。这说明在 AAEM 燃料电池中要更加注重水管理和催化剂层的问题[76-78]。

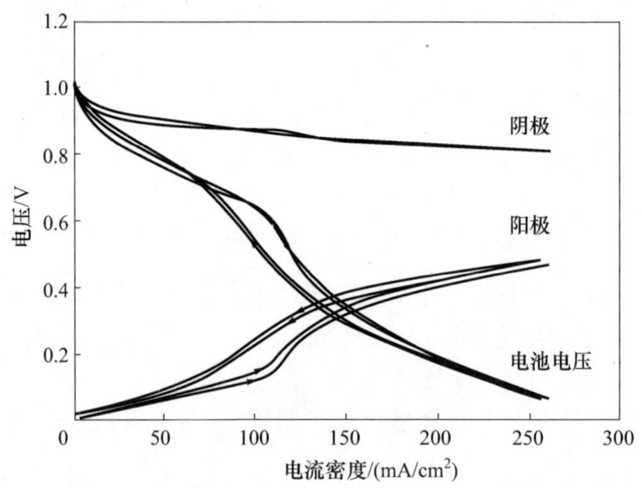


图 2.12　整个电池的极化曲线（$V_{cell}$）及阴极和阳极的极化曲线，使用贵金属 Pd 包覆 Pt 线作为参比电极。阳极极化明显大于阴极极化[76]

早在 2001 年，Agel 等[7]做了一些关于 AAEM 燃料电池的工作。在常温常压下的 $H_2/O_2$ 系统中，他们用掺杂 KOH 的膜，以泡沫镍负载 Pt/C 为电极，在电压为 0.6V 时获得的电流密度约为 $25mA/cm^2$。这项研究表明，电极与膜间要有良好的离子传导能力，若在电极与膜间使用氢氧化钾凝胶，则电压在 0.6V 时，电流密度增加到约 $60mA/cm^2$。

在 2008 年，Lu 等首次制成了完全非铂 AAEM 电池，使用铬修饰的镍作阴极，用银作阳极[80]。他们将季铵聚砜膜溶解在溶剂中，能够控制铸膜的厚度和浸渍的电极层。在 60℃、1atm① 下，使用湿润的 $H_2/O_2$ 获得的功率密度峰值为 $50mW/cm^2$。

Park 等表明，60℃时，在湿润的 $H_2$/空气条件下，使用高负载量的 Ag 作为电

① 1atm = 101.325kPa。

极(2.0mg/$cm^2$)可以产生类似于Pt阴极在胺化的聚砜膜燃料电池中的功率密度峰值30mW/$cm^2$(图2.13)[22]。他们也获得了约为1.05V的高开路电压,可能预示着更高的交换电流密度或更低的燃料交叉来减少电压损失(Varcoe也观察到类似的结果[8])。

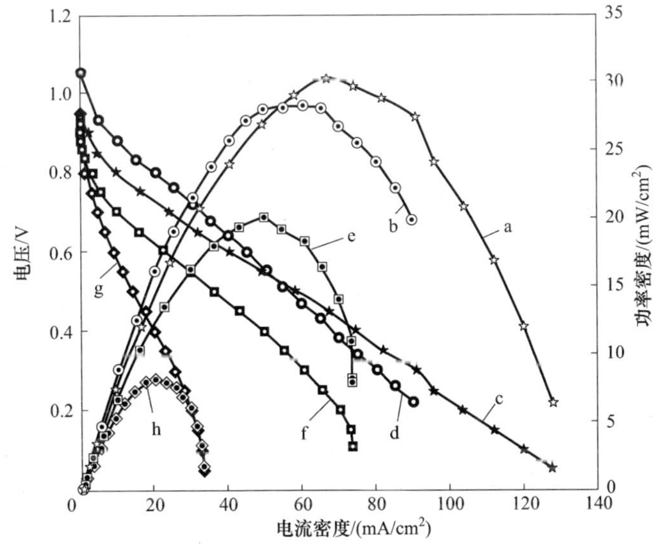


图2.13 在AAEM燃料电池中,以Pt为阳极(0.5mg/$cm^2$),分别以Ag作为阴极((g,h)0.5mg/$cm^2$,(e,f)1mg/$cm^2$,(a,c)2mg/$cm^2$)和以Pt作为阴极((b,d)0.5mg/$cm^2$)的极化曲线和能量密度[22]

2009年,Gu等将含有季鏻盐的离子交联聚合物溶液应用于商业AAEM中,再次证明电极和膜间良好离子传导的重要性[54]。观察到,AEM电池中添加与不添加离子交联聚合物溶液,其性能明显不同,在80℃时,$H_2/O_2$背压下电流密度峰值接近200mA/$cm^2$。Mamlouk等在他们的自制膜上也得到了相似的结果,总结出在催化剂层中最佳的离子交联聚合物含量主要取决于几个因素,如电极厚度和氧气分压。他们还指出,由于水淹问题,阳极层应比阴极层厚[81]。

在2010年,Piana等制成了性能最佳的AAEM燃料电池[82]。他们在50℃、$H_2$/空气(无二氧化碳)气氛中,使用商用膜并结合一种新的自制的离子交联聚合物溶液作为催化剂黏结剂。当使用铂作阳极和阴极时,功率密度峰值为400mW/$cm^2$,当使用公开的过渡金属负载碳作为阴极,功率密度值为200mW/$cm^2$。他们还证明了二氧化碳对系统的影响,在低电流密度下,由于二氧化碳与氢氧根反应致使性能大幅下降,随后电导率降低。然而,在高电流密度下,电化学半反应产生更多的氢氧根离子,有利于克服这一问题,这就是自动净化机制(图2.14)。极化曲线有明显变化表示在高电流密度下膜电导率增加(尽管此时操作电压下降),这已影响到AAEM燃料电池的调节。Varcoe表明,电池应该迅速调到高电流密度,然后通过降低负载回到开路电压来获得极化曲线(这种方法比增加开路电压的电流负载产生更好的性能)[78]。

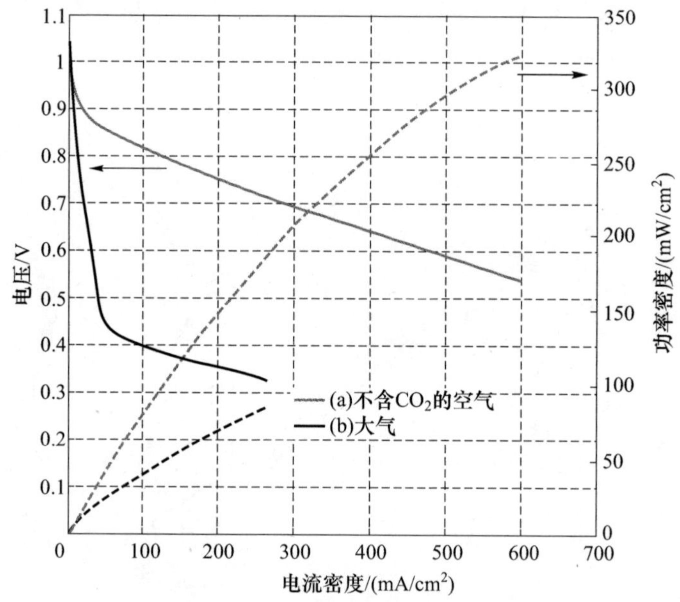


图 2.14 极化和能量密度曲线,用于说明 $CO_2$ 对膜性能的影响

不含 $CO_2$ 的空气(灰色)的性能很好,但在大气中(黑色)碳酸盐效应对膜的性能影响很大。当电流密度升高,这种效应减弱,如图所示黑色极化曲线的梯度变得与无 $CO_2$ 空气的极化曲线相似[82]

2012 年,Cao 等[51]通过化学法将氨基引入聚合物,开发了一种聚甲基乙烯基醚交替马来酸酐(PMVMA)膜。这种膜在 150℃ 以下均很稳定,且从室温到 60℃时离子电导率增加。在 35℃、$H_2/O_2$ 电池中获得功率密度峰值为 155mW/cm$^2$,且在较低温度下性能有很大改善(低温下欧姆电阻较高)(图 2.15)。

关于 AAEM 燃料电池的一些最具前景的工作来自碱性膜供应商之一 Tokuyama公司。不过他们仅发布了简短的且含有较少细节的会议记录,但这项研究工作显示出了固体碱性燃料电池的前景[9,50,56,63,83-85]。他们已开发出大量的季铵膜,电导率约为 40mS/cm,且可以控制膜厚度为 10~40μm[63]。使用 Pt/C 作为阳极和阴极,不含 $CO_2$ 的空气作为氧化剂,达到功率密度峰值为 325mW/cm$^2$,同时也证实了 Piana 等讨论的在高电流密度下,电池具有自净作用[9,83,86]。Tokuyama 也是第一个致力于提高 AAEM 电池操作温度研究的公司,使得电池性能有所改进,在 80℃时,开发的膜可稳定运行 200h[84]。2011 年,他们在没有阴极气流加湿时比较了电池性能,从而证实了一个非常有趣但表面上有悖常理的事实,即碱性介质中的 ORR 消耗水作为反应物,因此通常假定阴极需要良好的

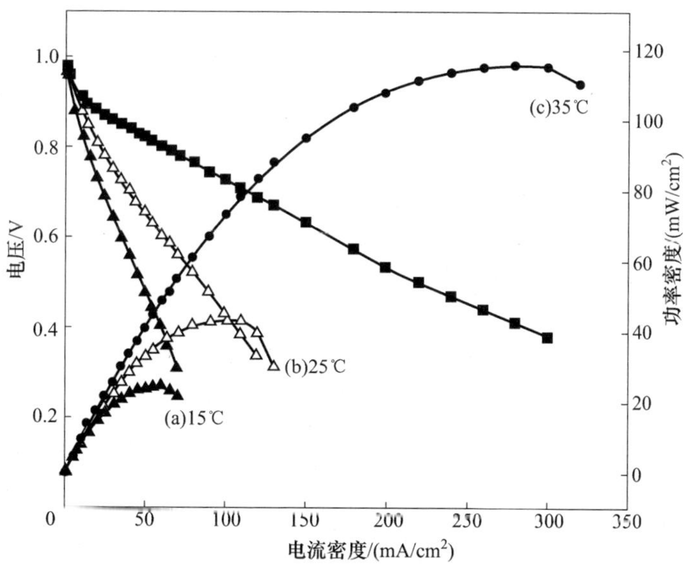


图2.15　温度对AAEM电池电阻和极化性能的影响[51]

润湿,但这项研究表明当阳极增湿、阴极不增湿时,水会从阳极流入阴极,从而补偿缺水的状况。他们还证实了不增湿气流时(即阴极、阳极均不增湿)的灾难性影响,阳极HOR产生2mol水不足以让膜充分润湿(图2.16)[50]。如果AAEM用于便携设备,如汽车,那么部分不增湿操作很重要。

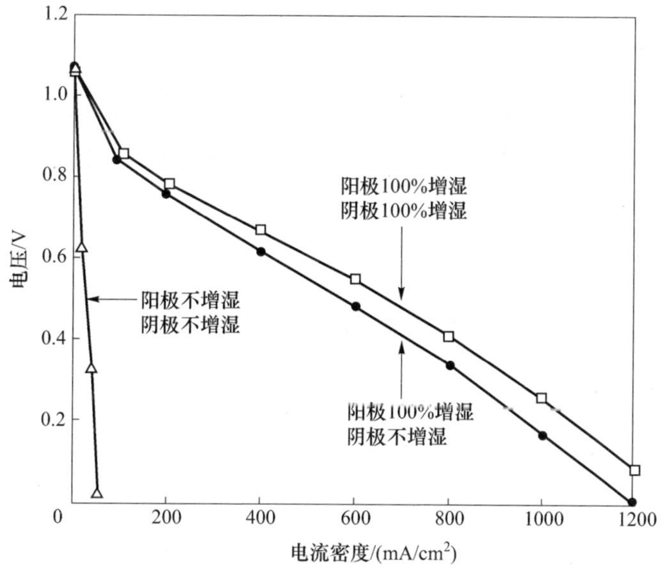


图2.16　增湿对Tokuyama公司AAEM电池的影响。只增湿阳极气流(●)与阳极、阴极均增湿(□)比较图。当两股气流均不增湿(△),性能下降明显[50]

## 2.4 本章小结

以氢作为燃料的固体碱性燃料电池是改善酸性 PEM 燃料电池性能和降低其成本的潜在途径。技术还处于开发的早期阶段，缺乏研究数据，在这些研究中技术和结果都不同，因此需要应用特定的方法对碱性情况（而不是试图将酸性 PEM 原理应用到可能是一种非常不同的技术上）进行进一步的研究，从而达到最新 PEM 燃料电池的性能水平。

（1）燃料电池的活化损失可由半反应的交换电流密度表示，且其为反应活化能的函数。

（2）在碱性介质中，ORR 较快，HOR 较慢，这意味着碱性燃料电池要得到更便宜的阴极催化剂必须同时考虑阳极催化剂。

（3）在酸性与碱性介质中的电化学半反应是不同的：

| | |
|---|---|
| $H_2 \rightleftharpoons 2H^+ + 2e^-$ | 阳极，酸性介质 |
| $H_2 + 2OH^- \rightleftharpoons 2H_2O + 2e^-$ | 阳极，碱性介质 |
| $\frac{1}{2}O_2 + 2H^+ + 2e^- \rightleftharpoons H_2O$ | 阴极，酸性介质 |
| $\frac{1}{2}O_2 + H_2O + 2e^- \rightleftharpoons 2OH^-$ | 阴极，碱性介质 |
| $H_2 + \frac{1}{2}O_2 \rightleftharpoons H_2O$ | 总反应 |

（4）碱性燃料电池与 PEM 燃料电池相比，由于其半反应不同使得水处理方式不同。

（5）固体碱性膜的使用避免了使用液体碱性电解质的危险，且减少了碳酸盐的影响。

（6）碱性膜研究不像酸性膜研究那么充分，且需要进一步研究以达到 PEM 电池的性能。但是通过开发更便宜的催化剂为低成本燃料电池提供了一条潜在途径。

## 参考文献

[1] Nørskov, J. K., Rossmeisl, J., Logadottir, A., Lindqvist, L., Kitchin, J. R., Bligaard, T., and Jónsson, H. (2004) Origin of the overpotential for oxygen reduction at a fuel-cell cathode. *The Journal of Physical Chemistry B*, **108**, 17886 – 17892.

[2] Wang, B. (2005) Recent development of non-platinum catalysts for oxygen reduction reaction. *Journal of Power Sources*, **152**, 1 – 15.

[3] Chen, S. and Kucernak, A. (2004) Electrocatalysis under conditions of high mass transport: investigation of hydrogen oxidation on single submicron Pt particles supported on carbon. *The Journal of Physical Chemistry B*, **108**, 13984 – 13994.

[4] Li, X. (2006) Principles of fuel cells. *Platinum Metals Review*, **50**, 200.

[5] Srinivasan, S., Enayetullah, M. A., Somasundaram, S., Swan, D. H., Manko, D., Koch, H., and Appleby, A. J. (1989) Recent advances in solid polymer electrolyte fuel cell technology with low platinum loading electrodes. *Journal of Power Sources*, **29**, 367 - 387.

[6] Appleby, A. J. (1993) Electrocatalysis of aqueous dioxygen reduction. *Journal of Electroanalytical Chemistry and Interfacial Electrochemistry*, **357**, 117 - 179.

[7] Agel, E., Bouet, J., and Fauvarque, J. F. (2001) Characterization and use of anionic membranes for alkaline fuel cells. *Journal of Power Sources*, **101**, 267 - 274.

[8] Varcoe, J. R., Slade, R. C. T., Wright, G. L., and Chen, Y. (2006) Steady-state DC and impedance investigations of $H_2/O_2$ alkaline membrane fuel cells with commercial Pt/C, Ag/C, and Au/C cathodes. *The Journal of Physical Chemistry B*, **110**, 21041 - 21049.

[9] Yanagi, H., Watanabe, S., Sadasue, K., Isomura, T., Inoue, H., and Fukuta, K. (2009) Improved performance of alkaline membrane fuel cells based on newly developed electrolyte materials. *ECS Meeting Abstracts*, **902**, 341 - 341.

[10] McLean, G. F., Niet, T., Prince-Richard, S., and Djilali, N. (2002) An assessment of alkaline fuel cell technology. *International Journal of Hydrogen Energy*, **27**, 507 - 526.

[11] Blizanac, B. B., Ross, P. N., and Markovic, N. M. (2007) Oxygen electroreduction on Ag(111): the pH effect. *Electrochimica Acta*, **52**, 2264 - 2271.

[12] Blizanac, B. B., Ross, P. N., and Marković, N. M. (2006) Oxygen reduction on silver low-index single-crystal surfaces in alkaline solution: rotating ring disk(Ag (hkl)) studies. *The Journal of Physical Chemistry B*, **110**, 4735 - 4741.

[13] Poynton, S. D., Kizewski, J. P., Slade, R. C. T., and Varcoe, J. R. (2010) Novel electrolyte membranes and non-Pt catalysts for low temperature fuel cells. *Solid State Ionics*, **181**, 219 - 222.

[14] Bidault, F., Brett, D. J. L., Middleton, P. H., Abson, N., and Brandon, N. P. (2009) A new application for nickel foam in alkaline fuel cells. *International Journal of Hydrogen Energy*, **34**, 6799 - 6808.

[15] Bidault, F., Brett, D. J. L., Middleton, P. H., Abson, N., and Brandon, N. P. (2010) An improved cathode for alkaline fuel cells. *International Journal of Hydrogen Energy*,**35**, 1783 - 1788.

[16] Yeager, E. (1984) Electrocatalysts for $O_2$ reduction. *Electrochimica Acta*, **29**, 1527 - 1537.

[17] Wroblowa, H. S. (1976) Electroreduction of oxygen: a new mechanistic criterion. *Journal of Electroanalytical Chemistry and Interfacial Electrochemistry*, **69**, 195.

[18] Marković, N. M. and Ross, P. N., Jr. (2002) Surface science studies of model fuel cell electrocatalysts. *Surface Science Reports*, **45**, 117 - 229.

[19] Grgur, B. N., Marković, N. M., and Ross, P. N. (1997) Temperature-dependent oxygen electrochemistry on platinum low-index single crystal surfaces in acid solutions. *Canadian Journal of Chemistry*, **75**, 1465 - 1471.

[20] Marković, N. M., Gasteiger, H. A., Grgur, B. N., and Ross, P. N. (1999) Oxygen reduction reaction on Pt(111): effects of bromide. *Journal of Electroanalytical Chemistry*, **467**, 157 - 163.

[21] Stamenkovic, V., Marković, N. M., and Ross, P. N., Jr. (2001) Structurerelationships in electrocatalysis: oxygen reduction and hydrogen oxidation reactions on Pt(111) and Pt(100) in solutions containing chloride ions. *Journal of Electroanalytical Chemistry*, **500**, 44 - 51.

[22] Park, J.-S., Park, S.-H., Yim, S.-D., Yoon, Y.-G., Lee, W.-Y., and Kim, C.-S. (2008) Performance of solid alkaline fuel cells employing anion-exchange membranes. *Journal of Power Sources*, **178**, 620 - 626.

[23] Bidault, F., Brett, D. J. L., Middleton, P. H., and Brandon, N. P. (2009) Review of gas diffusion cathodes for alkaline fuel cells. *Journal of Power Sources*, **187**, 39 - 48.

[24] Ernest, Y. (1986) Dioxygen electrocatalysis: mechanisms in relation to catalyst structure. *Journal of Molecular Catalysis*, **38**, 5 - 25.

[25] Alcaide, F., Brillas, E., and Cabot, P. L. (2005) Hydrogen oxidation reaction in a Pt-catalyzed gas diffusion electrode in alkaline medium. *Journal of the Electrochemical Society*, **152**, E319 - E327.

[26] Markovic, N., Gasteiger, H., and Ross, P. N. (1997) Kinetics of oxygen reduction on Pt (hkl) electrodes: implications for the crystallite size effect with supported Pt electrocatalysts. *Journal of the Electrochemical Society*, **144**, 1591 - 1597.

[27] Parthasarathy, A., Srinivasan, S., Appleby, A. J., and Martin, C. R. (1992) Electrode kinetics of oxygen reduction at carbonsupported and unsupported platinum microcrystallite/Nafion® interfaces. *Journal of Electroanalytical Chemistry*, **339**, 101 - 121.

[28] Srinivasan, S., Manko, D. J., Koch, H., Enayetullah, M. A., and Appleby, A. J. (1990) Recent advances in solid polymer electrolyte fuel cell technology with low platinum loading electrodes. *Journal of Power Sources*, **29**, 367 - 387.

[29] Song, C., Tang, Y., Zhang, J. L., Zhang, J., Wang, H., Shen, J., McDermid, S., Li, J., and Kozak, P. (2007) PEM fuel cell reaction kinetics in the temperature range of 23 - 120°C. *Electrochimica Acta*, **52**, 2552 - 2561.

[30] Hayden, B. E., Pletcher, D., Suchsland, J. - P., and Williams, L. J. (2009) The influence of support and particle size on the platinum catalysed oxygen reduction reaction. *Physical Chemistry Chemical Physics*, **11**, 9141 - 9148.

[31] Gasteiger, H. A. and Ross, P. N. (1996) Oxygen reduction on platinum low-index single-crystal surfaces in alkaline solution: rotating ring diskPt (hkl) studies. *The Journal of Physical Chemistry*, **100**, 6715 - 6721.

[32] Kinoshita, K. (1990) Particle size effects for oxygen reduction on highly dispersed platinum in acid electrolytes. *Journal of the Electrochemical Society*, **137**, 845 - 848.

[33] Gasteiger, H. A., Kocha, S. S., Sompalli, B., and Wagner, F. T. (2005) Activity benchmarks and requirements for Pt, Pt-alloy, and non-Pt oxygen reduction catalysts for PEMFCs. *Applied Catalysis B: Environmental*, **56**, 9 - 35.

[34] Sheng, W., Gasteiger, H. A., and ShaoHorn, Y. (2010) Hydrogen oxidation and evolution reaction kinetics on platinum: acid vs alkaline electrolytes. *Journal of the Electrochemical Society*, **157**, B1529 - B1536.

[35] Lima, F. H. B., Zhang, J., Shao, M. H., Sasaki, K., Vukmirovic, M. B., Ticianelli, E. A., and Adzic, R. R. (2006) Catalytic activity-d-band center correlation for the $O_2$ reduction reaction on platinum in alkaline solutions. *The Journal of Physical Chemistry C*, **111**, 404 - 410.

[36] Adzic, R. (1998) Recent advances in the kinetics of oxygen reduction.

[37] Yang, Y. -F., Zhou, Y. -H., and Cha, C. -S. (1995) Electrochemical reduction of oxygen on small palladium particles supported on carbon in alkaline solution. *Electrochimica Acta*, **40**, 2579 - 2586.

[38] Taylor, R. J. and Humffray, A. A. (1975) Electrochemical studies on glassy carbon electrodes: II. Oxygen reduction in solutions of high pH (pH > 10). *Journal of Electroanalytical Chemistry and Interfacial Electrochemistry*, **64**, 63 - 84.

[39] Appel, M. and Appleby, A. J. (1978) A ringdisk electrode study of the reduction of oxygen on active carbon in alkaline solution. *Electrochimica Acta*, **23**, 1243 - 1246.

[40] Jürmann, G. and Tammeveski, K. (2006) Electroreduction of oxygen on multiwalled carbon nanotubes modified highly oriented pyrolytic graphite electrodes in alkaline solution. *Journal of Electroanalytical Chemistry*, **597**, 119 – 126.

[41] Neyerlin, K. C., Gu, W., Jorne, J., and Gasteiger, H. A. (2007) Study of the exchange current density for the hydrogen oxidation and evolution reactions. *Journal of the Electrochemical Society*, **154**, B631 – B635.

[42] Gasteiger, H. A., Panels, J. E., and Yan, S. G. (2004) Dependence of PEM fuel cell performance on catalyst loading. *Journal of Power Sources*, **127**, 162 – 171.

[43] Bagotzky, V. S. and Osetrova, N. V. (1973) Investigation of hydrogen ionization on platinum with the help of microelectrodes. *Journal of Electroanalytical Chemistry*, **43**, 233 – 249.

[44] Cabot, P. -L., Alcaide, F., and Brillas, E. (2009) Hydrogen reaction at open circuit in alkaline media on Pt in a gas-diffusion electrode. *Journal of Electroanalytical Chemistry*, **626**, 183 – 191.

[45] Markovica, N. M., Sarraf, S. T., Gasteiger, H. A., and Ross, P. N. (1996) Hydrogen electrochemistry on platinum low-index single-crystal surfaces in alkaline solution. *Journal of the Chemical Society, Faraday Transactions*, **92**, 3719 – 3725.

[46] Schmidt, T. J., Ross, P. N., Jr., and Markovic, N. M. (2002) Temperature dependent surface electrochemistry on Pt single crystals in alkaline electrolytes: Part 2. The hydrogen evolution/oxidation reaction. *Journal of Electroanalytical Chemistry*, **524 – 525**, 252 – 260.

[47] Markovicć, N. M., Grgur, B. N., and Ross, P. N. (1997) Temperature dependent hydrogen electrochemistry on platinum low-index single-crystal surfaces in acid solutions. *The Journal of Physical Chemistry B*, **101**, 5405 – 5413.

[48] Bacon, F. T. (1960) The high pressure hydrogen-oxygen fuel cell. *Industrial & Engineering Chemistry*, **52**, 301 – 303.

[49] Brushett, F. R., Naughton, M. S., Ng, J. W. D., Yin, L., and Kenis, P. J. A. (2012) Analysis of Pt/C electrode performance in a flowingelectrolyte alkaline fuel cell. *International Journal of Hydrogen Energy*, **37**, 2559 – 2570.

[50] Isomura, T., Fukuta, K., Yanagi, H., Ge, S., and Wang, C. -Y. (2011) Impact of low cathode humidification on alkaline membrane fuel cell performance. *ECS Meeting Abstracts*, **1101**, 221.

[51] Cao, Y. C., Wang, X., and Scott, K. (2012) The synthesis and characteristic of an anion conductive polymer membrane for alkaline anion exchange fuel cells. *Journal of Power Sources*, **201**, 226 – 230.

[52] Merle, G., Wessling, M., and Nijmeijer, K. (2011) Anion exchange membranes for alkaline fuel cells: a review. *Journal of Membrane Science*, **377**, 1 – 35.

[53] Grew, K. N. and Chiu, W. K. S. (2010) A dusty fluid model for predicting hydroxyl anion conductivity in alkaline anion exchange membranes. *Journal of the Electrochemical Society*, **157**, B327 – B337.

[54] Gu, S., Cai, R., Luo, T., Chen, Z., Sun, M., Liu, Y., He, G., and Yan, Y. (2009) A soluble and highly conductive ionomer for high-performance hydroxide exchange membrane fuel cells. *Angewandte Chemie: International Edition*, **48**, 6499 – 6502.

[55] Chempath, S., Boncella, J. M., Pratt, L. R., Henson, N., and Pivovar, B. S. (2010) Density functional theory study of degradation of tetraalkylammonium hydroxides. *The Journal of Physical Chemistry C*, **114**, 11977 – 11983.

[56] Watanabe, S., Fukuta, K., and Yanagi, H. (2010) Determination of carbonate ion in MEA during the alkaline membrane fuel cell (AMFC) operation. *ECS Transactions*, **33**, 1837 – 1845.

[57] Couture, G., Alaaeddine, A., Boschet, F., and Ameduri, B. (2011) Polymeric materials as

anion-exchange membranes for alkaline fuel cells. *Progress in Polymer Science*, **36**, 1521 – 1557.

[58] Varcoe, J. R. and Slade, R. C. T. (2006) An electron-beam-grafted ETFE alkaline anion-exchange membrane in metalcation-free solid-state alkaline fuel cells. *Electrochemistry Communications*, **8**, 839 – 843.

[59] Beillard, M., Varcoe, J. R., Halepoto, D. M., Kizewski, J. P., Poynton, S. D., and Slade, R. C. T. (2008) Membrane and electrode materials for alkaline membrane fuel cells. University of Surrey, Guilford.

[60] Varcoe, J. R., Slade, R. C. T., Yee, E. L. H., Poynton, S. D., and Driscoll, D. J. (2007) Investigations into the ex situ methanol, ethanol and ethylene glycol permeabilities of alkaline polymer electrolyte membranes. *Journal of Power Sources*, **173**, 194 – 199.

[61] Slade, R. C. T. and Varcoe, J. R. (2005) Investigations of conductivity in FEP-based radiation-grafted alkaline anion-exchange membranes. *Solid State Ionics*, **176**, 585 – 597.

[62] Herman, H., Slade, R. C. T., and Varcoe, J. R. (2003) The radiation-grafting of vinylbenzyl chloride onto poly (hexafluoropropylene-*co*-tetrafluoroethylene) films with subsequent conversion to alkaline anion-exchange membranes: optimisation of the experimental conditions and characterisation. *Journal of Membrane Science*, **218**, 147 – 163.

[63] Yanagi, H. and Fukuta, K (2008) Anion exchange membrane and ionomer for alkaline membrane fuel cells (AMFCs). *ECS Meeting Abstracts*, **802**, 783 – 783.

[64] Varcoe, J. R. and Slade, R. C. T. (2005) Prospects for alkaline anion-exchange membranes in low temperature fuel cells. *Fuel Cells*, **5**, 187 – 200.

[65] Danks, T. N., Slade, R. C. T., and Varcoe, J. R. (2003) Alkaline anion-exchange radiation-grafted membranes for possible electrochemical application in fuel cells. *Journal of Materials Chemistry*, **13**, 712 – 721.

[66] Varcoe, J. R., Slade, R. C. T., and Lam How Yee, E. (2006) An alkaline polymer electrochemical interface: a breakthrough in application of alkaline anion-exchange membranes in fuel cells. *Chemical Communications*, 1428 – 1429.

[67] Hong, J.-H., Li, D., and Wang, H. (2008) Weak-base anion exchange membranes by amination of chlorinated polypropylene with polyethyleneimine at low temperatures. *Journal of Membrane Science*, **318**, 441 – 444.

[68] Kostalik, H. A., Clark, T. J., Robertson, N. J., Mutolo, P. F., Longo, J. M., Abruña, H. C. D., and Coates, G. W. (2010) Solvent processable tetraalkylammoniumfunctionalized polyethylene for use as an alkaline anion exchange membrane. *Macromolecules*, **43**, 7147 – 7150.

[69] Clark, T. J., Robertson, N. J., Kostalik, IV, H. A., Lobkovsky, E. B., Mutolo, P. F., Abruña, H. C. D., and Coates, G. W. (2009) A ring-opening metathesis polymerization route to alkaline anion exchange membranes: development of hydroxide-conducting thin films from an ammonium-functionalized monomer. *Journal of the American Chemical Society*, **131**, 12888 – 12889.

[70] Wan, Y., Peppley, B., Creber, K. A. M., and Bui, V. T. (2010) Anion-exchange membranes composed of quaternized-chitosan derivatives for alkaline fuel cells. *Journal of Power Sources*, **195**, 3785 – 3793.

[71] Varcoe, J. R. (2007) Investigations of the *ex situ* ionic conductivities at 30 degrees C of metal-cation-free quaternary ammonium alkaline anion-exchange membranes in static atmospheres of different relative humidities. *Physical Chemistry Chemical Physics*, **9**, 1479 – 1486.

[72] Ogumi, Z. (2002) Preliminary study on direct alcohol fuel cells employing anion exchange membrane. *Electrochemistry*, **70**, 980 – 983.

[73] Tanaka, M., Fukasawa, K., Nishino, E., Yamaguchi, S., Yamada, K., Tanaka, H., Bae, B.,

Miyatake, K. , and Watanabe, M. (2011) Anion conductive block poly(arylene ether)s: synthesis, properties, and application in alkaline fuel cells. *Journal of the American Chemical Society*, **133**, 10646 – 10654.

[74] Passalacqua, E. , Lufrano, F. , Squadrito, G. ,Patti, A. , and Giorgi, L. (2001) Nafion content in the catalyst layer of polymer electrolyte fuel cells: effects on structure and performance. *Electrochimica Acta*, **46**,799 – 805.

[75] Adams, L. A. , Poynton, S. D. , Tamain, C. ,Slade, R. C. , and Varcoe, J. R. (2008) A carbon dioxide tolerant aqueouselectrolyte-free anion-exchange membrane alkaline fuel cell. *ChemSusChem*, **1**, 79 – 81.

[76] Zeng, R. , Poynton, S. D. , Kizewski, J. P. ,Slade, R. C. T. , and Varcoe, J. R. (2010) A novel reference electrode for application in alkaline polymer electrolyte membrane fuel cells. *Electrochemistry Communications*, **12**, 823 – 825.

[77] Zeng, R. , Slade, R. C. T. , and Varcoe, J. R. (2010) An experimental study on the placement of reference electrodes in alkaline polymer electrolyte membrane fuel cells. *Electrochimica Acta*, **56**,607 – 619.

[78] Tamain, C. , Poynton, S. D. , Slade, R. C. T. Carroll, B. , and Varcoe, J. R. (2007) Development of cathode architectures customized for $H_2/O_2$ metal-cation-free alkaline membrane fuel cells. *The Journal of Physical Chemistry C*, **111**, 18423 – 18430.

[79] Zhou, J. , Unlu, M. , Vega, J. A. , and Kohl,P. A. (2009) Anionic polysulfone ionomers and membranes containing fluorenyl groups for anionic fuel cells. *Journal of Power Sources*, **190**, 285 – 292.

[80] Lu, S. , Pan, J. , Huang, A. , Zhuang, L. , and Lu, J. (2008) Alkaline polymer electrolyte fuel cells completely free from noble metal catalysts. *Proceedings of the National Academy of Sciences of the United States of America*, **105**, 20611 – 20614.

[81] Mamlouk, M. , Scott, K. , Horsfall, J. A. , and Williams, C. (2011) The effect of electrode parameters on the performance of anion exchange polymer membrane fuel cells. *International Journal of Hydrogen Energy*, **36**, 7191 – 7198.

[82] Piana, M. , Boccia, M. , Filpi, A. , Flammia,E. , Miller, H. A. , Orsini, M. , Salusti, F. ,Santiccioli, S. , Ciardelli, F. , and Pucci, A. (2010) $H_2$/air alkaline membrane fuel cell performance and durability, using novel ionomer and non-platinum group metal cathode catalyst. *Journal of Power Sources*, **195**, 5875 – 5881.

[83] Fukuta, K. , Inoue, H. , Chikashige, Y. , and Yanagi, H. (2010) Improved maximum power density of alkaline membrane fuel cells by the optimization of catalyst layer construction. *ECS Meeting Abstracts*, **1001**, 275 – 275.

[84] Isomura, T. , Fukuta, K. , Yanagi, H. , Ge, S. , and Wang, C. -Y. (2010) alkaline membrane fuel cell operated at elevated temperatures. *ECS Meeting Abstracts*, **1002**, 751 – 751.

[85] Oda, K. ,Kato, H. , Fukuta, K. , and Yanagi, H. (2010) Optimization of RRDE method for the evaluation of catalyst activity in alkaline solution. *ECS Meeting Abstracts*, **1001**, 192.

[86] Inoue, H. , Fukuta, K. , Watanabe, S. , Yanagi, H. (2009) *In-situ* observation of $CO_2$ through the self-purging in alkaline membrane fuel cell (AMFC). *ECS Transactions*, **19**, 23 – 27.

# 第3章

# 用于质子交换膜燃料电池的催化剂载体材料

Xin Wang, Shuangyin Wang

## 3.1 引　　言

质子交换膜燃料电池(PEMFC)作为一种环境友好型技术在固定电源、交通运输和移动设备应用方面受到了人们广泛关注[1-3]。在质子交换膜燃料电池中,氢气和各种有机小分子(SOM)可在阳极作为燃料,而氧气作为氧化剂用于阴极,氢气/有机小分子(SOM)与氧气直接电化学转化为水和二氧化碳($CO_2$)产生电力。阳极和阴极的电催化反应在催化剂的活性表面上发生。一般来说,质子交换膜燃料电池常以纳米结构铂和铂基合金用作电催化剂。在燃料电池中,只有发生在三相界面(电极-电解液-燃料/空气)上的电化学反应才对质子交换膜燃料电池的整体性能有作用。

决定燃料电池技术商业化应用的关键因素是它的成本,而其成本主要是与大量贵金属的使用有关。大量研究者一直致力于通过设计电催化剂以及提高整体效率以减少贵金属的使用量。从铂黑到碳载铂催化剂已经大大减少铂需求量。今天在电极上铂的典型负载量为0.4~0.8mg/cm$^2$,这已远远低于早期铂黑的负载量(25mg/cm$^2$)。美国能源部(DOE)设定2010年铂负载量减少到0.3mg/cm$^2$,2015年减少到0.2mg/cm$^2$。

众所周知,催化剂的活性主要取决于铂粒子尺寸大小及其在载体上的分散形式。人们已经发现通过使用理想的载体材料以及适当的制备方法可获得最佳分散形式和铂颗粒尺寸。

理想的催化剂载体应具有如下结构和性能:

(1) 高比表面积和良好的导电性能;

(2) 反应气接触电催化剂的通道;

(3) 良好的耐腐蚀性;

（4）在燃料电池运行条件下高的化学和电化学稳定性。

在碳载金属电催化剂中，金属与碳载体之间的电子相互作用对其电化学性能有重要的影响[4]。对于碳载铂电催化剂，碳能够促进电极 - 电解质界面的电子转移，从而加速电极过程。典型地，在碳基催化剂表面上电子从铂转移到氧物种同时在接触相界面上形成化学键或发生电荷转移过程，而这些被认为有利于提高电催化剂的活性和稳定性。金属催化剂与载体之间的电子相互作用的研究可通过各种物理、光谱和电化学方法从实验上实现。铂对碳基载体的供电子行为已通过电子自旋共振技术（ESR）和 X 射线光电子能谱（XPS）得到验证；铂与载体之间电子相互作用取决于它们之间的电子费米能级。人们认为由铂与载体之间电子相互作用引起的电子结构变化对电催化系统催化性能的增强和稳定性的提高有积极作用。然而，碳载催化剂的电子相互作用与电催化性能之间精确的定量关系至今仍没有建立[4]。

## 3.2　载体材料研究现状以及碳在燃料电池载体中的应用

由于其化学惰性、宽的电化学窗口和优异的电子迁移率，碳材料在作为燃料电池催化剂载体上有着良好的应用前景。炭黑因其巨大的比表面积和较低的价格成为最受欢迎和广泛使用的燃料电池电催化剂碳材料。炭黑包括许多类型，如乙炔黑、Vulcan XC-72 炭黑及 Ketjen 炭黑，这些通常是通过裂解碳氢化合物（如天然气或石油提炼过程中的石油馏分）制备。一般来说，低比表面积的炭黑（如乙炔黑）不能用于制备高度分散的载体催化剂。Antolini 等[5]发现炭黑的分散特性对载体电催化剂金属纳米粒子分散以及后续燃料电池上电催化反应有着重要的影响，Pt 粒径随着炭黑的比表面积增大而减小，高比表面积炭黑（如 Ketjen Black）能负载高度分散的催化剂纳米颗粒，然而，Ketjen Black 型载体催化剂在燃料电池运行时表现出高欧姆阻抗以及有限的传质效率，Vulcan XC-72（比表面积为 $250m^2/g$）则广泛应用于燃料电池催化剂载体。尽管炭黑粒子具有高表面积，但是炭黑作为电催化剂载体存在两个主要问题：①炭黑载体因其致密结构在质量传递中存在严重限制，从而导致极低的 Pt 利用率；②在燃料电池阴极，炭黑表面易发生电化学氧化成氧化物以及二氧化碳。随着炭黑被腐蚀，在炭黑表面的贵金属纳米颗粒（如 Pt）可能会脱离电极并有可能聚合成更大的粒子，这会降低 Pt 的表面积，从而导致燃料电池性能的衰减[6]。

大量的努力一直致力于寻求新的碳材料，主要包括：有序介孔碳（OMC）材料、碳纤维（CNF）、碳纳米管（CNT）、碳纳米角（CNH）、螺旋碳纳米纤维（CNC）、碳凝胶（CAG）、石墨烯。本章首先简要叙述应用于质子交换膜燃料电池催化剂各种可能有前景的可选载体。

## 3.3 新型碳材料作为燃料电池的电催化剂载体材料

### 3.3.1 介孔碳作为燃料电池的电催化剂载体材料

在炭黑表面存在大量的微孔(小于2nm)会导致燃料电池内部燃料传输受阻并使催化剂活性降低。此外,由于炭黑微孔孔隙之间低连通性导致导电性较低。另外,介孔碳(MPC)(2~50nm)被用于燃料电池电催化剂载体。根据结构和形貌,介孔碳材料可为两种类型:一种为有序介孔碳,通常通过模板方法制备;另一种是无序介孔碳(DOMC),这类介孔碳孔结构不规则,导电性低且孔径分布较宽。这两类介孔碳相比,有序介孔碳材料因其高比表面积、高导电性以及孔结构中易传质等特点更适合作为燃料电池电催化剂载体。在有序介孔碳材料上负载的电催化剂在燃料电池电极反应中已表现出优良的性能。有序介孔碳负载电催化剂在燃料电池上拥有良好的催化性能这一点很好理解,这是因为有序介孔碳孔结构的有序性,其作为电催化剂载体能够生成高度均匀分散的金属纳米粒子(电催化剂)从而加快传质。典型的电催化反应,如燃料电池上的电极反应,只会发生在特定的纳米区域,称为三相界面(TPB),是反应物、产物、电解质和电子均可接触和进入的区域。对于传统炭黑作为载体材料来说,由于本身存在很多微孔致使反应物、产物和电解质不易接触,因此很难形成三相界面。另外,大孔碳(孔径大于50nm)因比表面积低以及导电性较低,因此不适合作为催化剂载体材料。考虑介孔碳高表面积、大孔容和高度连通的孔道结构,有序介孔碳(OMC)可使催化剂纳米颗粒均匀分散并有良好的传质(反应物/产物和电解质)[5,7]。

OMC的制备涉及使用具有特殊孔结构的有序介孔二氧化硅(OMS)模板[7]。如图3.1所示,通过浸渍法将合适的碳前驱体(碳源,如蔗糖、糠醇、乙炔气、吡咯、丙烯腈)填充到模板剂孔道内,之后进行碳化得到硅-碳复合物,再使用氢氟酸或氢氧化钠的乙醇-水溶液去除模板得到介孔碳。所制备的介孔碳(OMC)的结构很大程度上取决于使用的模板结构。Chang等[7]已对应用于燃料电池催化剂载体的有序介孔碳(OMC)的合成进行了综述。棒状、管状介孔碳结构可分别通过将碳前驱体填充进模板剂孔道内和将碳源涂抹到模板剂孔壁形成薄膜制备。为了得到结构清晰的OMC,模板剂应该具有三维相互连通的孔道结构。另外,前躯体的碳化应该严格限制在有序介孔二氧化硅模板介孔孔道内,并且孔道须填充足够多的碳前驱体。因此,在热解过程前,应该通过采用酸聚合催化剂将碳源转换为交联聚合物[5,7]。

根据采用不同介孔二氧化硅及硅铝酸盐作为模板,许多介孔碳制备技术已经开发出来[5,7]。自采用具有双连续立方相Ia3d空间群的MCM-48作为硬模板

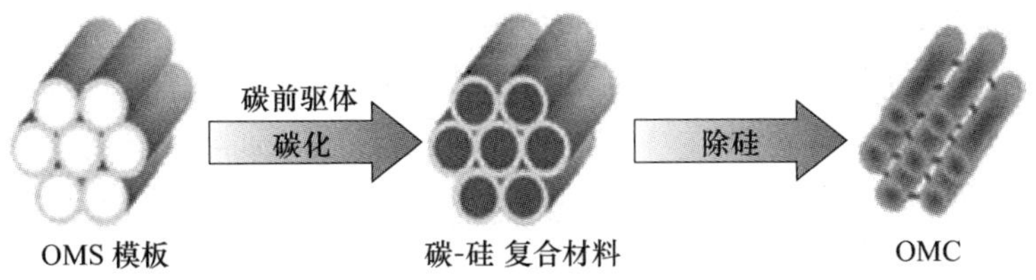


图 3.1　有序介孔二氧化硅(OMS)为模板制备有序介孔碳材料(OMC)合成过程示意图[7]

剂制备出有序介孔碳材料(被命名为 CMK-1)被首次报道之后,各种具有不同孔道结构的 OMC 均有报道。对于这些 OMC,均一的介孔是相互连通的,在 $2\theta<5$ 时会出现不同的 X 射线衍射峰。同时,它们还拥有大的表面积和高孔容。其它的结构参数如孔直径、粒子形态和大小以及碳骨架的微观结构可通过制备过程的合理的实验设计进行调整。Joo 等[8]通过采用有序介孔二氧化硅作为模板成功制备了高度有序、孔径均一可调的刚性纳米多孔碳阵列,去除模板剂后得到部分有序的石墨骨架。制备的 OMC 材料被证明能负载高分散的铂纳米粒子,超过其他常见的微孔碳材料,包括炭黑、木炭、活性炭纤维。在这个载体材料上,电催化剂 Pt 的粒子尺寸可制造到 3nm 并且高度分散,显著提高了燃料电池氧还原反应(ORR)电催化活性[7]。

Ding 等[9]利用 SBA-15 为硅模板成功制备出 CMK-3 有序介孔碳(图 3.2)。将负载有 Pt 纳米粒子的 CMK-3 催化剂应用于燃料电池的氧还原反应,结果发现这种催化剂(Pt/CMK-3)的氧还原电催化活性远高于商业催化剂。通过复制介孔二氧化硅 SBA-15(⊥)模板,Mou 和他的合作者制备出短通道垂直于薄膜的有介孔碳薄膜,并在上面沉积了 Pt – Ru 纳米颗粒。结果证明所设计结构的优点是对甲醇氧化电催化活性有增强作用,短程纳米通道的薄膜碳结构可增加纳米电催化剂的利用效率。Liu 等[11]直接利用 SBA-15 介孔氧化硅作为模板,糠醇、均三甲苯为初始碳源,乙酰丙酮铂和乙酰丙酮钌为共金属源及碳源,在介孔碳材料

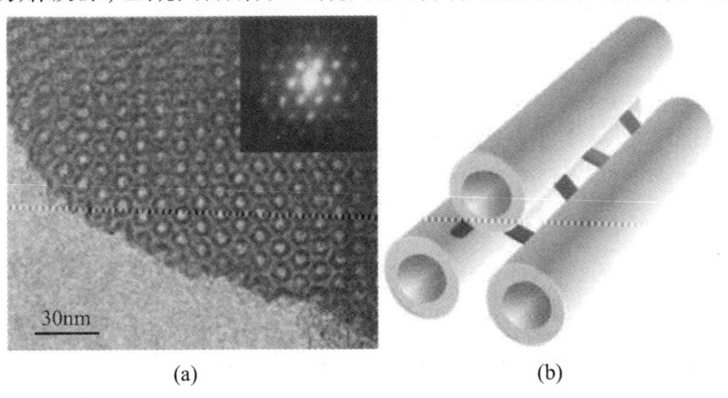


图 3.2　有序介孔二氧化硅 SBA-15 为模板制备有序介孔碳[8]

上制备出高分散、高稳定性、粒径为 2 ~ 3nm 的 Pt - Ru 纳米颗粒（Pt - Ru - CMM）。透射电镜图（图 3.3）显示 Pt - Ru - CMM 具有独特的长程有序介孔阵列，同时发现所有负载型 Pt - CMM、Pt - Ru - CMM 及 Ru - CMM 催化剂、单金属（Pt 和 Ru）Pt - Ru 合金均匀分散镶嵌在碳棒表面。

图 3.3 （(a)，(b)）$Pt_{100}$ - CMM；（(c)，(d)）$Pt_{50}Ru_{50}$ - CMM；（(e)，(f)）$Ru_{100}$ - CMM 的 TEM 图

通过 X 射线吸收光谱法（XAS）证明相对于典型商业电催化剂而言，Pt - Ru - CMM 的高电化学活性归因于高度合金状态的 Pt - Ru 纳米粒子。结果发现 $Pt_{50}Ru_{50}$ - CMM 样品展现出最好的电催化性能和持久稳定性，适用于直接甲醇燃料电池（DMFC）阴极电催化剂。

Yu 等[12]报道了通过诱导苯酚和甲醛不同的聚合过程产物作为碳源，采用胶体晶体模板制备具有有趣形貌变化的独特多孔碳复制体。在这个工作中，只需简单地改变碳前躯体酸催化缩合过程的酸起始位点就能控制形貌。特别地，这些高度有序多孔碳作为催化剂载体能显著改善燃料电池甲醇氧化性能。Zhao 等[13]在不负载任何催化物种的纯介孔 SBA-15 硅模板孔道内用化学气相沉积法（CVD）裂解苯制备石墨化有序介孔碳（GMPC）。结果表明 CVD 方法因高度结晶热解碳的渗透有利于形成高度有序介孔碳石墨孔壁和低的碳收缩。介孔碳作为 Pt 纳米粒子催化剂载体在室温甲醇氧化催化性能测试结果表明介孔碳 Pt 纳米粒子催化性能优于商业 Pt/C（E-TEK）。Joo 等[14]报道了有序介孔碳（OMC）的石墨特性对直接甲醇燃料电池用 OMC 负载的催化剂性能的影响。以菲和蔗糖为碳源，介孔硅为模板采用纳米模板法制备出两种具有六方介孔结构 OMC 样品。结构表征显示两种 OMC 样品具有大的表面积和独特的介孔结构，然而，以菲为碳源制备出的 OMC 展现出更低的薄层阻抗。Pt 纳米粒子在两个 OMC 均具有很高的分散度。在 DMFC 单电池的测试中，OMC 负载的 Pt 催化剂比商业 Pt/C 催化剂表现出更高的电催化性能，这可能归因于 OMC 的高比表面积和独特介孔网络结构。特别地，通过降低 OMC 阻抗，OMC 负载的 Pt 催化剂展现出更优的催化性能[14]。此外，Yan 等[15]发现高稳定性石墨化介孔碳可通过在 2600℃ 热处理聚合物模板得以制备，将石墨化介孔碳基 Pt 催化剂（Pt/GMPC）电化学稳定性与炭黑基 Pt/C 催化剂（（Pt/XC-72）进行对比，Pt/GMPC 电化学稳定性优于 Pt/XC-72。

具有高孔容和高表面积的碳气凝胶被开发作为 Pt 载体，用于 $H_2$/空气质子交换膜燃料电池阴极膜电极。研究表明，具有不同初始孔大小的碳气凝胶表现出有类似的动力学活性，但却有不同的扩散极化损失：碳气凝胶孔隙越大，传质极化越高[16,17]。然而，由于碳气凝胶无定形特性，催化剂的化学稳定性是有限的。

氮掺杂碳材料在催化活性和稳定性方面展现出良好的催化剂载体性质[18,19]。具有三维扩展高度有序孔阵列的石墨化氮化碳作为直接甲醇燃料电池阴极 Pt－Ru 合金催化剂载体已经被报道。纳米结构 $C_3N_4$ 比商业炭黑 Vulcan XC-72 能量密度高 73%～83%。同有序介孔碳（OMC）相比，这种有序大孔 $C_3N_4$ 虽然比表面积较低，但是由于它更高的石墨化以及骨架氮加强电子转移速率，因而具有更高的电导率和更好的电催化性能[20]。

极高稳定性导电掺硼金刚石作为质子交换膜燃料（PEMFC）电池催化剂载体具有吸引力。掺硼金刚石作为催化剂载体对 ORR[21] 以及甲醇电化学氧化展现出优越的稳定性[22]。然而，掺杂金刚石作为电催化载体仍然存在一些问题，即低的电导率，低的比表面积和金属粒子分散不均匀。此外，难以实现在金刚石粉末上掺杂硼均匀且可控[23]。

### 3.3.2　石墨纳米纤维作为燃料电池电催化剂载体

石墨纳米纤维（GNF）因具有良好的石墨化结构及高的导电性在作为载体方面的应用引起了广泛的关注[6,24]。许多不同种类的石墨纳米纤维如薄片、带和鱼骨型被用作燃料电池电催化剂载体材料。GNF 通常由在金属表面裂解含碳气体制备。在石墨纳米纤维（GNF）结构中，基底平面只在边缘地区暴露。另外，不同的石墨纳米纤维具有不同的空腔。由于其独特的结构，石墨纳米纤维（GNF）可以直接用于负载金属纳米粒子而不需要任何预处理，这通常会对石墨纤维完美结构有影响，这是由于 GNF 的薄片及鱼骨型结构可作为金属沉积潜在的活性位点。许多不同的合成方法已经被开发用于石墨纳米纤维上负载金属纳米电催化剂，在燃料电池方面展现出潜在应用前景。Gangeri 等[24]采用等体积浸渍方法在石墨纳米纤维沉积 Pt。相对于商业 Pt/C 性能而言，在碳纳米纤维负载 Pt(Pt/GNF)的性能可作为质子交换膜燃料电池电极材料的替代材料。纳米碳纤维在两种不同类型的微形貌碳载体(毡和布)通过化学气相沉积生长，之后在这些纳米/微米碳载体上沉积 Pt。分析测试结果，特别是极化曲线表明：①Pt/GNF 材料比商业材料的电催化性能更好；②Pt/GNF 材料传质损失最低。电化学测试表明，基于 Pt 纳米簇及碳纳米结构的新电极材料在燃料电池应用上可能会展现出广泛的前景。此外，纳米纤维石墨微观结构对于影响负载型催化剂电催化性能影响的研究肯定会具有吸引力。Bessel 等[25]以甲醇电化学氧化作为探针反应发现了石墨纳米纤维负载型 Pt 催化剂作为电极在燃料电池应用上的潜力。许多不同类型的石墨纳米纤维被应用并且在这些材料和 Vulcan carbon (XC-72)上负载 Pt 纳米颗粒性能进行了对比。薄片及带状型石墨纳米纤维负载 5%(质量分数)Pt 的纳米催化剂对反应物主要在边缘暴露，表明 Vulcan carbon (XC-72)负载约为 25%(质量分数)Pt 的催化性能。此外，同传统催化剂体系相比，石墨纳米纤维负载型金属颗粒更耐 CO 中毒。这种改善的性能被认为是与金属颗粒被分散在高度裁剪的石墨纳米纤维结构上采用了特定晶体取向有关。

Lukehart 等[26]成功制备出 Pt - Ru 石墨纳米纤维复合材料，这种材料在直接甲醇燃料电池上展现出相对高的性能。采用含 Pt 和 Ru 金属单一分子前驱体在鱼骨型石墨纳米碳纤维上多步沉积和活化分解提供了 GNF 载体上富含 Pt - Ru 合金纳米簇 Pt - Ru/GNF 复合材料，这种纳米复合材料总金属含量为 42%(质量分数)(Pt/Ru 原子比例约为 1:1)，且通过 TEM 图片直接测的合金簇的平均粒径为 7nm(图 3.4)。制备的复合材料的 XRD 和电化学分析及储存在常压

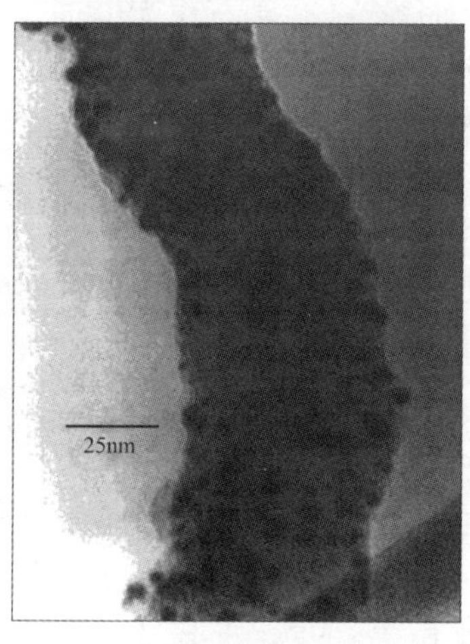


图 3.4　Pt - Ru/GNF 纳米颗粒的亮场 TEM 图[26]

条件下表明存在少量的 Ru 金属及氧化物金属物种。将这种纳米复合材料作为直接甲醇燃料电池阳极催化剂和没使用载体且具有类似比表面积和催化颗粒尺寸的 Pt – Ru 胶体相比,结果显示 Pt – Ru/GNF 纳米复合材料的性能提高了 50% 。

最近,Kang 等[27]成功制备了表面改性石墨纳米纤维作为燃料电池阴极催化剂载体。利用催化生长碳纳米纤维高温石墨化法合成了 GNF。与碳纳米管相比,石墨化碳纳米纤维展现出暴露的石墨化边缘。石墨化后,在 GNF 上相邻石墨边缘将很大程度上结合,导致 GNF 表面重构。制备的 GNF 展示出高度石墨化和与 Pt 纳米粒子的强相互作用, 从而 DMFC 阴极性能得到提高。TEM 图像如图 3.5 所示,原始 GNF 展现出典型的鱼骨状结构,纳米纤维的轴向石墨化程度降低。石墨化后,在 GNF 上的石墨层结合良好,(图 3.5(c)及(d)),结果表明更高的石墨化,此外,从图 3.5(d)可知,在原始 GNF 表面所暴露的石墨边缘很大程度上变成了许多的环,这表明了在石墨化过程中的结构的重构。

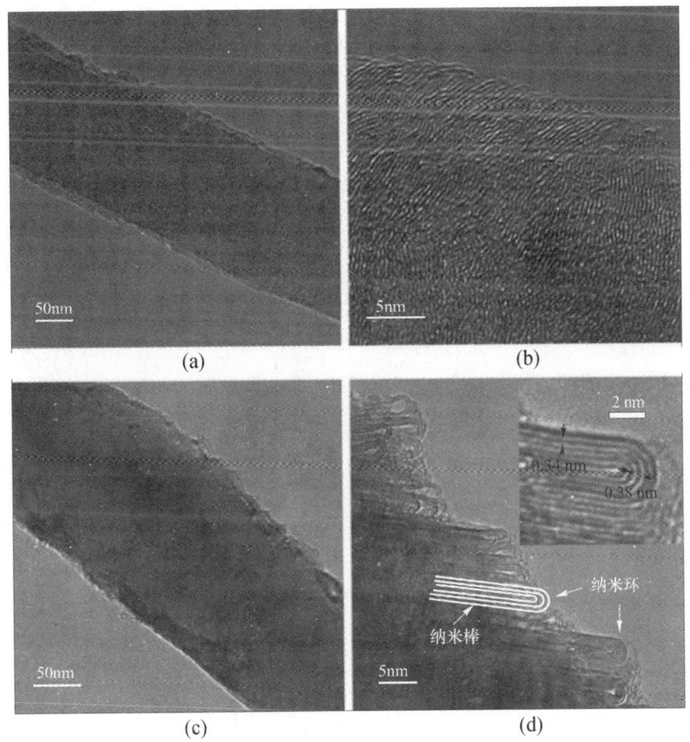


图 3.5 ((a),(b))GNF 和((c),(d))表面重构的 GNF 的 TEM 及 HRTEM 图片,(d)中的插图为形成的纳米环的放大图[27]

由于石墨边缘具有较高表面能的碳悬空键,相邻石墨为了降低表面能压缩边缘形成这些纳米环。由于石墨纳米纤维表面的重建,3nm 大小的 Pt 纳米粒子仍然可以在高度石墨化的 GNF 载体上均匀分散,且无需任何进一步的预处理。

与 Pt/CNF 和 Pt/Vulcan 催化剂相比，Pt/GNF 催化剂在单 DMFC 性能测试中展现出更好的性能，这可能是因为 GNF 高石墨化程度产生更高的导电性及 Pt/GNF 疏水界面增强。GNF 高石墨化程度同时也提高了 Pt/GNF 在长期使用中的稳定性[27]。

### 3.3.3 碳纳米管作为燃料电池的载体材料

碳纳米管因具有独特的电子及结构性质，在很多应用研究中引起广泛的关注，如超导体、储氢、场发射及非均相催化剂。由于具有高表面积、优良的电导率和高的化学稳定性，碳纳米管被广泛研究作为燃料电池 Pt 和 Pt 合金催化剂的载体材料。由于原始碳纳米管的化学惰性，为了锚定和沉积催化纳米颗粒，对碳纳米管石墨表面进行活化是必要的[28-33]。在碳纳米管上负载的金属纳米粒子的沉积、分布和大小很大程度上取决于碳纳米管表面处理及表面性质。Pt 纳米粒子的活性也显著受碳纳米管与 Pt 相互之间本质作用和碳纳米管初始性质的影响。有许多综述报道了碳纳米管的合成及作为一种新型催化剂载体的特征。碳纳米管的合成涉及碳源物质在气或固相中催化分解[5,33]。制备碳纳米管典型的技术包括 CVD、电弧放电及激光蒸发合成。制备的碳纳米管通常有高相对分子质量并形成强疏水表面，这通常导致碳纳米管上金属分散不匀及限制碳纳米管负载型电催化剂整体电催化性能[5,6,33]。金属纳米电催化剂的电化学活性强烈依赖于碳载体材料上电催化剂的分散和粒子尺寸。另外，碳纳米管负载金属纳米电催化剂的分散、分布和颗粒大小主要受碳纳米管表面性质的影响，这可通过不同的功能化方法改善。对碳纳米管的表面功能化进行了广泛的研究活动。基本上，碳纳米管表面功能化可分为两类，即共价和非共价改性。碳纳米管的共价表面改性包括材料表面的永久改性，这样功能化活性基团后可与另一个分子形成共价键。非共价表面改性并没有涉及一个分子与碳纳米管表面之间的化学键的形成，这些作用力包括范德华力、静电力、氢键以及其他的吸引力[30-33]。

作为最常用的功能化方法，剧烈氧化过程通常被用于碳纳米管功能化，例如经在浓硫酸及浓硝酸中回流在碳纳米管管壁及管节产生官能团。由于这种酸处理，碳纳米管表面的完整芳香共轭环结构可被摧毁。相应地，碳纳米管可被许多官能团功能化，如羟基、羧基和羰基，这些官能团与金属离子及金属纳米粒子有强的相互作用及锚定能力。Yu 等[34]用混合酸（$HNO_3-H_2SO_4$）处理碳纳米管。他们推导了 Pt 在碳纳米管上的沉积机理，如图 3.6 所示。当碳纳米管用混合酸（$HNO_3-H_2SO_4$）回流，石墨层表面与氧化剂反应产生了高密度的表面官能团。当 Pt 离子被引入到体系中时，它们可以相互作用及通过离子交换或者配位与表面官能团结合并提供成核前驱体。采用氢气还原表面 $Pt^{2+}$ 后，Pt 金属纳米颗粒在碳纳米管表面上可获得良好分散沉积[33,34]。Xu 等[35]报道了处理方式对 Pt 纳米粒子沉积的影响。他们发现，硫酸和硝酸的混合溶液回流之后，用过氧化氢

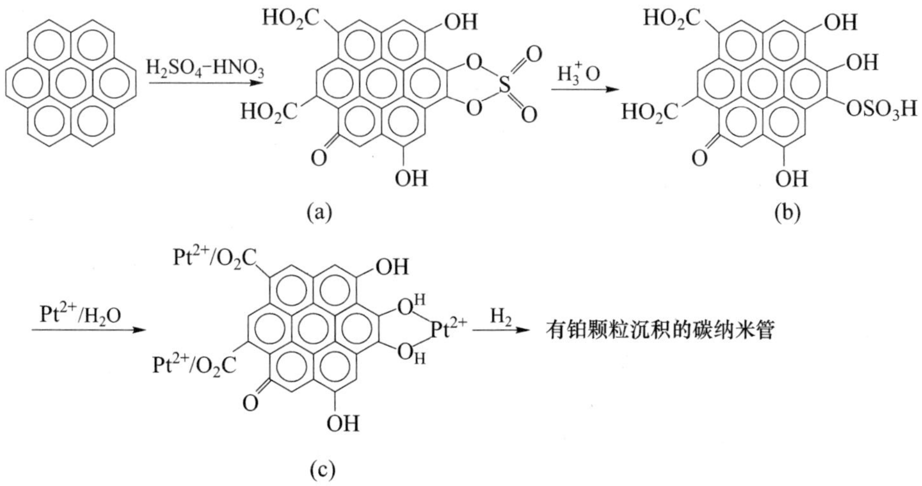

$H_2SO_4$-$HNO_3$ →（a）$H_3^+O$ →（b）

$Pt^{2+}/H_2O$ →（c）$H_2$ → 有铂颗粒沉积的碳纳米管

图3.6　碳纳米管的功能化和Pt纳米粒子的沉积[34]

溶液浸渍是在CNT表面上沉积Pt纳米粒子的有效预处理方法（称为混合过程）。采用这种方法，在碳纳米管上可获得3nm尺寸大小的Pt纳米粒子。同其他纯酸处理相比，采用混合处理过程得到的纳米粒子展现出最好的催化性能，这表明它们在碳纳米管上尺寸更小、分布更加均匀[33,35]。

采用类似的酸处理过程（硫酸氧化），超声化学辅助处理也被开发出来作为碳纳米管表面官能化有效的方法[33,36]。例如，Xing等[36]发现Pt纳米粒子在超声化学处理过的碳纳米管上可均匀沉积，通过图3.7透射电镜图可证明。在质子交换膜燃料电池中，所制备的Pt/CNT电催化剂比炭黑负载的催化剂具有更高的催化活性。

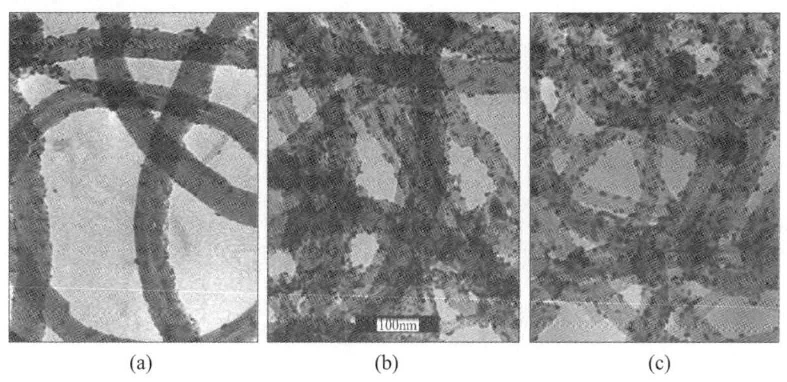


(a)　(b)　(c)

图3.7　超声化学处理碳纳米管沉积Pt纳米粒子的TEM图（图(a)～(c)中Pt负载量分别为10%（质量分数）、20%（质量分数）、30%（质量分数））[36]

尽管共价官能团化已经被广泛开发，但是由于化学氧化法在碳纳米管表面引入了大量的缺陷降低了碳纳米管材料的导电性及抗腐蚀性能。碳或碳纳米管载体的腐蚀已被确认为在燃料电池测试中 Pt 电催化剂电化学活性表面损失和耐久性降低的主要原因之一[30-33]。因此,对于碳纳米管而言,有必要开发一个更好、更有效的功能化方法,这种方法不仅可以引入高密度和均匀表面官能团,而且结构性破坏很少或几乎没有。Hsin 等[37]报道了一种采用 PVP 原位聚合官能团化碳纳米管作为载体沉积 Pt 和 Pt - Ru 纳米颗粒。其他人在吸附 Au 凝胶纳米颗粒前用聚苯乙烯磺酸钠(PSS)及聚二烯丙基二甲基氯化铵(PDDA)包覆碳纳米管。但是为了达到一个更好的聚合电解质吸附,在聚电解质包覆之前,碳纳米管通常需要用酸处理,这就会在碳纳米管上不可避免地引入表面缺陷及导致结构毁坏。

最近,碳纳米管的非共价键官能团化因其能够对纳米粒子和碳纳米管混合物的性质进行调控同时仍然保留碳纳米管几乎所有的内在属性而引起特别关注。Wang 等[38]采用含芘的分子作为交联剂将半导体与金属纳米粒子在多壁碳纳米管上进行复合,但是实验中纳米粒子的集聚体的形成表明纳米颗粒在碳纳米管上的分散不令人满意。Correa-Duarte 等[39]采用 PSS 和 PDDA 多层修饰的碳纳米管作为模板负载 Si 包覆的 Au 纳米粒子。Sacher 等[40]通过与苄硫醇之间 π—π 键的结合成功制备了巯基官能团化的碳纳米管。如图 3.8 所示,官能团化的碳纳米管表面通过形成 Pt—S 键与 Pt 纳米粒子产生强相互作用,从而有一个高 Pt 纳米粒子负载量(高分散和窄的粒径分布)。同其他催化剂相比,所制备的 Pt/MWCNT 展现出更高的电催化活性和耐 CO 中毒。这种有前景的制备方法可拓宽至在碳纳米管上复合其他贵金属催化剂用于燃料电池[32]。

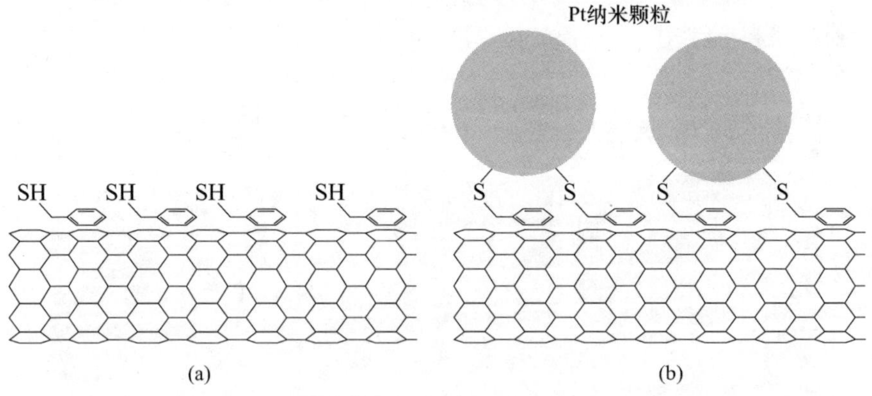


图 3.8　苄硫醇功能化的 CNT 及后续沉积 Pt[41]

Wang 等[30]采用聚合电解质官能团技术功能化碳纳米管。带正电的水溶性聚季铵盐 PDDA 在水溶液中被用于包覆 CNT。这非共价键功能化不但在碳纳米管上产生高密度和均匀分散的表面官能团,而且不破坏多壁碳纳米管的完美表面结构性质并保留其原始的性质。因为碳管表面带正电荷,大量的带负电荷的 Pt 前体可以通过静电相互作用锚定到 MWCNT 表面。随后在这些高度功能化的碳纳米管用乙二醇还原得到均匀分布和高密度的Pt纳米粒子。

Wang等[31]的进一步工作表明这种官能团化方法可以在碳纳米管表面得到高密度且均匀引入的官能团,这可以在碳纳米管上允许高负载量的金属纳米粒子。电催化活性与Pt纳米粒子在碳纳米管上相互联结之间的关系已经得到验证。结果发现,Pt/CNT催化剂的电催化活性基本与Pt纳米粒子在碳纳米管上相互联结有关。Pt纳米粒子相互联结数量是影响电催化活性的关键因素,相互联结的Pt纳米粒子比独立分散的纳米粒子具有更高活性。高联结的Pt纳米粒子的高催化活性被认为与电子边界活性的增加有关,这显著提高了Pt纳米粒子的电催化活性。另外,相互联结的Pt纳米粒子能够显著削弱含氧物种(如$CO_{ad}$和$OH_{ad}$)的化学吸附,从而提高CO、甲醇氧化和氧还原的电催化性。增加相互联结的Pt纳米粒子可以降低粒子电子传输的界面阻抗[31,41]。

类似地,含有各种特征官能团的聚合电解质作为交联剂锚定Pt纳米粒子被用于功能化碳纳米管作为Pt电催化剂载体。Pt纳米粒子与碳纳米管之间交联剂在甲醇氧化电催化活性的影响已被研究,如图3.9所示[42]。结果发现,离子状态的聚苯乙烯磺酸钠(PSS)及聚丙烯酸钠(PAA)对在Pt纳米粒子负载碳纳米管上的甲醇电氧化有利,X射线光电子能谱(XPS)证明这主要是通过将聚阴离子的电荷转移到Pt位置和含氧物种的供应修饰它们的电子结构。来自聚阴离子电荷转移到Pt位置,Pt周围的电子密度增加会引起Pt 5d带部分填充,从而导致d带中心减速和减弱含氧物种(CO等)的化学吸收作用。在Pt纳米粒子上化学吸收作用减弱将促进甲醇电氧化。然而,聚阳离子会对Pt纳米粒子的电子结构和化学吸收作用性质有不利影响。另外,长期稳定性测试表明阳离子功能化的碳纳米管作为Pt载体通过Pt纳米粒子与碳纳米管的静电吸附强相互作用会增强其稳定性[42]。

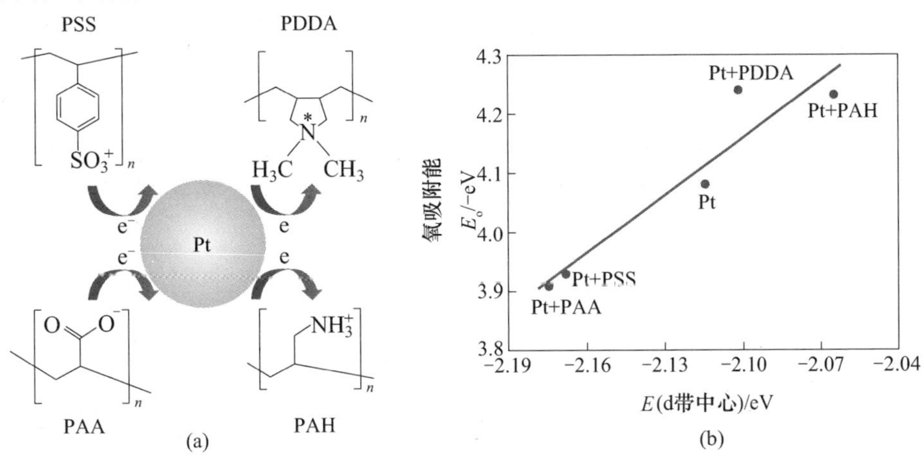


图3.9 (a)聚合阴离子(PSS和PAA)和聚合阳离子(PDDA和PAH)的带电基团对Pt NP的供电子-接受电子行为的可能影响以及(b)O吸附能与Pt片d带中心之间的关系[42]

最近,双官能化分子已经被用于改性电催化剂载体碳纳米管表面[32]。它涉及1-氨基芘分子(1-AP)在多壁碳管表面的吸附。芘官能团(本质上高度芳香性)被认为通过π键堆叠与石墨底面产生强相互作用力。类似的作用,1-氨基芘分子(1-AP)的芘官能团可与多壁碳纳米管的管壁产生强相互作用,从而把1-氨基芘分子固定在多壁碳纳米管上。当溶液pH值控制在弱酸性时,锚定在碳纳米管上的1-氨基芘的氨基官能团被质子化并带微弱的正电荷。这会与带负电的$PtCl_6^{2-}$产生静电作用力,之后同带正电荷的$Ru^{3+}$在1-氨基芘分子功能化的多壁碳纳米管后续自组装。在乙二醇作还原剂条件下通过微波辅助多元醇处理还原Pt-Ru前驱体,在多壁碳纳米管上形成了Pt-Ru纳米颗粒。这些表面官能团也可能作为锚定位直接沉积还原的金属纳米粒子,这些金属纳米粒子通常带负电荷。强化学酸主要是用于在多壁碳纳米管上产生碳氧活性位,与酸氧化多壁碳纳米管不同,1-氨基芘分子(1-AP)功能化处理保持了多壁碳纳米管的完整性和电子结构。如图3.10所示,尺寸分布窄的金属纳米粒子均匀沉积在1-AP-CNT上,这归因于碳纳米管表面独特均匀分布的官能团。粒子的平均尺寸为2nm,并没有集聚体发生,甚至高Pt-Ru负载量条件下也是如此。然而,在酸氧化的碳纳米管上,Pt-Ru纳米粒子因强酸氧化引起较差的官能团分布,趋于形成大颗粒的集聚体。Pt-Ru电催化剂密度可以通过调整金属前体的加料浓度有效控制。因此,同酸处理的多壁碳纳米管(AO-MWCNT)相比,在1-氨基芘分子(1-AP)功能化的碳纳米管上的Pt-Ru电催化剂有更高的电化学比表面、更好的活性以及在酸性溶液中对甲醇电化学氧化更高的稳定性[32]。

基于以上例子可知,碳纳米管的非共价功能化可以轻易有效地使碳纳米管具有高密度的特殊官能团。这些官能团可作为活性位锚定金属前驱体或者金属纳米粒子,这会产生分散度好、分布窄以及负载量可控的金属纳米粒子/CNT电催化剂。除了物理作用,即影响分散、尺寸和催化剂纳米颗粒负载量,特定官能团可能会与催化剂纳米粒子相互作用并影响其内在电化学活性。现在,大部分调控活性的策略局限于Pt与其他金属的调整,然而,很少有人关注能够联结金属纳米粒子和载体材料的锚定官能团,这也许同等重要,这个方向是值得研究的。因此,对于采用尺寸形貌可控Pt基纳米粒子及其他纳米催化剂修饰碳纳米管而言,非共价功能化策略是一个很有吸引力的方法。

螺旋碳纳米纤维及碳纳米管构成一个新类别碳纳米材料,明显区别于其他形式的碳。螺旋碳纳米纤维的结构类似于多壁碳纳米管(螺旋形状除外)。同Vulcan XC-72型炭黑基催化剂相比,螺旋碳纳米纤维基催化剂展现出更好的电催化性能。尤其是螺旋碳纳米纤维上负载的Pt-Ru合金催化剂拥有更好的结晶度及大比表面积,同螺旋碳纳米纤维负载其他催化剂相比,展现出优越的电催化性能[43]。在富勒烯纳米簇电化学沉积Pt后,富勒烯($C_{60}$)膜电极也可作为甲醇氧化催化剂载体[44]。

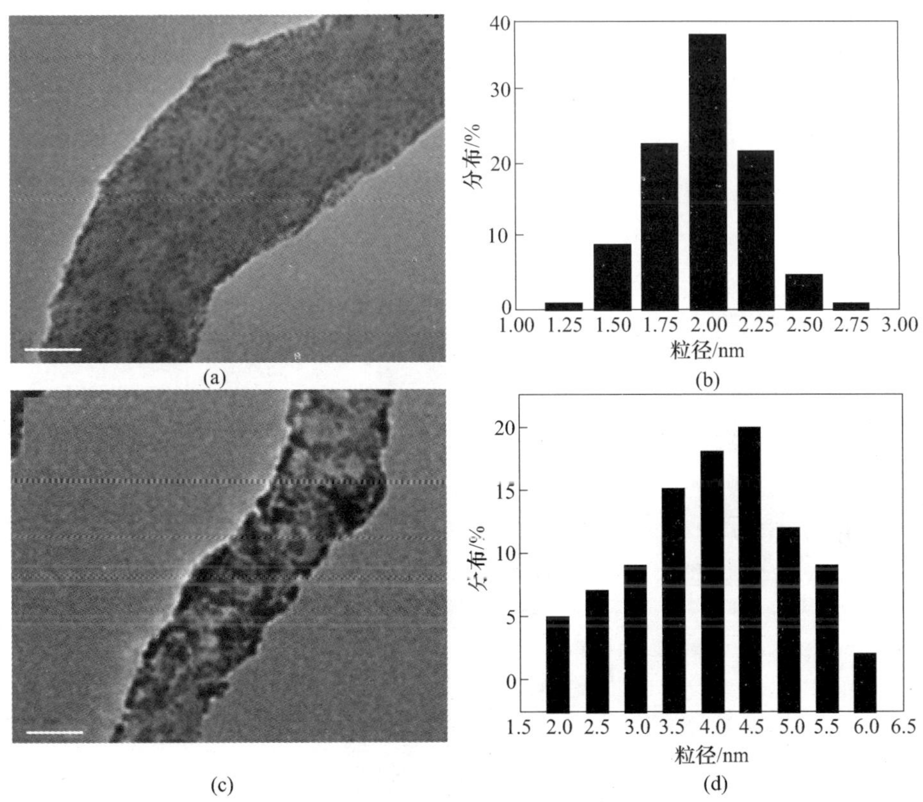


图3.10 ((a),(b))1-AP-MWCNT和((c),(d))AO-MWCNT上Pt-Ru纳米颗粒TEM图和分布直方图(Pt-Ru负载量为40%(质量分数))[32]

### 3.3.4 石墨烯作为燃料电池的载体材料

石墨烯是一种理想的催化剂载体,主要是由于其高导电性、优良的力学性能、高比表面积、独特的石墨基底平面结构和潜在的生产成本低[45,46]。Yoo等[47]在石墨烯阵列(GNS)上沉积Pt亚纳米簇,显著改善了Pt纳米簇电催化剂的性质并且对甲醇氧化反应具有极高的活性。在所测试的催化剂样品中,Pt/GNS电催化剂对CO氧化具有不同的特征。结果发现,在石墨烯阵列(GNS)形成低于0.5nm的Pt纳米粒子,从而获得特定的电子结构Pt,改善其催化活性。Sharma等[48]报道了采用微波辅助多元醇还原法在还原氧化石墨烯上负载Pt电催化剂。因为在边缘以及基底两侧存在多种氧官能团(O-基因),可设想在还原石墨烯阵列表面锚定良好分散的纳米簇。这个构想可通过高倍TEM验证(图3.11(b))。值得注意的是这些锚定在RGO上的Pt NP也可阻止层和单个阵列构建功能分离之间的π—π键堆叠。这个体系在作为甲醇电氧化阳极材料方面有潜在的用途。与商业化的碳基 Pt 电催剂相比，Pt/RGO 展现出优越的抗 CO 中毒性、高电化学活性比表面以及对甲醇氧化反应高质量催化活性，分别增加了 110% 、134% 、和 60% 。结果发现，在还原氧化石墨烯上高密度的氧官能团在去除相邻 Pt 位置上的碳物种起着重要的作用，表明石墨烯上的氧物种与 Pt 纳米粒子存在协同作用。目前，微波辅助合成 Pt/RGO 提供了一种制备优越电催化活性及抗 CO 中毒性电催化剂的方法，这在能源相关应用中具有重要的意义[45]。

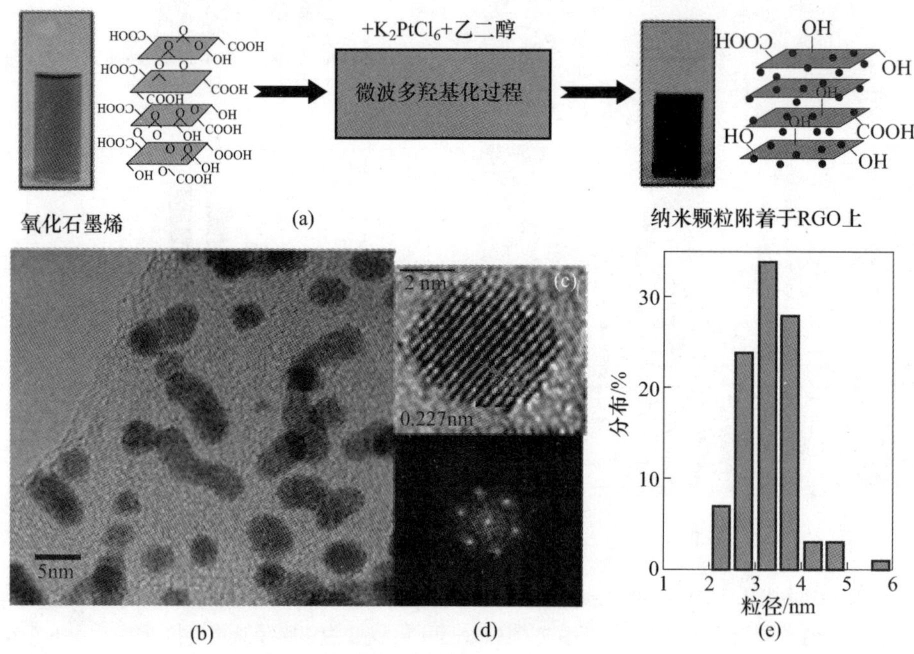


图 3.11 (a)Pt/RGO 复合物制备示意图；(b)Pt/RGO 的 TEM 图；(c)HRTEM 图；(d)单个 Pt NP 的 FFT；(e)在 Pt/RGO 上 Pt NP 尺寸分布[48]

最近，Lin 等[49]报道了在石墨烯上沉积金属氧化物和金属纳米粒子的新方法以及形成了稳定的金属 - 金属氧化物 - 石墨烯三元结合体应用于燃料电池电催化剂(图 3.12)。他们首先直接在功能化的石墨烯阵列上合成锡铟氧化物(ITO)纳米晶，从而形成锡铟氧化物 - 石墨烯杂化物。之后，沉积 Pt 纳米粒子，形成一个独特的三联点结构(Pt - ITO - 石墨烯)。他们的 DFT 计算证明 Pt 纳米粒子在金属氧化物 - 石墨烯结合体上沉积从热力学上来说有优势且稳定。石墨烯上的缺陷和功能官能团有助于提高催化剂的稳定性。Pt - ITO - 石墨烯纳米复合材料在质子交换膜燃料电池应用上可作为潜在的氧还原电催化剂。ITO - 石墨烯混合基质具有金属氧化物和石墨烯阵列所期望的性质。石墨烯阵列功能化作为基体可提供高比表面积以及显著地增加导电性。ITO 纳米粒子从本质上能够分散及保护石墨烯不被腐蚀，从而改善基体耐用性。在纳米复合材料中

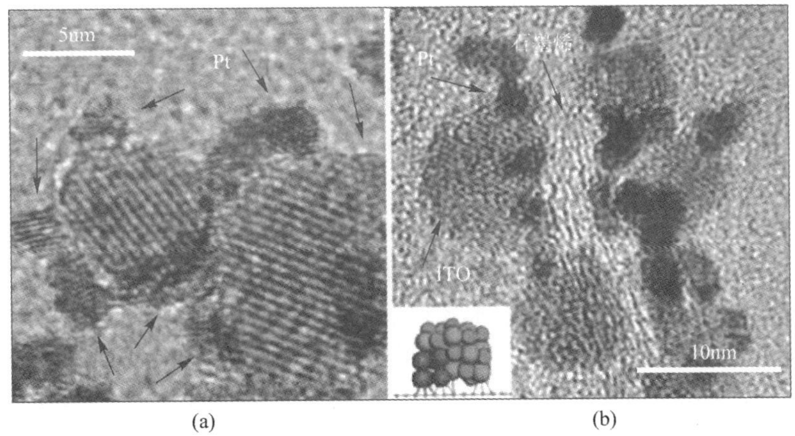


(a) (b)

图3.12 Pt-ITO-石墨烯的TEM图:(a)石墨烯平面;(b)石墨烯截面[49]

独特的三元结合结构具有高表面积、良好的金属分散以及良好的导电性较适合应用于质子交换膜燃料电池。电化学测试结果表明,ITO-石墨烯复合物基体负载的Pt的性能,尤其是耐用性不但比石墨系列负载的Pt更好而且比其他广泛使用的碳材料负载Pt电催化剂都好(如导电碳黑Vulcan XC-72以及碳纳米管)[49]。

Wang等[45]在非共价功能化石墨烯载体上同时使用Pt和Au纳米颗粒组装制备出DFAFC用高效甲酸氧化电催化剂。基于带正电PDDA碳纳米管的非共价功能化也是一个不破坏石墨烯电子特性的石墨烯功能化非常有效的方法。不同比例的Pt和Au纳米粒子在PDDA功能化石墨烯上同时自组装具有高均匀性和可控密度和组成,在PDDA功能化石墨烯上(或Pt-Au/PDDA-G)形成Pt和Au纳米粒子电催化剂用于甲酸(HCOOH)氧化反应。Pt和Au混合纳米粒子在PDDA功能化石墨烯自组装的原理如图3.13所示。结果表明Pt-Au/PDDA-G电催化剂对甲酸(HCOOH)氧化展现出卓越的电化学活性,在Pt1Au8/PDDA-G(Pt:Au原子比为1:8)上甲酸氧化质量电流密度是Pt/PDDA-G催化剂的32倍[45]。

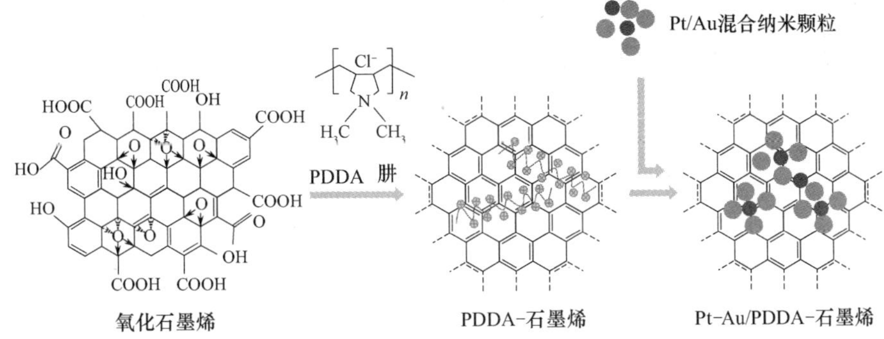


图3.13 PDDA官能化石墨烯上的Pt和Au纳米颗粒自组装[45]

### 3.3.5 掺氮碳材料

掺氮碳材料(如碳纳米管及石墨烯)被认为是 Pt 催化剂的良好载体。碳纳米管掺杂其他元素(如氮)可能是改善它们电气和力学性能的特别有趣的方式。引入氮元素改善碳纳米管的结构,会产生:①高比表面积;②高密度缺陷;③化学活性异质位及④窄管(随着氮的掺入壁的数目减小)[33]。

掺杂氮原子不仅可提供金属离子锚定位也可作为燃料电池反应化学活性位点。氮掺杂碳纳米管(N-CNT)上的氮位点被报道与金属有很强的键能,从而在金属/N-CNT 材料上产生很好的分散度。在碳纳米管掺杂氮诱导表面改性能够在许多催化应用中加强碳载体催化剂反应活性和选择性[33]。因此,它应该有可能避免功能化过程使用强酸处理,从而相对很容易在 N-CNT 沉积金属催化剂[33]。

氮功能化碳纳米管载体通过与氮位点 $sp^2$ 杂化轨道平面上的孤对电子形成键从而决定了 Pt 粒子尺寸。在未经处理的炭黑上这些 N 位点比氧位点电负性更负。在催化过程中的假设 Pt 也许与吡啶位结合力更强,从而在未经处理的炭黑上可在一定程度上防止 Pt 粒子烧结。吡啶型氮功能化对 Pt 离子增强的供电子能力也许有利于提高甲醇氧化动力学[18,19]。

最近,N-CNT 被报道在碱性条件下对 ORR 有较好的催化活性(图 3.14)[51]。N-CNT 的 ORR 活性及形貌和性质直接与碳氮前驱体类型及助催化剂的使用有关系。这种电催化活性直接归因于氮掺入到碳纳米管(CNT)石墨结构,从而增强了结构及电子性能。最近,Chen 等[52]利用氮掺杂碳纳米管作为铂纳米颗粒载体材料及阐明了氮前驱体溶液对 N-CNT 生长的影响。与未掺杂的 CNT 相比,采用含氮丰富的乙二胺前驱体溶液(ED-CNT)制备的 N-CNT 被

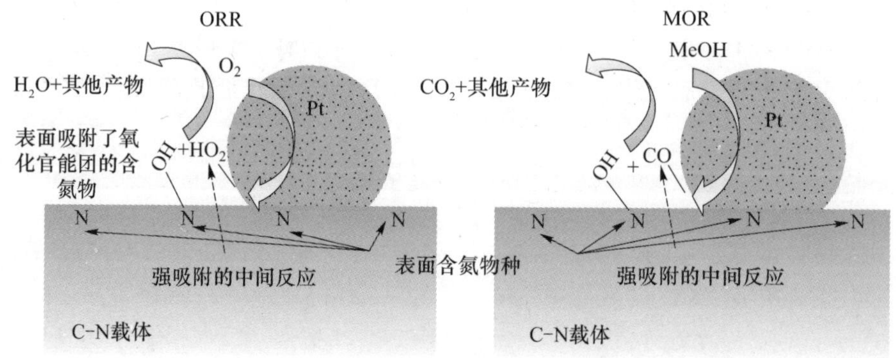


图 3.14 低密度 Pt 催化剂条件下在 C/Pt 催化剂颗粒表面附近的嗜氧性 C—N 的双功能 ORR 和 MOR 机理示意图(不按比例)。表面含氧物种的吸附促进了强烈吸收中间反应物种的反应,否则阻塞催化剂活性位点,从而增加了电化学反应的净转换频率[50]

发现对 ORR 有卓越的催化性能。与未掺杂的 Pt/CNT 相比,当用作铂纳米颗粒载体时,Pt/N－CNT 对 ORR 展现出更高的电催化活性,性能的提高归因于不同氮掺杂产生的独特结构和电子的增强。结果发现 Pt/N－CNT 作为质子交换膜燃料电池阴极催化剂的性能高于 Pt/CNT。Pt/N－CNT 及 Pt/CNT 在转压法制备的单电池 $H_2/O_2$ 膜电极系统中极化和能量密度曲线如图 3.15所示。Pt/ED－CNT 的峰电流密度是 1.04W/cm$^2$,比 Pt/CNT(0.89W/cm$^2$)增加了约 16.9%。在电池电压为 0.6V 时,Pt/ED－CNT 展现出的极限电流为 1.55 A/cm$^2$,比 Pt/CNT(1.25A/cm$^2$)高 24%。Pt/ED－CNT 与 Pt/CNT 相比,在燃料电池条件下展现出更高的 ORR 催化性能。

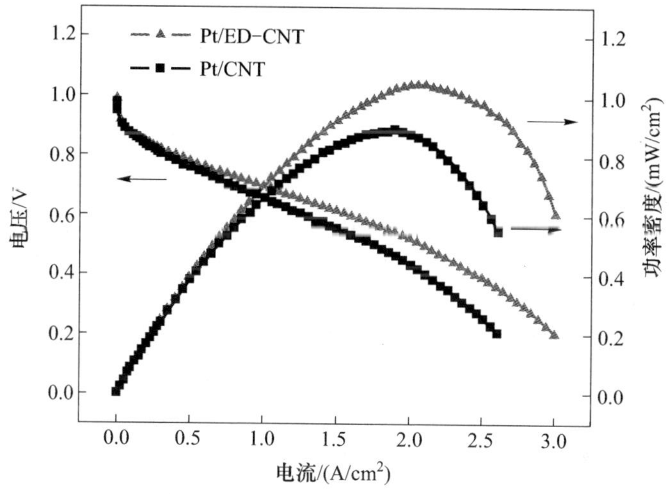


图 3.15　单电池 $H_2/O_2$ 全氟磺酸 112 PEM 体系商业 Pt/C 负载量为 0.2mg$_{Pt}$/cm$^2$ 作为阴极催化剂,Pt/ED－CNT 及 Pt/CNT 负载量为 0.2mg$_{Pt}$/cm$^2$ 作为阴极催化剂膜电极极化和能量曲线[52]

Ramaprabhu 等研究了氮掺杂石墨烯纳米片作为 Pt 催化剂载体应用于 ORR[53]。实验中通过使用氮等离子体处理石墨烯制备氮掺杂石墨烯纳米片,用 Pt/N－石墨烯制得膜电极作为 ORR 催化剂展现出最大功率密度为 440mW/cm$^2$,而 Pt/石墨烯制得膜电极最大功率密度为 390mW/cm$^2$。Pt/N－石墨烯作为 ORR 催化剂改善的性能归因于氮掺杂后在相邻碳原子的五边形及六边形的形成及活性的增强。据报道,氮掺杂在石墨烯堆叠引起无序,以及这些无序结构可以作为 Pt 粒子沉积锚定位点[53]。

## 3.4　导电金属氧化物作为载体材料

除了传统的碳材料之外,许多不同的导电金属氧化物,包括 $TiO_x$、$WO_x$、

$SnO_2$、$IrO_2$、ITO 等也可以用作电催化剂载体。在燃料电池环境中,导电氧化物具有抗腐蚀性、很好的热稳定性及电化学稳定性。金属氧化物用作载体材料具有潜在的强金属-载体交互作用(SMSI),这除了稳定外还可以调节活性。金属-金属氧化催化剂被认为可通过双功能机理起作用[54]。

二氧化钛($TiO_2$)纳米管在甲醇氧化中的促进作用可理解为基于强金属-载体交互作用(SMSI)以及 OH 在 Ti 离子位点吸附从而促进了 CO 在 Pt 位点的氧化,否则金属位点就会中毒,致使这金属位不适合用于甲醇氧化。这些假设的图画模型如图 3.16 所示[55]。

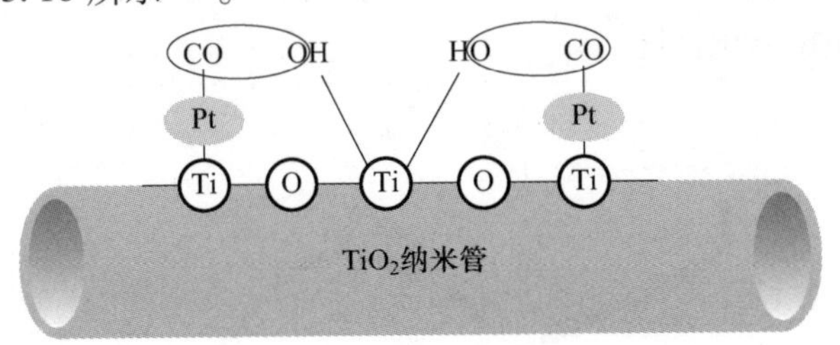


图 3.16 $TiO_2$ 纳米管负载的 Pt 催化剂甲醇氧化过程中 CO 有毒中间体的去除可能机理[55]

为了提高电子电导率,亚化学计量的 $TiO_2$($Ti_4O_7$)以及铌掺杂 $TiO_2$ 被用作载体,它们的氢氧化反应及氧还原反应活性在 PEMFC 条件下同 Vulcan XC-72 负载的 Pt 进行了对比[56-60]。

结果发现,$Pt/TiO_2$ 催化剂比碳基 Pt 催化剂展现出更高的催化活性,因此能够改善氧还原活性。Drew 等[61]发现,$TiO_2$ 和碳纤维负载的 Pt-Ru 对甲醇氧化表现出促进作用。设计的纳米结构的氧化钛被用于燃料电池电催化剂载体。Shanmugam 和 Gedanken[62]制备碳包覆锐钛矿型二氧化砝 $TiO_2$@C 核-壳结构负载 Pt 纳米颗粒用于 ORR 和 MOR。粒子有一个微弱的碳壳包覆的黑色二氧化钛核,复合粒子的尺寸为 20~40nm。Pt 在 CCT 上的平均尺寸为 5~8nm。从高倍透射电镜(HRTEM)研究中清晰地看到在锐钛型 $TiO_2$ 核周围存在几层碳层。然而,在 CCT 上壳层的厚度为 2~4nm 变化。结果发现,随着温度的升高,碳含量和碳层的厚度也会改变。同商业 Pt/C 电催化剂相比,所制备的电催化剂对 ORR 和 MOR 表现出更高的催化活性及耐久性[63]。

例如,Kowal 等[63]成功制备出三元 $PtRhSnO_2/C$ 电催化剂,在室温下能有效地打断乙醇 C—C 键,发生氧化反应生成 $CO_2$。分析结果表明,其催化活性取决于电催化剂三组分之间的协同作用。通过强吸附水[64,65]并与 Pt 和 Rh 沉积表面相互作用,$SnO_2$ 形成金属—OH 能阻止 Pt 和 Rh 活性位与 $H_2O$ 反应,从而使其有利于甲醇氧化。$SnO_2$ 和水提供 OH 物种在 Rh 活性位氧化游离的 CO 以及 Pt 促进乙醇脱氢。它同时也能改善 Rh 电子结构以便对乙醇、中间产物及产品提供合适的键,这有利于 C—C 键断裂从而乙醇氧化。DFT 计算结果证明乙醇在 $PtRh/SnO_2$ 上的氧化通过氧化金属杂环结构实现,这能够以合适速率促进 C—C 键直接断开。Pt - Ru 对甲醇氧化的高活性以及乙醇氧化低活性似乎是由于 Ru 形成 RuOH 倾向与水在 $E>0{:}0V$ 时相互作用。这个反应不能用 SnOH 抑制 OH—OH 排斥,这是由于与 RhOH 形成了微弱的键,RuOH 不吸附乙醇及不能把 C—C 键断裂,此工作证明二氧化锡对甲醇氧化的重要作用。

氧化钨($WO_{3-x}$)作为燃料电池催化剂的载体材料[66-69]引起了研究人员的兴趣。据报道,在 $WO_3$ 负载 Pt 纳米粒子催化剂展现出卓越的抗 CO 性能以及高催化性能。例如,$Pt/WO_3$ 基电极对氧还原反应在磷酸中电催化活性据报道是 Pt/C 的两倍[69]。Sun 等[69]在碳纸上通过 CVD 法成功生长了氧化钨纳米线($W_{18}O_{49}$ NW)。通过简单的还原过程,尺寸分布在 2 ~ 4nm 的分散度极好的 Pt 纳米粒子沉积在 $W_{18}O_{49}$ NW 表面。制备的 $Pt/W_{18}O_{49}$ NW/碳纸复合物形成了三维电极结构。同传统的 Pt/C 电催化剂相比,$Pt/W_{18}O_{49}$ NW/碳纸复合物在单电池质子交换膜燃料电池中对氧还原反应展现出更高的电催化活性以及更佳的抗 CO 毒性。$Pt/Ti_{0.7}W_{0.3}O_2$ 被提出作为另一个有前途的高稳定性及抗 CO 毒性质子交换膜燃料电池电催化剂。初始测试证明 $Pt/Ti_{0.7}W_{0.3}O_2$ 比 Pt/C 及 PtRu/C 催化剂更稳定。500 次循环后,新催化剂的 CV 的集成电荷库仑的损失只有 5%,然而,商业化的 E - TEK PtRu/C 催化剂损失超过 30%[70]。Suzuki 等[71]报道了采用硫酸氧化锆作为 Pt 载体用于 PEMFC,$Pt/S-ZrO_2$ 的电催化活性低于 Pt/C。

## 3.5 金属碳化物及金属氮化物作为催化剂载体

碳化钨(WC)由于相近的费米能级表现出 Pt 类似的催化性质,碳化钨电子态密度与贵金属铂相似[72]。其低廉的价格和良好的抗 CO 毒性使它成为贵金属催化剂有趣的替代者。因此,在过去几十年,碳化钨已经被测试作为替代 PEMFC Pt 电催化剂。

碳化钨被认为比碳载体热稳定性及电化学稳定性更好。然而,它在酸性电解质中的稳定性并不令人满意,主要因为其在硫酸中容易被腐蚀[73-75]。碳化钨不适合作为酸性燃料电池阴极催化剂或载体,这是由于它在酸性和氧化条件下低的抗腐蚀性能。二硼化钛($TiB_2$)展现出许多优越的性质,包括高熔点、高硬度、良好的电子特性以及高热传导性,在酸性介质中极好的热稳定性及抗腐蚀性。$Pt/TiB_2$ 的稳定性大约是商业 Pt/C 稳定性的 4 倍[76]。最近,报道了氮化钛负载的 Pt 应用于 PEM 燃料电池展现出比商业 Pt/C 催化剂更好的催化性能,但是 TiN 作为载体的耐久性并不清楚[77,78]。有必要进一步研究 TiN 作为催化剂载体,尤其是评价它的耐久性[79]。

## 3.6 导电聚合物作为燃料电池载体材料

最近,许多研究者使用导电聚合物(CP)作为电催化剂载体并取得了可喜的成果[80-82]。通过把导电聚合物和金属纳米粒子适当地结合,可制备出具有更高表面积和甲醇氧化活性的新型电催化剂。导电聚合物因其高比表面积、低电阻和高稳定性得到了很多关注。考虑到它们导电和稳定的三维结构,导电聚合物可以作为低温燃料电池催化剂合适的载体。导电聚合物/金属纳米颗粒复合物在电化学过程中通过聚合物骨架给电子电荷流。多孔结构和高比表面的 CP 被作为骨架与贵金属催化剂配合应用于燃料电池一些重要的电化学反应如氢和甲醇氧化和氧还原[81]。将金属颗粒与多孔聚合物阵列结合的主要原因是为了提高活性面积从而改善催化效率。另一个原因是同碳基负载铂电极严重 CO 中毒的问题相比,聚合物负载 Pt 纳米颗粒因其 CO 物种吸附具有更高的抗 CO 中毒性。的确,催化剂在醇氧化过程中由于 $H_2$ 或 CO 类似中间物种的强烈吸附 CO 中毒,这阻塞了 Pt 活性位,从而降低了催化剂活性[81,82]。此外,CP 不仅是电子导电,而且是质子导电材料,因此,它们可以替代燃料电池电极催化剂层的全氟磺酸并能增强性能。在这个例子中,理论上在燃料电池反应中只有两相界面是电子和离子转移所需的,而与之相比,当碳作为载体时需要三相界面,因此催化剂的总体利用率显著增加了(图 3.17)[82]。

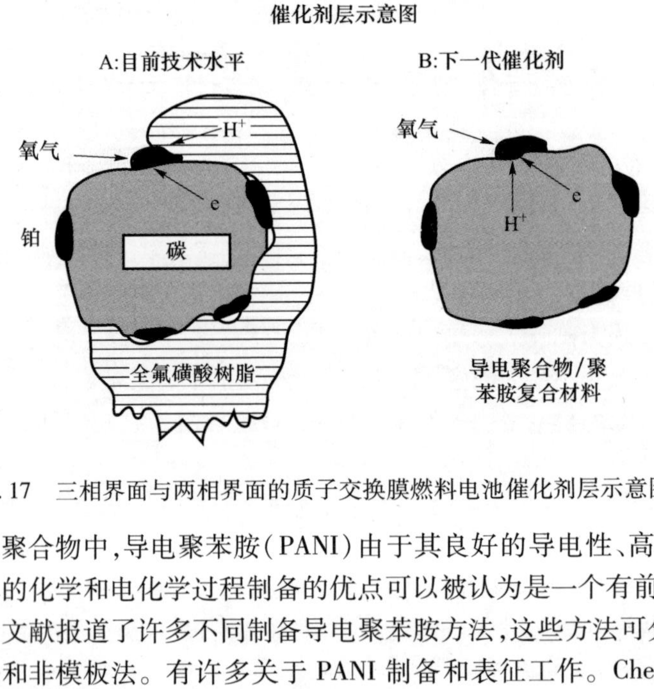

图 3.17 三相界面与两相界面的质子交换膜燃料电池催化剂层示意图[82]

在导电聚合物中,导电聚苯胺(PANI)由于其良好的导电性、高环境稳定性及通过简单的化学和电化学过程制备的优点可以被认为是一个有前途的催化剂载体材料。文献报道了许多不同制备导电聚苯胺方法,这些方法可分为两类,即模板辅助法和非模板法。有许多关于 PANI 制备和表征工作。Chen 等[83]制备了新型聚苯胺纳米纤维(PaniNF)负载的 Pt 纳米电化学催化剂用于 DMFC。不使用模板或功能掺杂剂通过一个可伸缩的界面聚合制备直径为 60nm 聚苯胺纳米纤维。采用乙二醇还原法制备了 PaniNF 负载的 Pt 电催化剂(Pt/PaniNF)及炭黑负载 Pt 电催化剂(Pt/C)。Pt 纳米粒子沉积到 PaniNF 比 Pt 纳米粒子沉积到炭黑具有更小直径和窄粒度分布。Pt/PaniNF 催化剂展现出更高的电化学活性表面(ECSA)及比 Pt/C 更高的甲醇氧化反应催化活性。

作为可取代聚苯胺导电聚合物之一的聚乙酰苯胺(PAANI)成功地被用作燃料电池催化剂载体[84]。PEDOT/PSS 也被报道作为载体材料,Pt/PEDOT/PSS 有与商业碳基催化剂相比拟的氧还原活性。然而,与碳材料相比,PEDOT/PSS 的长期稳定性以及导电性需要改善[85]。Choi 等在聚 N-乙烯基咔唑(PVK)和聚(9-(4-乙烯基 - 苯基)咔唑)(P4VPCz)上电化学沉积 Pt - Ru 催化剂。由于 PVK 低的导电性能,碳负载的 Pt - Ru 在 DMFC 上比 Pt - Ru/PVK 展现出更好的性能。

## 3.7　导电聚合物接枝碳纳米材料

Pt 催化剂分散在高比表面积导电载体上,促进了电子从外部通道传递到催化剂,但不能传递质子。质子传递材料如全氟磺酸经常被加入以便使质子从催化剂传递到膜表面。因此,质子传导可能极大地取决于催化剂层的厚度。在催化剂设计上的最新进展已经开始解决这个局限。同电子传导类似,薄的催化剂层有助于减少质子传导引起的电阻损失。另外,碳具有相对疏水性,且反应气体、水和碳制成的固体电极表面之间的边界接触导致燃料电池中的高电接触电阻和欧姆功率损失,从而导致燃料电池的效率降低。并且,接枝在碳表面的导电聚合物影响质子电导率和催化剂载体表面化学是可调的,如图 3.18 所示。然而,这些改变必须被认真执行以减少对催化剂其他重要特性的任何不利影响。同时,由于铂溶解及碳载体腐蚀,催化剂可能会失去稳定性。

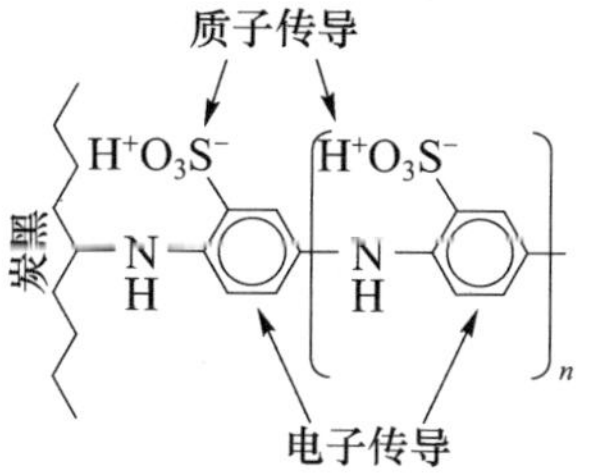

图 3.18　炭黑表面改性催化剂展现的更高的质子和电子传导[89]

在碳载体上铂粒子烧结会显著降低活性表面积。催化剂的烧结可通过加强金属 - 载体相互作用减少。碳载体表面聚合物接枝减少了金属粒子烧结。导电接枝聚合物碳材料有助于均匀分散,以及通过锚定聚合物上的杂原子(如 N、O、S 等)有助于金属粒子的稳定。聚合物骨架中的杂原子(硫或氮)作为路易斯碱能够有效地锚定 Pt 纳米粒子及阻止金属晶体粒子(如 Pt)团聚[87,88]。

## 3.8 三维纳米薄膜作为燃料电池载体材料

三维纳米薄膜(NSTF)催化剂载体已经被很好地开发,主要由基于二萘嵌苯二甲酰亚胺复合物片段的有序纳米尺度有机晶须组成[90,91]。这片段材料产生了数密度的单层定向结晶晶须,其矩形截面约 30 ~ 50nm,平均长度为 0.5 ~ 1μm。Debe 等[90,91]早期的工作证明了与传统的碳基分散 Pt 催化剂相比,NSTF 催化剂具有更高的稳定性和耐用性。而且有机晶须载体外延生长影响后续成核和 Pt 晶须增长,从而使得 Pt(111)晶面数量最大化[92]。除了作为优异的催化剂载体外,NSTF 载体对于燃料电池应用方面是一个好的基底,用于广泛潜在的快速筛查衬层材料。衬层材料展示出用于 NSTF 改善催化剂载体晶须的维度和形貌或者替代传统高比表面积碳载体。NSTF 技术应用于研究衬层材料,是因为其可以通过溅射沉积制备许多不同的元素及其化合物,因为它们高纵横比和表面积从而可后续制备高比表面积铂基催化剂用于实际条件下燃料电池测试[90,91]。

## 3.9 总结与展望

各种类型的载体材料,包括碳材料(即炭黑、有序介孔碳、石墨纳米纤维、碳纳米管和石墨烯)、导电金属氧化物、导电聚合物,都被广泛应用于燃料电池。催化剂载体材料用于提高金属纳米粒子负载量和分散度,从而提高贵金属催化剂的利用效率和耐久性。载体材料的性质对燃料电池的催化活性、耐久性、成本有很大的影响。燃料电池电催化剂的合适载体材料条件:高电化学接触表面积;良好的电子导电性;合适的质量传输孔道;燃料电池操作条件下良好的电化学和热稳定性。最常用的催化剂载体是炭黑。本章中讨论炭黑作为载体材料的几点不足,因此发展可替代燃料电池载体材料在过去 10 年引起了研究人员的广泛关注。

有序介孔碳材料(OMC)由于其高表面积、高导电性以及能促进传质的孔道成为了一种很有前途的候选者。OMC 负载的电催化剂在 PEMFC 电极反应中已经展现卓越的性能。在 OMC 负载电催化剂观察到的改善的燃料电池性能是可以理解的,主要是由于 OMC 因其高比表面积和相互连通的孔道结构会产生均匀分散金属粒子和增强的传质特性。石墨纳米纤维因其良好的石墨结构、低阻抗和高导电性在作为载体材料的研究中吸引了研究人员广泛的兴趣。与碳纳米管结构不同,在 GNF 没有中空腔和只在边缘有暴露的底面。由于其独特的结构,

GNF可以直接用于金属纳米粒子载体而无需任何预处理,这通常会破坏石墨纤维完整结构,因为GNF的底面和鱼脊结构存在潜在的反应官能团,通常用作沉积金属。由于它们的高表面积、良好的电子导电性及高化学稳定性,碳纳米管(CNT)广泛地用作燃料电池中Pt和Pt合金催化剂载体材料。然而,碳纳米管负载的Pt或Pt合金纳米粒子沉积、分布和尺寸取决于碳纳米管的表面处理和表面性质。Pt纳米粒子与碳纳米管相互作用和碳纳米管内在本质属性显著影响Pt纳米粒子的活性。由于原始碳纳米管是化学惰性的,因此有必要活化碳纳米管石墨表面以便锚定和沉积催化纳米粒子。许多功能化方法已经被开发出来,包括共价和非共价功能化。非共价方法可以在不破坏CNT完整结构的情况下在CNT表面引入多种官能团。石墨烯因其独特的电子和机械性质已经吸引了越来越多研究者的研究兴趣。同碳纳米管相比,石墨烯和化学改性的石墨烯阵列具有更高导电性和比表面积及有许多潜在应用。本章也综述了其他不同种类的载体材料如金属氧化物、导电聚合物等。

尽管已经开发了负载燃料电池纳米电催化剂的各种载体材料,然而,制备的催化剂性在活性、耐久性以及成本方面距离燃料电池商业化研究目标仍然很远,需要进行更广泛的研究工作。

## 参考文献

[1] Borup, R., Meyers, J., Pivovar, B., Kim, Y. S., Mukundan, R., Garland, N., Myers, D., Wilson, M., Garzon, F., Wood, D., Zelenay, P., More, K., Stroh, K., Zawodzinski, T., Boncella, J., Mcgrath, J. E., Inaba, M., Miyatake, K., Hori, M., Ota, K., Ogumi, Z., Miyata, S., Nishikata, A., Zyun, S., Uchimoto, Y., Yasuda, K., Kimijima, K.-I., and Iwashita, N. (2007) Scientific aspects of polymer electrolyte fuel cell durability and degradation. *Chemical Reviews*, 3904 - 3951.

[2] Markovic, N. M. and Ross, P. N. (2002) Surface science studies of model fuel cell electrocatalysts. *Surface Science Reports*, **45** (4 - 6), 121 - 229.

[3] Rajalakshmi, N., Pandiyan, S., and Dhathathreyan, K. S. (2008) Design and development of modular fuel cell stacks for various applications. *International Journal of Hydrogen Energy*, **33**, 449 - 454.

[4] Yu, X. and Ye, S. (2007) Recent advances in activity and durability enhancement of Pt/C catalytic cathode in PEMFC: Part I. Physico-chemical and electronic interaction between Pt and carbon support, and activity enhancement of Pt/C catalyst. *Journal of Power Sources*, **172** (1), 133 - 144.

[5] Antolini, E. (2009) Carbon supports for low-temperature fuel cell catalysts. *Applied Catalysis B: Environmental*, **88** (1 - 2), 1 - 24.

[6] Shao, Y. Y., Liu, J., Wang, Y., and Lin, Y. H. (2009) Novel catalyst support materials for PEM fuel cells: current status and future prospects. *Journal of Materials Chemistry*, **19** (1), 46 - 59.

[7] Chang, H., Joo, S. H., and Pak, C. (2007) Synthesis and characterization of mesoporous carbon for fuel cell applications. *Journal of Materials Chemistry*, **17** (30), 3078 - 3088.

[8] Joo, S. H., Choi, S. J., Oh, I., Kwak, J., Liu, Z., Terasaki, O., and Ryoo, R. (2001) Ordered nanoporous arrays of carbon supporting high dispersions of platinum nanoparticles. *Nature*, **412** (6843), 169 -

172.

[9] Ding, J. , Chan, K. -Y. , Ren, J. , and Xiao, F. -S. (2005) Platinum and platinum – ruthenium nanoparticles supported on ordered mesoporous carbon and their electrocatalytic performance for fuel cell reactions. *Electrochimica Acta*, **50** (15), 3131 – 3141.

[10] Lin, M. -L. , Huang, C. -C. , Lo, M. -Y. , and Mou, C. -Y. (2008) Well-ordered mesoporous carbon thin film with perpendicular channels: application to direct methanol fuel cell. *Journal of Physical Chemistry C*, **112** (3), 867 – 873.

[11] Liu, S. -H. , Yu, W. -Y. , Chen, C. -H. , Lo, A. -Y. , Hwang, B. -J. , Chien, S. -H. , and Liu, S. -B. (2008) Fabrication and characterization of well-dispersed and highly stable PtRu nanoparticles on carbon mesoporous material for applications in direct methanol fuel cell. *Chemistry of Materials*, **20** (4), 1622 – 1628.

[12] Yu, J. -S. , Kang, S. , Yoon, S. B. , and Chai, G. (2002) Fabrication of ordered uniform porous carbon networks and their application to a catalyst supporter. *Journal of the American Chemical Society*, **124** (32), 9382 – 9383.

[13] Su, F. , Zeng, J. , Bao, X. , Yu, Y. , Lee, J. Y. , and Zhao, X. S. (2005) Preparation and characterization of highly ordered graphitic mesoporous carbon as a Pt catalyst support for direct methanol fuel cells. *Chemistry of Materials*, **17** (15), 3960 – 3967.

[14] Joo, S. H. , Pak, C. , You, D. J. , Lee, S. -A. , Lee, H. I. , Kim, J. M. , Chang, H. , and Seung, D. (2006) Ordered mesoporous carbons (OMC) as supports of electrocatalysts for direct methanol fuel cells (DMFC): effect of carbon precursors of OMC on DMFC performances. *Electrochimica Acta*, **52** (4), 1618 – 1626.

[15] Shanahan, P. V. , Xu, L. , Liang, C. , Waje, M. , Dai, S. , and Yan, Y. S. (2008) Graphitic mesoporous carbon as a durable fuel cell catalyst support. *Journal of Power Sources*, **185** (1), 423 – 427.

[16] Du, H. , Li, B. , Kang, F. , Fu, R. , and Zeng, Y. (2007) Carbon aerogel supported Pt – Ru catalysts for using as the anode of direct methanol fuel cells. *Carbon*, **45** (2), 429 – 435.

[17] Marie, J. , Chenitz, R. , Chatenet, M. , Berthon-Fabry, S. , Cornet, N. , and Achard, P. (2009) Platinum supported on resorcinol-formaldehyde based carbon aerogels for PEMFC electrodes: influence of the carbon support on electrocatalytic properties. *Journal of Power Sources*, **190**, 423 – 434.

[18] Maiyalagan, T. , Viswanathan, B. , and Varadaraju, U. V. (2005) Nitrogen containing carbon nanotubes as supports for Pt: alternate anodes for fuel cell applications. *Electrochemistry Communications*, **7**, 905 – 912.

[19] Maiyalagan, T. (2008) Synthesis and electro-catalytic activity of methanol oxidation on nitrogen containing carbon nanotubes supported Pt electrodes. *Applied Catalysis B: Environmental*, **80** (3 – 4), 286 – 295.

[20] Kim, M. , Hwang, S. , and Yu, J. S. (2007) Novel ordered nanoporous graphitic $C_3N_4$ as a support for Pt – Ru anode catalyst in direct methanol fuel cell. *Journal of Material Chemistry*, **17**, 1656 – 1659.

[21] Roustom, B. El. , Sine, G. , Foti, G. , and Comninellis, C. (2007) A novel method for the preparation of bi-metallic (Pt – Au) nanoparticles on boron doped diamond (BDD) substrate: application to the oxygen reduction reaction. *Journal of Applied Electrochemistry*, **37**, 1227 – 1236.

[22] Fischer, A. E. and Swain, G. M. (2005) Preparation and characterization of borondoped diamond powder: a possible dimensionally stable electrocatalyst support material. *Journal of the Electrochemical Society*, **152** (9), B369 – B375.

[23] Salazar-Banda, G. R. , Eguiluz, K. I. B. , and Avaca, L. A. (2007) Boron-doped diamond powder as catalyst support for fuel cell applications. *Electrochemistry Communications*, **9** (1), 59 – 64.

[24] Gangeri, M., Centi, G., Malfa, A. L., Perathoner, S., Vieira, R., Pham-Huu, C., and Ledoux, M. J. (2005) Electrocatalytic performances of nanostructured platinumcarbon materials. *Catalysis Today*, **102 – 103**, 50 – 57.

[25] Bessel, C. A., Laubernds, K., Rodriguez, N. M., and Baker, R. T. K. (2001) Graphite nanofibers as an electrode for fuel cell applications. *The Journal of Physical Chemistry B*, **105** (6), 1115 – 1118.

[26] Steigerwalt, E. S., Deluga, G. A., Cliffel, D. E., and Lukehart, C. M. (2001) A PtRu/graphitic carbon nanofiber nanocomposite exhibiting high relative performance as a direct-methanol fuel cell anode catalyst. *The Journal of Physical Chemistry B*, **105** (34), 8097 – 8101.

[27] Gan, L., Du, H., Li, B., and Kang, F. (2011) Surface-reconstructed graphite nanofibers as a support for cathode catalysts of fuel cells. *Chemical Communications*, **47** (13), 3900 – 3902.

[28] Girishkumar, G., Hall, T. D., Vinodgopal, K., and Kamat, P. V. (2006) Single wall carbon nanotube supports for portable direct methanol fuel cells. *Journal of Physical Chemistry B*, **110** (1), 107 – 114.

[29] Li, W., Liang, C., Zhou, W., Qiu, J., Zhou, Z., Sun, G., and Xin, Q. (2003) Preparation and characterization of multiwalled carbon nanotube-supported platinum for cathode catalysts of direct methanol fuel cells. *Journal of Physical Chemistry B*, **107** (26), 6292 – 6299.

[30] Wang, S., Jiang, S. P., and Wang, X. (2008) Polyelectrolyte functionalized carbon nanotubes as a support for noble metal electrocatalysts and their activity for methanol oxidation. *Nanotechnology*, **19** (26), 265601.

[31] Wang, S. Y., Jiang, S. P., White, T. J., Guo, J., and Wang, X. (2009) Electrocatalytic activity and interconnectivity of Pt nanoparticles on multiwalled carbon nanotubes for fuel cells. *Journal of Physical Chemistry C*, **113** (43), 18935 – 18945.

[32] Wang, S. Y., Wang, X., and Jiang, S. P. (2008) PtRu nanoparticles supported on 1-aminopyrene-functionalized multiwalled carbon nanotubes and their electrocatalytic activity for methanol oxidation. *Langmuir*, **24** (18), 10505 – 10512.

[33] Saha, M. S. and Kundu, A. (2010) Functionalizing carbon nanotubes for proton exchange membrane fuel cells electrode. *Journal of Power Sources*, **195** (19), 6255 – 6261.

[34] Yu, R., Chen, L., Liu, Q., Lin, J., Tan, K.-L., Ng, S. C., Chan, H. S. O., Xu, G.-Q., and Hor, T. S. A. (1998) Platinum deposition on carbon nanotubes via chemical modification. *Chemistry of Materials*, **10** (3), 718 – 722.

[35] Xu, C., Chen, J., Cui, Y., Han, Q., Choo, H., Liaw, P. K., and Wu, D. (2006) Influence of the surface treatment on the deposition of platinum nanoparticles on the carbon nanotubes. *Advanced Engineering Materials*, **8** (1 – 2), 73 – 77.

[36] Xing, Y. (2004) Synthesis and electrochemical characterization of uniformly-dispersed high loading Pt nanoparticles on sonochemically-treated carbon nanotubes. *The Journal of Physical Chemistry B*, **108** (50), 19255 – 19259.

[37] Hsin, Y. L., Hwang, K. C., and Yeh, C.-T. (2007) Poly(vinylpyrrolidone)-modified graphite carbon nanofibers as promising supports for PtRu catalysts in direct methanol fuel cells. *Journal of the American Chemical Society*, **129** (32), 9999 – 10010.

[38] Li, X. L., Liu, Y. Q., Fu, L., Cao, L. C., Wei, D. C., and Wang, Y. (2006) Efficient synthesis of carbon nanotube-nanoparticle hybrids. *Advanced Functional Materials*, **16** (18), 2431 – 2437.

[39] Correa-Duarte, M. A., Sobal, N., LizMarzán, L. M., and Giersig, M. (2004) Linear assemblies of silica-coated gold nanoparticles using carbon nanotubes as templates. *Advanced Materials*, **16** (23 – 24), 2179 – 2184.

[40] Yang, D. Q., Hennequin, B., and Sacher, E. (2006) XPS demonstration of interaction between benzyl mercaptan and multiwalled carbon nanotubes and their use in the adhesion of Pt nanoparticles. *Chemistry of Materials*, **18** (21), 5033-5038.

[41] Wang, S. (2010) Nanostructured electrocatalysts for proton exchange membrane fuel cells (PEMFCs). Ph. D. thesis, Nanyang Technological University, Singapore.

[42] Wang, S., Yang, F., Jiang, S. P., Chen, S., and Wang, X. (2010) Tuning the electrocatalytic activity of Pt nanoparticles on carbon nanotubes via surface functionalization. *Electrochemistry Communications*, **12** (11), 1646-1649.

[43] Park, K.-W., Sung, Y.-E., Han, S., Yun, Y., and Hyeon, T. (2004) Origin of the enhanced catalytic activity of carbon nanocoil-supported PtRu alloy electrocatalysts. *Journal of Physical Chemistry B*, **108** (3), 939-944.

[44] Lee, G., Shim, J. H., Kang, H., Nam, K. M., Song, H., and Park, J. T. (2009) Monodisperse Pt and PtRu/C60 hybrid nanoparticles for fuel cell anode catalysts. *Chemical Communications*, (33), 5036-5038.

[45] Wang, S., Wang, X., and Jiang, S. P. (2011) Self-assembly of mixed Pt and Au nanoparticles on PDDA-functionalized graphene as effective electrocatalysts for formic acid oxidation of fuel cells. *Physical Chemistry Chemical Physics*, **13** (15), 6883-6891.

[46] Ha, H. W., Kim, I. Y., Hwang, S. J., and Ruoff, R. S. (2011) One-pot synthesis of platinum nanoparticles embedded on reduced graphene oxide for oxygen reduction in methanol fuel cells. *Electrochemical and Solid-State Letters*, **14**(7), B70-B73.

[47] Yoo, E., Okata, T., Akita, T., Kohyama, M., Nakamura, J., and Honma, I. (2009) Enhanced electrocatalytic activity of Pt subnanoclusters on graphene nanosheet surface. *Nano Letters*, **9** (6), 2255-2259.

[48] Sharma, S., Ganguly, A., Papakonstantinou, P., Miao, X., Li, M., Hutchison, J. L., Delichatsios, M., and Ukleja, S. (2010) Rapid microwave synthesis of CO tolerant reduced graphene oxide-supported platinum electrocatalysts for oxidation of methanol. *The Journal of Physical Chemistry C*, **114** (45), 19459-19466.

[49] Kou, R., Shao, Y., Mei, D., Nie, Z., Wang, D., Wang, C., Viswanathan, V. V., Park, S., Aksay, I. A., Lin, Y., Wang, Y., and Liu, J. (2011) Stabilization of electrocatalytic metal nanoparticles at metal-metal oxide-graphene triple junction points. *Journal of the American Chemical Society*, **133** (8), 2541-2547.

[50] Zhou, Y., Neyerlin, K., Olson, T. S., Pylypenko, S., Bult, J., Dinh, H. N., Gennett, T., Shao, Z., and O'Hayre, R. (2010) Enhancement of Pt and Pt-alloy fuel cell catalyst activity and durability via nitrogen-modified carbon supports. *Energy and Environmental Science*, **3** (10), 1437-1446.

[51] Gong, K. P., Du, F., Xia, Z. H., Durstock, M., and Dai, L. M. (2009) Nitrogen-doped carbon nanotube arrays with high electrocatalytic activity for oxygen reduction. *Science*, **323** (5915), 760-764.

[52] Higgins, D. C., Meza, D., and Chen, Z. (2010) Nitrogen-doped carbon nanotubes as platinum catalyst supports for oxygen reduction reaction in proton exchange membrane fuel cells. *The Journal of Physical Chemistry C*, **114** (50), 21982-21988.

[53] Imran Jafri, R., Rajalakshmi, N., and Ramaprabhu, S. (2010) Nitrogen doped graphene nanoplatelets as catalyst support for oxygen reduction reaction in proton exchange membrane fuel cell. *Journal of Materials Chemistry*, **20**, 7114-7711.

[54] Lv, R., Cui, T., Jun, M.-S., Zhang, Q., Cao, A., Su, D. S., Zhang, Z., Yoon, S.-H., Miyawaki, J., Mochida, I., and Kang, F. (2011) Open-ended, N-doped carbon nanotube-graphene hybrid nano-

structures as high-performance catalyst support. *Advanced Functional Materials*, **21**,999 – 1006.

[55] Maiyalagan, T. , Viswnathan, B. , and Varadaraju, U. V. (2006) Electro-oxidation of methanol on $TiO_2$ nanotube supported platinum electrodes. *Journal of Nanoscience and Nanotechnology*, **6** (7), 2067 – 2071.

[56] Ioroi, T. , Siroma, Z. , Fujiwara, N. ,Yamazaki, S. -I. , and Yasuda, K. (2005) Sub-stoichiometric titanium oxidesupported platinum electrocatalyst for polymer electrolyte fuel cells. *Electrochemistry Communications*, **7** (2), 183 – 188.

[57] Huang, S. -Y. , Ganesan, P. , and Popov, B. N. (2010) Electrocatalytic activity and stability of niobium-doped titanium oxide supported platinum catalyst for polymer electrolyte membrane fuel cells. *Applied Catalysis B*: Environmental, **96** (1 – 2), 224 – 231.

[58] Beak, S. , Jung, D. , Nahm, K. S. , and Kim, P. (2010) Preparation of highly dispersed Pt on $TiO_2$-modified carbon for the application to oxygen reduction reaction. *Catalysis Letters*, **134** (3 – 4), 288 – 294.

[59] Lee, J. -M. , Han, S. -B. , Kim, J. -Y. , Lee, Y. -W. , Ko, A. -R. , Roh, B. , Hwang, I. , and Park, K. -W. (2010) $TiO_2$@ carbon core-shell nanostructure supports for platinum and their use for methanol electrooxidation. *Carbon*, **48**, 2290 – 2296.

[60] He, D. , Yang, L. , Kuang, S. , and Cai, Q. (2007) Fabrication and catalytic properties of Pt and Ru decorated $TiO_2$*n*CNTs catalyst for methanol electrooxidation. *Electrochemistry Communications*, **9**, 2467 – 2472.

[61] Drew, K. , Girishkumar, G. , Vinodgopal, K. , and Kamat, P. V. (2005) Boosting fuel cell performance with a semiconductor photocatalyst: $TiO_2$/Pt – Ru hybrid catalyst for methanol oxidation. *The Journal of Physical Chemistry B*, **109** (24), 11851 – 11857.

[62] Shanmugam, S. and Gedanken, A. (2007) Carbon-coated anatase $TiO_2$ nanocomposite as a high-performance electrocatalyst support. *Small*, **3** (7), 1189 – 1193.

[63] Kowal, A. , Li, M. , Shao, M. , Sasaki, K. ,Vukmirovic, M. B. , Zhang, J. , Marinkovic, N. S. , Liu, P. , Frenkel, A. I. , and Adzic, R. R. (2009) Ternary Pt/Rh/$SnO_2$ electrocatalysts for oxidizing ethanol to $CO_2$. *Nature Materials*, **8** (4), 325 – 330.

[64] Pang, H. L. , Lu, J. P. , Chen, J. H. , Huang, C. T. , Liu, B. , and Zhang, X. H. (2009) Preparation of $SnO_2$ – CNTs supported Pt catalysts and their electrocatalytic properties for ethanol oxidation. *Electrochimica Acta*, **54**, 2610 – 2615.

[65] Du, C. , Chen, M. , Cao, X. , Yin, G. , and Shi, P. (2009) A novel CNT@ $SnO_2$ core-sheath nanocomposite as a stabilizing support for catalysts of proton exchange membrane fuel cells. *Electrochemistry Communications*, **11** (2), 496 – 498.

[66] Hobbs, B. S. and Tseung, A. C. C. (1969) High performance, platinum activated tungsten oxide fuel cell electrodes. *Nature*, **222** (5193), 556 – 558.

[67] Maiyalagan, T. and Viswanathan, B. (2008) Catalytic activity of platinum/tungsten oxide nanorod electrodes towards electrooxidation of methanol. *Journal of Power Sources*, **175** (2), 789 – 793.

[68] Saha, M. S. , Banis, M. N. , Zhang, Y. , Li, R. , Sun, X. , Cai, M. , and Wagner, F. T. (2009) Tungsten oxide nanowires grown on carbon paper as Pt electrocatalyst support for high performance proton exchange membrane fuel cells. *Journal of Power Sources*, **192** (2), 330 – 335.

[69] Raghuveer, V. and Viswanathan, B. (2005) Synthesis, characterization and electrochemical studies of Ti-incorporated tungsten trioxides as platinum support for methanol oxidation. *Journal of Power Sources*, **144**, 1 – 10.

[70] Subban, C. V., Zhou, Q., Hu, A., Moylan, T. E., Wagner, F. T., and Disalvo, F. J. (2010) Sol-gel synthesis, electrochemical characterization, and stability testing of $Ti_{07}W_{0.3}O_2$ nanoparticles for catalyst support applications in proton-exchange membrane fuel cells. *Journal of the American Chemical Society*, **132** (49), 17531 – 17536.

[71] Suzuki, Y., Ishihara, A., Mitsushima, S., Kamiya, N., and Ota, K. -I. (2007) Sulfatedzirconia as a support of Pt catalyst for polymer electrolyte fuel cells. *Electrochemical and Solid-State Letters*, **10** (7), B105 – B107.

[72] Chhina, H., Campbell, S., and Kesler, O. (2007) Thermal and electrochemical stability of tungsten carbide catalyst supports. *Journal of Power Sources*, **164** (2), 431 – 440.

[73] Wang, Y., Song, S., Maragou, V., Shen, P. K., and Tsiakaras, P. (2009) High surface area tungsten carbide microspheres as effective Pt catalyst support for oxygen reduction reaction. *Applied Catalysis B: Environmental*, **89** (1 – 2), 223 – 228.

[74] Ganesan, R., Ham, D. J., and Lee, J. S. (2007) Platinized mesoporous tungsten carbide for electrochemical methanol oxidation. *Electrochemistry Communications*, **9** (10), 2576 – 2579.

[75] Nie, M., Shen, P. K., Wu, M., Wei, Z., and Meng, H. (2006) A study of oxygen reduction on improved Pt-WC/C electrocatalysts. *Journal of Power Sources*, **162** (1), 173 – 176.

[76] Yin, S., Mu, S., Lv, H., Cheng, N., Pan, M., and Fu, Z. (2010) A highly stable catalyst for PEM fuel cell based on durable titanium diboride support and polymerstabilization. *Applied Catalysis B: Environmental*, **93** (3 – 4), 233 – 240.

[77] Ottakam Thotiyl, M. M., Ravikumar, T., and Sampath, S. (2010) Platinum particles supported on titanium nitride: an efficient electrode material for the oxidation of methanol in alkaline media. *Journal of Materials Chemistry*, **20** (47), 10643 – 10651.

[78] Musthafa, O. T. M. and Sampath, S. (2008) High performance platinized titanium nitride catalyst for methanol oxidation. *Chemical Communications*, (1), 67 – 69.

[79] Avasarala, B. and Haldar, P. (2011) On the stability of TiN-based electrocatalysts for fuel cell applications. *International Journal of Hydrogen Energy*, **36** (6), 3965 – 3974.

[80] Huang, S. -Y., Ganesan, P., and Popov, B. N. (2009) Development of conducting polypyrrole as corrosion-resistant catalyst support for polymer electrolyte membrane fuel cell (PEMFC) application. *Applied Catalysis B: Environmental*, **93** (1 – 2), 75 – 81.

[81] Antolini, E. and Gonzalez, E. R. (2009) Polymer supports for low-temperature fuel cell catalysts. *Applied Catalysis A: General*, **365** (1), 1 – 19.

[82] Qi, Z., Lefebvre, M. C., and Pickup, P. G. (1998) Electron and proton transport in gas diffusion electrodes containing electronically conductive protonexchange polymers. *Journal of Electroanalytical Chemistry*, **459** (1), 9 – 14.

[83] Chen, Z., Xu, L., Li, W., Waje, M., and Yan, Y. (2006) Polyaniline nanofibre supported platinum nanoelectrocatalysts for direct methanol fuel cells. *Nanotechnology*, **17** (20), 5254.

[84] Jiang, C. and Lin, X. (2007) Preparation of three-dimensional composite of poly(*N*-acetylaniline) nanorods/platinum nanoclusters and electrocatalytic oxidation of methanol. *Journal of Power Sources*, **164** (1), 49 – 55.

[85] Tintula, K. K., Pitchumani, S., Sridhar, P., and Shukla, A. K. (2010) PEDOT-PSSA as an alternative support for Pt electrodes in PEFCs. *Bulletin of Materials Science*, **33** (2), 157 – 163.

[86] Choi, J. -H., Park, K. -W., Lee, H. -K., Kim, Y. -M., Lee, J. -S., and Sung, Y. -E. (2003) Nanocomposite of PtRu alloy electrocatalyst and electronically conducting polymer for use as the anode in a direct

methanol fuel cell. *Electrochimica Acta*, **48** (19), 2781 – 2789.

[87] Zhu, Z.-Z., Wang, Z., and Li, H.-L. (2008) Functional multi-walled carbon nanotube/polyaniline composite films as supports of platinum for formic acid electrooxidation. *Applied Surface Science*, **254** (10), 2934 – 2940.

[88] Selvaraj, V. and Alagar, M. (2007) Pt and Pt – Ru nanoparticles decorated polypyrrole/multiwalled carbon nanotubes and their catalytic activity towards methanol oxidation. *Electrochemistry Communications*, **9** (5), 1145 – 1153.

[89] He, C., Desai, S., Brown, G., and Bollepalli, S. (2005) PEM fuel cell catalysts: cost, performance, and durability. *Electrochemical Society Interface*, **14** (3), 41 – 44.

[90] Debe, M. K., Schmoeckel, A. K., Vernstrom, G. D., and Atanasoski, R. (2006) High voltage stability of nanostructured thin film catalysts for PEM fuel cells. *Journal of Power Sources*, **161** (2), 1002 – 1011.

[91] Garsuch, A., Stevens, D. A., Sanderson, R. J., Wang, S., Atanasoski, R. T., Hendricks, S., Debe, M. K., and Dahn, J. R. (2010) Alternative catalyst supports deposited on nanostructured thin films for proton exchange membrane fuel cells. *Journal of the Electrochemical Society*, **157** (2), B187 – B194.

# 第4章

# 低温直接醇类燃料电池阳极催化剂

Wenzhen Li

## 4.1 引　　言

低温聚合物电解质燃料电池可将燃料中所存储的化学能(如氢气和醇类等)直接转化为电能,并具有低排放和高能效等优点,广受瞩目,成为目前极具应用前景的电化学能源设备[1-8];同时其反应过程不受卡诺循环限制,所以具有极高的能量利用效率:占电能中的40% ~50%和总能量(电能+产生的热量)中的80% ~85%[6]。氢气由于其氧化的动力学过程非常迅速且产物只有水,因此被认为是在汽车应用方面的最佳燃料;然而,由于氢气本身只是一个能量载体而非自然资源,在生产、运输和储存方面都面临着巨大的技术挑战[9]。与氢气相比,液体醇类燃料则拥有更明显的优势:不但具有更高能量密度和热力学能量转换效率,类似的电动势(热力学位能),同时还能避免氢气在生产和存储方面面临的问题。许多醇类都可以从可再生生物质原料中大量获取[10-12]:例如,甲醇可从发酵农产品、生物质中提取出来;乙醇可从农产品发酵中得到;乙二醇可以通过对木质素进行多相催化加氢来大量获取;丙三醇则是生物质柴油生产的主要副产物。它们价格低廉且来源丰富,并可以利用现有液体燃料的基础设施进行储存。因此,低温直接乙醇燃料电池(DAFC)作为一种清洁可持续的便携式电子设备,在运输系统中也有很好的应用前景。

表4.1给出了部分醇类燃料在标准状况下的热力学性质。虽然醇类的电动势比氢气稍微低一些,但是甲醇、乙醇、乙二醇和甘油的热力学能量转化效率都高达97% ~99%,高于氢气(83%);同时醇类的质量和体积能量密度也比氢气高。

表4.1 标准状况下醇类选择性氧化的电子转移数($N_e$)、电势($E^o$)、体积能量密度($W_e$)和热力学能量转换效率($\varepsilon_{rev}$)

| 燃料 | $N_e$ | $E^o$/V | $W_e$/(kW·h/L) | $\varepsilon_{rev}$/% |
|---|---|---|---|---|
| 氢气 | 2 | 1.23 | 2.6(液氢) | 83 |
| 甲醇 | 6 | 1.18 | 4.8 | 97 |
| 乙醇 | 12 | 1.15 | 6.3 | 97 |
| 乙二醇 | 10 | 1.22 | 5.9 | 99 |
| 丙三醇 | 14 | 1.22 | 6.3 | 99 |

目前直接醇类燃料电池与氢气质子交换膜燃料电池相比,尽管醇类在热力学中有明显的优势,但其输出功率密度和效率却不高,大大制约了其应用。可以显著减少醇类渗透和降低离子电导率的高分子聚合膜,以及需要显著提高的醇类缓慢的反应动力学性能,是直接醇类燃料电池在广泛应用的过程中面临的巨大挑战。例如,公认催化性能最好的PtRu/C催化剂在电流密度为0.5A/$cm^2$时,甲醇氧化的过电位大于0.3V,这比氢气氧化的过电位(仅为0.02V)高得多[4];乙醇的氧化过程也有相似的较慢的动力学,更严重的是,乙醇中的C—C在低温下很难被破坏(如小于90℃),主要的产物为乙醛和乙酸,而目前所用的PtSn催化剂,产生的$CO_2$低于10%[6,13],乙醇氧化的不彻底也导致了乙醇利用率和能量转换效率较低。

如何将小有机分子中所储存的化学能直接高效地转化为电能,这是科学家们的一个长期目标。近年来,为了开发出高性能的直接醇类燃料电池阳极催化剂,研究者对醇类电催化氧化机理进行了大量的探索。本章主要致力于研究低温直接醇类燃料电池的阳极催化材料。首先,介绍目前已知的在酸性和碱性介质中电氧化醇类(甲醇、乙醇、乙二醇和丙三醇)性能最高的阳极催化剂;其次,回顾一些催化剂的制备方法和新型的碳载体材料;最后,讨论该领域未来所面临的机遇和挑战。

## 4.2 直接甲醇燃料电池阳极催化剂:二元和三元催化剂性能的提高

### 4.2.1 直接甲醇燃料电池工作原理

基于质子交换膜的直接甲醇燃料电池包括以下几个部件:阳极、质子交换膜和阴极。在阳极中,甲醇被氧化为$CO_2$;在阴极中,氧气与质子反应生成水:

$$CH_3OH + H_2O \longrightarrow CO_2 + 6H^+ + 6e^- \quad +0.05V\ 相对于\ SHE\ (阳极) \tag{4.1}$$

$$\frac{3}{2}O_2 + 6H^+ + 6e^- \longrightarrow 3H_2O \quad +1.23V \text{ 相对于 SHE（阴极）} \tag{4.2}$$

总反应为

$$CH_3OH + 3/2O_2 \longrightarrow CO_2 + 2H_2O \quad +1.18V \tag{4.3}$$

### 4.2.2 甲醇电化学氧化的催化反应机理

在过去的几十年间，研究者在小有机分子的低温电化学氧化领域进行了大量的探索[14-38]。电化学研究通常利用光谱、质谱、物理化学工具以及理论计算（如密度泛函数计算）等表征方式[21,27,28]研究醇类在氧化过程中电极表面的吸附物种和活性中间体，进而探索醇类的氧化反应途径。

小分子醇类的电氧化一般都使用贵金属 Pt 作为催化剂；然而，在低温条件下，甲醇氧化所产生的 CO 类中间产物很容易与 Pt 反应使其中毒。经过研究，研究者发现 Pt 基二元或者三元催化剂具有优异的抗中毒能力，如 $Pt-M_1$，$Pt-M_1-M_2$（M = Ru，Sn 等），这主要归功于其双官能团效应：

$$Pt + CH_3OH \longrightarrow Pt-(CH_3OH)_{ads} \tag{4.4}$$

$$Pt-(CH_3OH)_{ads} \longrightarrow Pt-(CO)_{ads} + 4H^+ + 4e^- \tag{4.5}$$

$$M + H_2O \longrightarrow M-(H_2O)_{ads} \tag{4.6}$$

$$M-(H_2O)_{ads} \longrightarrow M-(H_2O)_{ads} + H^+ + e^- \tag{4.7}$$

$$Pt-(CO)_{ads} + M-(H_2O)_{ads} \longrightarrow Pt + M + CO_2 + H^+ + e^- \tag{4.8}$$

当甲醇吸附在 Pt 活性点上之后，甲醇脱氢产生 $CO_{ads}$。式(4.6)和式(4.7)为水在金属 M 上的活化。Pt 和 M 共同作用将 $CO_{ads}$ 氧化为 $CO_2$（式(4.8)）。一直以来，式(4.6)～式(4.8)被认为是反应的速率控制步骤。值得一提的是，Wieckowski 等使用密度泛函数理论和单晶模型催化剂，计算出了甲醇脱氢过程所需要的能量，并且阐明了甲醇脱氢是通过下面步骤进行的[38]：

$$\begin{aligned} CH_3OH_{ads} &\longrightarrow CH_2OH_{ads} + H_{ads} \longrightarrow CHOH_{ads} + 2H_{ads} \\ &\longrightarrow COH_{ads} + 3H_{ads} \longrightarrow CO_{ads} + 4H_{ads} \end{aligned} \tag{4.9}$$

内在机理表明 M 可以对 Pt 的电子结构进行修饰，改变了氧化物种的吸附甚至是甲醇的解离吸附。Pt 活性位点上吸附的 CO 的稳定性由两个效应共同决定：电子从 CO 充满的 5σ 分子轨道转移到 Pt 的空 dσ 轨道，反馈电子由金属的 dπ 轨道再转移到 CO 的 $2\pi^*$ 反键轨道。σ-型键的产生增强了 P-型键，反之亦然。而在 Pt－M 合金中，发生了对 Pt 空电子态密度的修饰，就 CO 分子轨道的能量而言费米能级发生移动。这种协同效应的产生减弱了 Pt—CO 键，从而促进了甲醇氧化的动力学。

在所有的 Pt 基合金中，PtRu 是甲醇电氧化过程最有希望的催化剂[17,18,39-41]。PtRu 合金的组分和结构都会对催化剂的性能有显著的影响，据报道,40%~60%(原子分数)Ru 对甲醇氧化有最佳的催化活性。在早期的一些研究中,十分注重 PtRu 合金的合成:根据双官能团机理,要提升 CO 的氧化必须使 Ru 的位点更加接近 Pt 的位点(式(4.10));另外,Pt-Ru 更近距离的相互作用能提升 Pt 的电子效应,使 $CO_{ads}$ 更容易被去除。随着研究的深入,人们发现提高甲醇氧化活性的不仅仅是 PtRu 合金:Rolison 等发现如果 Ru 以含水氧化物的形式存在,则其对甲醇的氧化活性将会大幅的提高[42-44];Ren 等也展示了 $RuO_xH_y$ 的含量越高,直接甲醇燃料电池的性能越好[45],$RuO_xH_y$ 展现出的优异性能是因为其电子、质子和表面 OH 官能团的导电性能。虽然哪种形态 Ru 的性能最好仍有争论,但令人鼓舞的是 PtRu 基三元和四元催化剂可以进一步提高甲醇的氧化活性。实验和高通量的实验方案已经被提出,并在研究过程发现增进性能的元素有 W, Mo, Ir, Os, Ni, Co, V, Rh 等。Reddington 等证明了 $Pt_{44}Ru_{41}Os_{10}Ir_5$ 在 600 多种 Pt-Ru-Os-Ir 四元催化剂中,无论是半电池测试还是单电池测试中都是性能最佳的复合物[33]。Kim 等发现 PtRuSn 比 PtRu 的催化性能更优,MOR 活性的增强是由于作为水性催化剂的 Ru 和可以修饰 Pt 中电子的 Sn 协同作用的结果[46]。研究结果表明,第三和第四种元素的量应该低于一定的量,否则它们的存在将会使催化剂的活性降低。总的来说,Pt-Sn 催化剂表现出了较高的 CO 氧化活性,但其甲醇氧化性能不高,可能是因为在 Pt-Sn 表面甲醇吸附以及脱氢作用有所减少。与纯 Pt 相比,二元催化剂 Pt-W、Pt-Ni 和 Pt-Co 的 MOR 活性虽然得到了一定程度的提高,但仍不如 PtRu 催化剂[47-49]。

直接甲醇燃料电池技术在过去几十年间已经取得了许多成果。以氧气和空气为原料的质子交换膜直接甲醇燃料电池的峰值功率密度已经分别达到了 $500mW/cm^2$ 和 $300mW/cm^2$。当直接甲醇燃料电池的电压为 0.5V,温度接近或者高于 100℃,Pt 的载量为 $1\sim2mg/cm^2$ 时,其功率密度达到了 $200mW/cm^2$;然而,在环境温度和鼓入空气的操作模式下,功率密度只有 $10\sim40mW/cm^2$[4]。锂离子电池占据了当前全球便携式电子产品市场 60 亿美元的市场份额,但直接甲醇燃料电池由于其高能量密度,成为最有可能取代锂离子电池的产品。作为对比,以 Pd/多壁纳米管为阳极的碱性膜基直接甲醇燃料电池的峰值功率密度要低得多,只有约 $80mW/cm^2$[50];因此,碱性膜在直接 $C_{2+}$ 乙醇燃料电池中拥有更多的优势,在下面的章节中有详细探讨。

## 4.3 直接乙醇燃料电池阳极催化剂:破坏 C—C 键实现完全 12 电子传递氧化

乙醇是一种生物可再生分子,它可以通过甘蔗、玉米、谷物、棉花和许多种类的纤维素废料和农作物的光合作用制得。使用乙醇作为燃料有一个较大优势:它能减少大气中 $CO_2$ 的含量,因为植物可以吸收 $CO_2$ 并将其作为原料来生产乙醇[51]。目前使用的乙醇燃料是将变性乙醇与汽油进行混合,例如最近出现在美国中西部加油站的E85就是85%的乙醇与汽油的混合物。然而,由于所有的内燃机都受到卡诺循环的限制,从原理上看,通过直接乙醇燃料电池产生电能是一种更有效的乙醇利用方式[6]。

### 4.3.1 质子交换膜直接乙醇燃料电池的原理

在酸性条件下,阳极、阴极以及总反应如下:

$$C_2H_5OH + 3H_2O \longrightarrow 2CO_2 + 12H^+ + 12e^- \quad +0.085V \text{ 相对于 SHE(阳极)} \tag{4.10}$$

$$3O_2 + 12H^+ + 12e^- \longrightarrow 6H_2O \quad +1.23V \text{ 相对于 SHE(阴极)} \tag{4.11}$$

$$C_2H_5OH + 3O_2 \longrightarrow 2CO_2 + 3H_2O \quad +1.15V\text{(总反应)} \tag{4.12}$$

在标准状况下,热力学可逆能量效率为97%,这一数据被定义为电能产生的比率。然而,在工作条件下,电池电压一般都低于平衡电势,因此实际的能量效率比较低。例如,对于直接乙醇燃料电池来说,当它在0.55V和100mA/cm$^2$的状态下工作,$CO_2$和乙酸的选择性分别为80%和20%时,其效率为

$$\varepsilon_{cell} = \varepsilon_F \times \varepsilon_E \times \varepsilon_{rev} = (0.2 \times 4/12 + 0.8 \times 1) \times (0.55/1.15) \times 0.97 = 40\%$$

潜在的效率为$\varepsilon_E = 48\%$ (0.55/1.15)。法拉第效率$\varepsilon_F$与产品的分布有关(催化剂的选择性):产物为$CO_2$时法拉第效率为100% (12/12);而醋酸为产物时,则只发生了4电子的转移,法拉第效率为33% (4/12)。虽然高电流密度与乙醇完全氧化之间并没有必要的联系,但提高阳极催化剂对于$CO_2$的选择性将会增加直接乙醇燃料电池的整体效率和燃料的利用率。

### 4.3.2 反应机理和乙醇电氧化的催化剂

乙醇的完全电化学氧化是一个复杂的12电子转移反应,在其过程中会生成大量的反应中间体。Pt一般用作乙醇在酸性介质中电氧化的催化剂,其氧化过程主要包括两个关键步骤,分别为C—H键和C—C键的断裂。

基于半电池和单电池测试、原位红外光谱和色谱研究,图4.1所示为乙醇在Pt基催化剂上的氧化机理[52]:步骤1为乙醇通过O或者C吸附的分离吸附[53,54],在低于0.6V时形成了吸附的乙醛(步骤2);乙醛按照步骤3进行解吸并通过步骤4中所示的双官能团机理与吸附的OH反应生成醋酸,这一步经常发生在电极电势大于0.6V时且C—C键并未断裂;吸附的$CH_3CHOH$可以进一步脱氢(步骤5)并与所吸附的OH直接生成醋酸(步骤6)[6];但是,醋酸究竟是由乙醛还是$CH_3COH$生成(步骤5、6),目前还有争议。

在红外光谱中能清晰地看到0.3V RHE时的吸附峰,证明了Pt可以破坏C—C键,使得CO的吸附能在相较对低的阳极电势下进行[55]。之后的反应有两种不同路径:步骤7和8或步骤9和10。在第一种途径中,乙醇被C—H键吸

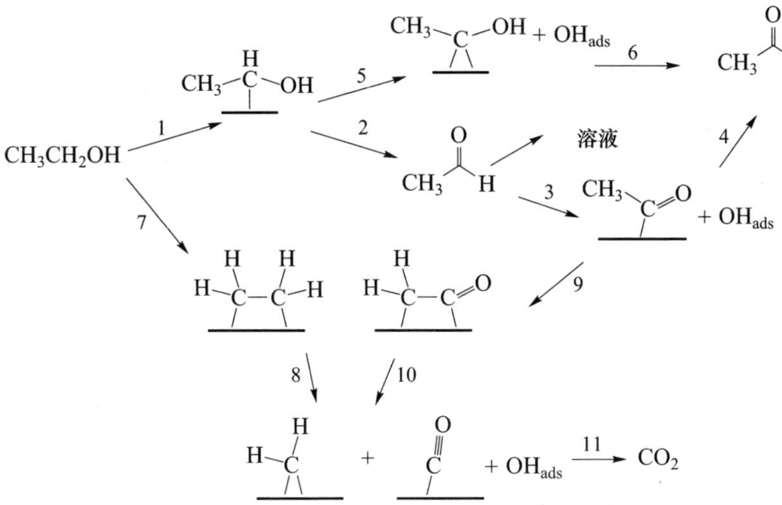

图 4.1 乙醇氧化的反应机理[52]

附之后分裂为两个碳原子;第二种途径则认为乙醛吸附后其中间产物的 C—H 键被捕获。CO 等与吸附的 OH 发生反应生成 $CO_2$,如步骤 11 所示。当电势低于 0.4V 时只检测到极少量的 $CH_4$,可能发生了如下反应[53,54]

$$Pt-(COCH_3)_{ads}+Pt \longrightarrow Pt-(CO)_{ads}+Pt-(CH_3)_{ads}, \quad E>0.3V\ SHE 时 \tag{4.13}$$

$$Pt-(CH_3)_{ads}+Pt-(H)_{ads} \longrightarrow 2Pt+CH_4, \quad E<0.4V\ SHE 时 \tag{4.14}$$

有趣的是在以 Pt/C 为阳极催化剂时,HPLC 只检测到了乙酸、乙醛和 $CO_2$[56],然而根据电极电位,在半电池电解过程中检测到了乙醛、乙酸、$CO_2$ 以及微量的 $CH_4$。同时还发现当电压低于 0.35V 时,经过长时间的电解,Pt 催化剂上只检测到了乙醛,而乙酸在该电势范围内并没有被检测到[6],这意味着乙醇的分解产物分布取决于电能的输入状况。

在酸性电解液中,Pt 基催化剂比其他铂族金属表现出了更好的电氧化活性。然而,Pt 自身很容易被多种 $C_1$、$C_2$ 中间产物毒化。研究者对于一些二元和三元的 Pt 基催化剂包括 Ru、Sn、Pb、Pd 等已经进行了深入的探讨,期待能提高其乙醇氧化的动力学性能。其中,Sn 基催化剂是最有前景的一类[2,7,57-59]:Xin 等已经检测了 Pt-M(M=Ru, Sn, Pd 和 W)等催化剂在单电池中的性能,并且发现这几种催化剂的活性顺序为 PtSn > PtRu > PtPd > Pt,W 和 Mo 也与 $Pt_1Ru_1$ 催化剂一起组成合金催化剂,但其电氧化活性与 PtSn 催化剂相比稍显逊色[60-65];Lamy 等研究了 PtSn/C(90∶10, 50∶50),PtRu/C(90∶10, 80∶10)和 $Pt_{86}Sn_{10}Ru_4$/C的性能,并证明 PtSn 的活性要比 Pt 高得多[52,56,66,67]。但是,与 Pt

相比，PtSn 催化剂上的电氧化产物分布已经发生了改变，醋酸的产量增加而乙醛和 $CO_2$ 的产量则减少；而且 Sn 的存在似乎可以将水分子激活，在较低的电势下通过双官能团机理将乙醛等氧化为醋酸；与此同时，Sn 还稀释了临近的 Pt 原子浓度，这就减少了乙醇解离吸附的概率，直接导致 $CO_2$ 产量的升高。Sn 的作用还可能包括一些 CO 氧化过程中电子方面的影响（配位效应）[68]。Wang 等利用原位红外光谱和在线质谱仪研究了 Pt/C、PtRu/C 和 $Pt_3Sn/C$，并发现 Ru 和 Sn 的添加并没有促进 C—C 键的断裂，$CO_2$ 总的产量与现在相比只降低了 2%[13]，由此证明，之前报道的 PtSn/C 催化剂上高的乙醇氧化电流密度是由于 $C_2$ 产量的提高所导致，而不是由乙醇完全氧化生成 $CO_2$ 所致。

最近的研究热点聚焦在发现可同时破坏 C—C 键的新型催化剂组分和结构，通过实现完全电氧化来增强（或至少是保持）电氧化活性。Rh 和 Pt 的添加可以促进 C—C 的断裂，然而它们整体的电氧化活性还是要比 PtSn 差一些[69,70]。Adzic 等最近发现三元合金 $Pt/Rh/SnO_2$ 纳米结构催化剂既可以使 C—C更易断裂，还能提高催化剂的电氧化动力学[71]。$PtRhSnO_2/C$ 催化剂的电氧化比活度要比 $PtSnO_2/C$ 和 PtRu/C 高很多。从图 4.2(b) 可以看出，$PtRhSnO_2/C$ 催化剂电氧化反应的起始电位与 PtRu/C 相比负移了 0.18V (0.33V - 0.15V 相对于 SHE)。图 4.2(c)和(d)中，当电势为 0.78V 时，Pt(111) 电极在 2342$cm^{-1}$处出现了 $CO_2$ 可能的非对称弹性振动特征峰，但在 $RhSnO_2/Pt(111)$ 电极上这个峰向上偏移了 0.3V，表明乙醇中 C—C 键断裂。根据傅里叶变换计算，脱氢作用和 C—C 键的断裂（16 和 17 步）对于乙醇的完全降解和氧化都是至关重要的。

$$^*CH_3CH_2OH \longrightarrow {}^*CH_3CH_2O + H^* \tag{4.15}$$

$$\longrightarrow {}^*CH_2CH_2O + 2H^* \tag{4.16}$$

$$\longrightarrow {}^*CH_2 + {}^*CH_2O + 2H^* \tag{4.17}$$

乙醇在 $RhPt/SnO_2$ 催化剂上的降解可以通过特定的构象实现（—$CH_2$ CHO）。

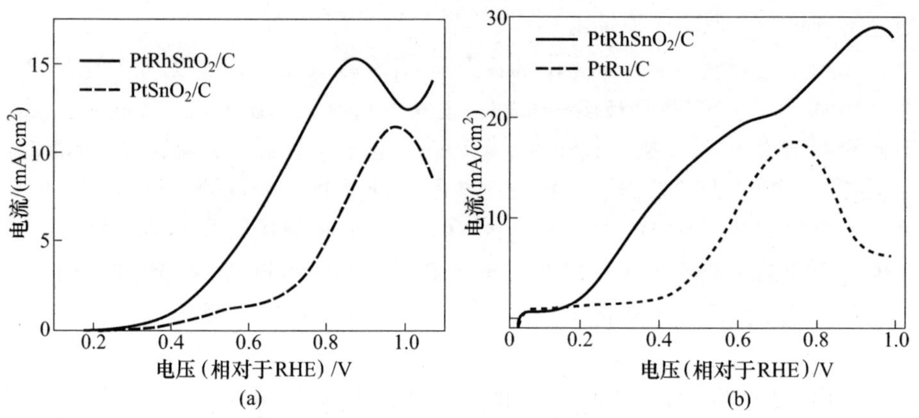

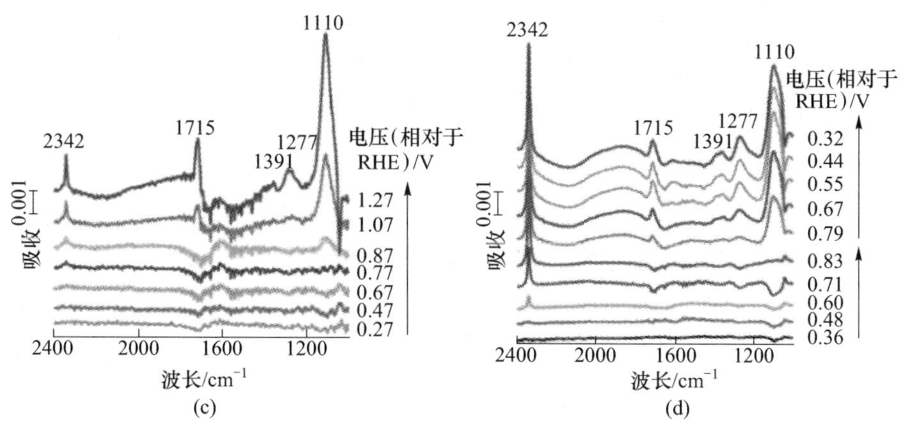


图 4.2 $PtRhSnO_2/C$ 和其他几种催化剂的用于比较乙醇氧化活性的电流－电压曲线；测定条件为 0.1M $HClO_4$ +0.2M 乙醇，50mV · $s^{-1}$；催化剂组成：(a) $PtRhSnO_2/C$ 为 30nmol Pt、8nmol Rh 和 60nmol $SnO_2$，$PtSnO_2/C$ 为 30nmol Pt 和 60nmol $SnO_2$；(b) $PtRhSnO_2/C$ 为 25nmol Pt、5nmol Rh 和 20 nmol $SnO_2$，PtRu/C 为 25nmol Pt 和 25nmol Ru。(c) Pt(111) 电极和 (d) $PtRhSnO_2/C$ 在 0.1M $HClO_4$ +0.2M 乙醇溶液中的电化学氧化原位红外反射吸收光谱

式(4.16)和式(4.17)这两个步骤，可通过掺入 Rh 等拥有更多活跃 d 带的元素与 Pt 一起构成合金来实现。

### 4.3.3 阴离子交换膜直接乙醇燃料电池

氧还原和乙醇氧化的动力学在 pH 值较高时都会有显著的提升，这是由于在碱性溶液中离子和电荷都更容易迁移[72]。阴离子交换膜直接乙醇燃料电池直接以乙醇为燃料，在阳极中，乙醇与 $OH^-$ 反应生成 $CO_2$（完全氧化）；在阴极，氧气与 $H_2O$ 和电子反应生成 $OH^-$。式(4.18)～式(4.20)分别为阳极、阴极、总反应和其理论电势：

$$C_2H_5OH + 12OH^- \longrightarrow 2CO_2 + 9H_2O + 12e^- \quad -0.75V\ 相对于\ SHE(阳极) \tag{4.18}$$

$$3O_2 + 6H_2O + 12e^- \longrightarrow 12OH^- \quad +0.40V\ 相对于\ SHE(阴极) \tag{4.19}$$

$$C_2H_5OH + 3O_2 \longrightarrow 2CO_2 + 3H_2O \quad +1.15V(总反应) \tag{4.20}$$

目前所使用的阴离子交换膜主要由季铵碱类高聚物组成[73]，研究发现其拥有很好的热力学和化学稳定性。Tokuyama A201 等商用阴离子交换膜的 $OH^-$ 电导率高达 38mS/cm[74]，这种膜能在 80℃条件下使用，结构不会发生变化。在这样的低温条件下反应，乙醇不会分解为那些不希望得到的产物；同时使用阴离子交换膜可以消除在液体碱性燃料电池中由电解液的碳酸化而引起的一系列严重问题。阴离子交换膜燃料电池与聚合物电解质膜燃料电池相比,价格上也有优势:阴离子交换膜的价格非常低廉;由于不易腐蚀的工作环境,所使用的催化剂为廉价的非铂族金属,研究发现 Ag 和 Fe/Co - N 都有拥有很强的氧还原活性和稳定性并可以用做阴离子交换膜燃料电池的阴极催化剂[72,75,76]。种种优点使得阴离子交换膜在近些年得到了更多的研究和关注。

### 4.3.4 阴离子交换膜直接乙醇燃料电池的阳极催化剂

研究表明,当 pH 值较高时,Pd 是所有已知的单金属催化剂中 EOR 活性最高的[5,7]。Pd 在地壳中的含量非常丰富,约为 Pt 的 200 倍,Pd 的价格也要比 Pt 低许多。Pd 的 EOR 活性与 pH 值密切相关[77,78]。Liang 等使用循环伏安法提出了在高 pH 值下 EOR 反应的机理[79]:在 Pd 的表面 α-C 首先被激活脱去两个氢原子,进一步破坏 O—H 形成—$CH_3CO$;乙酰基与吸附的—OH 进一步反应生成乙酸盐,这一步为反应的速控步。要想达到较高的氧化活性,乙醇和羟基的浓度之间需要有一个微妙的平衡,因为溶液中任何一种物质浓度的提高都可能会影响到所有物质的吸附,进而降低 EOR 活性。

$$Pd + OH^- \longrightarrow Pd - (OH)_{ads} + e^- \tag{4.21}$$

$$Pd - (CH_3CH_2OH)_{ads} + 3OH^- \longrightarrow Pd - (CH_3CO)_{ads} + 3H_2O + 3e^- \tag{4.22}$$

$$Pd - (CH_3CO)_{ads} + Pd - (OH)_{ads} \longrightarrow Pd - (CH_3COOH)_{ads} + Pd \tag{4.23}$$

研究者对 Pd 族模型进行密度泛函理论计算,结果表明在没有 $OH^-$ 辅助时脱氢反应很难发生;$OH^-$ 存在时,α-C 和羟基中的 H 都很容易参与到乙醇的氧化过程中,从而生成乙醛[78];然而,在循环伏安扫描中并没有发现它的氧化峰。

研究发现二元 Pd - M (M = Ru, Au, Sn, Cu 等)催化剂对 EOR 活性的提高有一定帮助:Chen 等发现 Pd - Ru 催化剂对甲醇、乙醇和乙二醇的催化活性都要比 Pd 更高一些,最佳比例为 1:1 [80];PdAu 和 PdSn 催化剂的抗毒化能力也比 Pt 更强[81]。各种氧化物(如 NiO、$CeO_2$ 等)的添加对碳载 Pd 催化剂性能的影响也被广泛地研究。在它们中间,NiO 表现出了最高的峰电流密度。氧化物之所以有效,是因为 $OH_{ads}$ 很容易在氧化物的表面生成,$OH_{ads}$ 的存在可以促进 Pd 表面 CO 等有毒中间体转化为 $CO_2$ 和其他的产物[82]。有趣的是,三元 Pd - Ni - Zn 合金催化剂表现出最优的 EOR 催化活性(在半电池中的比活性测试大于 3600A/$g_{Pd}$)且反应的稳定性也极佳[5,83,84]。Pd 基催化剂可以加速 EOR 的动力学过程,但是在高 pH 值的介质中 C—C 键是很难被破坏的,尤其是对于乙醇和丙醇等初级醇,产物仅为羧酸盐[85]。尽管多元醇(如丙三醇)的 C—C 可被破坏形成碳酸盐,但这只占到很少的一部分,主要的产物还是各种糖类[86,87]。

目前为止，阴离子交换膜直接乙醇燃料电池（AEM-DEFC）与质子交换膜直接乙醇燃料电池（PEM-DEFC）相比，具有更优越的性能[5]。例如，以 Pd-Ni-Zn/C 为阳极催化剂、Fe – Co – N/C 为阴极催化剂的 AEM-DEFC 在 80℃ 和 2atm 的 $O_2$ 背压条件下峰值功率密度可高达 $200mW/cm^2$[83]；而在 PEM-DEFC 中，在 130℃ 下，即使用 Pt 和 Pt 基催化剂仍会有很多的醇类不能被氧化，而性能最好的以 PtSn 为阳极催化剂的 PEM – DEFC 峰值功率密度也只有 $50 \sim 70mW/cm^2$[60,64]。然而，当前的 AEM – DEFC 仍面临许多挑战：把碱溶液与醇类进行混合以提供更多的 $OH^-$ 来提升反应的动力学；开发更高效的阳离子交换离聚物、构建有序的电极结构以及反应长期稳定性的检测是开发高效、耐用的 AEM – DEFC 必须要完成的研究任务。

## 4.4　直接多元醇燃料电池阳极催化剂：热电联产以及得到更高价值的化学品

### 4.4.1　多元醇电化学氧化概述

众所周知，在碱性电解液中，醇类氧化反应的催化活性会有显著提升[2,88]。近期，Koper 等在研究过程中发现醇类氧化去质子化的第一步为碱催化过程，而去质子化的第二步则与电极材料（Au 或者 Pt）吸附 $H_\beta$ 的能力密切相关[87]。密度泛函数理论计算进一步表明电极材料所吸附的 OH 等官能团，都能大大提升醇类氧化过程中的许多步骤反应速率。例如：当没有 $OH^-$ 存在时，去质子化第一步反应在 Au 和 Pt 催化剂上所需的活化能分别为 204kJ/mol 和 116kJ/mol；然而当反应过程中有 $OH^-$ 存在时，活化能则会降低一个数量级（在 Au 和 Pt 上分别为 22kJ/mol 和 18kJ/mol）[89]。这些工作表明醇类在碱性电解液中表现出高催化活性的主要原因是碱催化，而不是催化剂与 $OH^-$ 的相互作用。

由于在 pH 值较高的介质中，醇类氧化反应动力学会显著增强，因此近年来使用生物质醇类为燃料的阴离子交换膜燃料电池引起了人们的广泛关注。越来越多的研究结果表明在金属催化剂上，$C_{2+}$ 醇中 C—C 键低温下很难断裂，尤其是在高 pH 值的介质中[5]。例如，乙醇氧化的主要产物为乙醛和乙酸（或乙酸盐）。这将使直接乙醇燃料电池的法拉第效率降为 17% ~ 33%[6]。使用多元醇为原料将会是很好的选择，因为在多元醇中每个碳原子均连有一个羟基（—OH），它们可以被完全氧化为羰基（—CO）或羧基（—COOH），这样即使 C—C键不断裂它仍可以产生更多的电子，法拉第效率也会得到明显提升；另外，乙二醇和丙三醇拥有更高的能量密度（乙二醇和丙三醇的能量密度分别为

5.2kW·h/kg和5.0kW·h/kg,甲醇和乙醇的能量密度则分别为6.1kW·h/kg和8.0kW·h/kg),属于不易燃且无毒的燃料。

乙二醇的两个羟基官能团在C—C键不断裂时可以完全氧化为碳酸盐,其法拉第效率为80%:

$$CH_2OH—CH_2OH + 14OH^- \longrightarrow 2CO_3^{2-} + 10H_2O + 10e^- \quad (4.24)$$

$$CH_2OH—CH_2OH + 10OH^- \longrightarrow (COO—COO)^{2-} + 8H_2O + 8e^- \quad (4.25)$$

丙三醇中3个羟基被完全氧化生成二羟丙二酸盐,2个C—C都没有被破坏,其法拉第效率达到了71.5%:

$$CH_2OH—CHOH—CH_2OH + 20OH^- \longrightarrow 3CO_3^{2-} + 14H_2O + 14e^- \quad (4.26)$$

$$CH_2OH—CHOH—CH_2OH + 12OH^- \longrightarrow (COO—CO—COO)^{2-} + 10H_2O + 10e^- \quad (4.27)$$

此外,多元醇不完全氧化时会生成一些珍贵的化学品,如二羟基丙酮(一种贵重的鞣剂)和羟基丙酮酸(一种风味物质,可以作为DL-丝氨酸、丙醇二酸和丙酮二酸合成过程中的原材料,都是新型高聚物和药物合成过程的中间体)。因此,以阴离子交换膜燃料电池这一平台为基础,研究多元醇的热电联产并得到更高价值的化学品,不仅在开发电化学电源方面有很强的吸引力,在生物质能源的转化和利用方面也有广阔的发展前景。

### 4.4.2 乙二醇电氧化催化剂及其反应机理

乙二醇中由于其两个相邻羟基官能团的存在,使得其氧化过程要比乙醇复杂得多。对乙二醇在碱性介质中的电氧化的研究最早从20世纪70年代中期开始。由于乙二醇的完全氧化需要10个电子,因此在其氧化过程中会有多种不同反应同时进行,产生各类反应活性物种和反应中间体。图4.3所示为乙二醇氧化的反应机理[5]:方框中的化合物均是利用高效液相色谱在直接乙醇燃料电池的阳极室内检测到的,其中乙醛酸在电池(使用$Pt_{0.45}Pd_{0.45}Bi_{0.1}/C$催化剂)内0.58V下运行360min后就几乎检测不到了[90];圆圈中的化合物则是在半电池中使用红外光谱检测到的,乙醇醛和乙二醛都是反应的中间产物。乙二醇的氧化有两条路径,即催化剂中毒和未中毒。由于在催化剂上尤其是Pd的表面草酸的氧化速度非常缓慢,未中毒路径的最终产物为草酸,根据pH值的不同,它可由乙醇酸(glycolate)或乙醛酸(glycoxalate)可以进一步氧化而生成;由于乙醇酸的进一步氧化可使C—C键断裂开,中毒路径则会生成碳酸盐等$C_1$产物。外加电压对C—C键的断裂起到了重要的作用,电压低于400mV时Pt催化剂上C—C键不能断裂,而500mV下乙二醇的氧化过程中C—C会发生裂解,这一结果会导致催化剂CO中毒[91]。

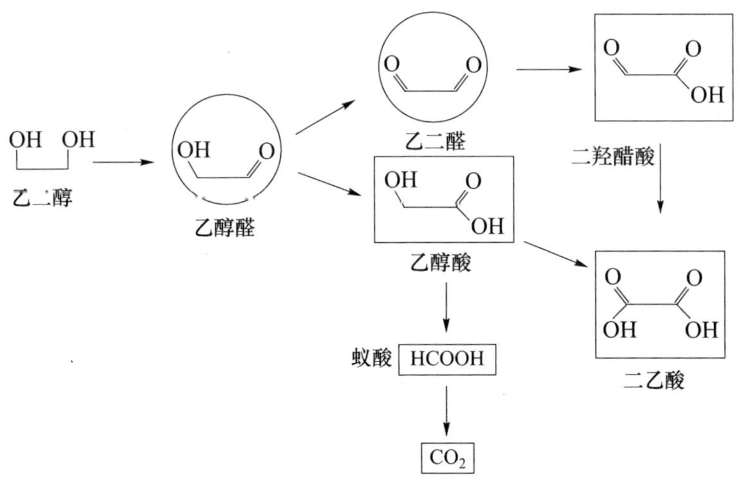


图 4.3　乙二醇氧化的反应机理[5]

在催化剂 Pt 和 Au 表面，乙二醇的电化学氧化动力学和产物分布有很大的差异。循环伏安曲线显示以 Pt 为催化剂时的起始电位更正，但是以 Au 为催化剂时的峰值电流却要比 Pt 大一些，这一结果说明其反应路径是不同的。Weaver 等研究了在碱性电解质中 Pt 和 Au 表面的乙二醇电氧化路径：发现 Au 对连续形成部分氧化的 $C_2$ 溶液（生成草酸盐和碳酸盐过程中的阶段性产物）起到了重要作用，而 Pt 则可以通过一系列的化学吸附中间产物将乙二醇氧化为碳酸盐[92]。在 pH 值较高时，Au 对醛类和醇类有很好的电化学氧化性能。碱性溶液中，乙二醇的电化学氧化在 Pd 及 Pt 电极上相比并未发生太大变化：随着乙醇酸的消耗，草酸酯和碳酸盐的含量不断增加，乙醇酸、草酸酯和碳酸盐在同一电势下生成。由于要想只产生羧酸盐产物，Pd/Pt 催化剂表面需要覆盖一层高浓度的 OH，因此低 pH 值理论上更有利于 C—C 键的断裂。

为了提高乙二醇的电氧化活性，人们对以下方法进行了探索研究：使用二元或三元催化剂，或将贵金属与其他金属构成合金结构，或使用杂质金属在贵金属表面进行修饰[90,93-95]。对 Pt - M（M = Bi, Cd, Cu, Pb, Re, Ru, Ti）催化剂的研究结果表面，Pb 和 Bi 均可提高乙二醇的氧化电流密度并接近扩散的极限值，这可归因于电催化的双官能团理论。Coutanceau 等研究了 Pt、Pt - Pd 和 Pt - Pd - Bi合金催化剂分别在碱性电解液和阴离子交换膜直接乙二醇燃料电池中的氧化活性，发现 Bi 的添加可以使起始电位提高 70mV，Pt - Pd - Bi 没能改变起始电位而是提升了电流密度；乙二醇在 Pt/C 催化剂上转化为了乙醇酸、草酸和甲酸，在 PtPdBi/C 催化剂上并没有生成甲酸，而是检测到了微量的乙醛酸；作者认为 Bi 的存在稀释了 Pt 表面的原子，有助于 OH 类物质的吸附，从而抑制了

C—C键的断裂[90]。Pd 的作用则是通过改变化学吸附物种的组成来限制 Pt 的毒化,研究发现,纳米结构的 Pd-(Ni-Zn)/C 和 Pt-(Ni-Zn-P)/C 比 Pd/C 的活性更高,达到了 3300A/gPd,并且能够改变氧化产物的分布,得到了乙醇酸、草酸和碳酸盐的混合物;在 Pd/C 电极上则更多的是生成乙醇酸;这一结果表明 Pd-(Ni-Zn)/C 可以促进 C—C 键的断裂,使氧化进行得更完全[5]。

### 4.4.3 丙三醇电化学氧化的机理

丙三醇是生物质柴油生产过程中的副产物,可以通过该途径大量生产[10,11]。由于丙三醇有可能被用于燃料电池,所以人们研究了它的电化学氧化的性能。在碱性介质中,Pt、Pd 和 Au 对丙三醇表现出了不同的作用方式。Pt/C 的起始电位比 Pd/C 和 Au/C 要低 150 mV,但峰值电流却高一些;Au/C 的起始电位较高但 GOR 活性电位区域却很广,这是因为 Au 的还原电位较高;PdAu/C 合金催化剂(Pd-Au 的物质的量比分别为 0.3:0.7, 0.5:0.5)比商业 Au/C 和 Pd/C 的起始电位要低一些,但与 Pt/C 相比还是略高[96]。

丙三醇的氧化产物包括甘油酸、丙醇二酸、乙醇酸、甲酸、草酸和 $CO_2$ 等,其分布取决于催化剂的组成、结构和操作电压(阳极过电势)等[5,50,86-88,96]。例如:以 Pd/CNT 为阳极催化剂的阴离子交换膜直接丙三醇燃料电池可提供 8.4h 的稳定电流,产生 3070C 的电流,转化效率达到 28.6%;消耗 4mmol 的丙三醇可产生 27% 甘油酸、23% 丙醇二醇盐、4% 乙醇酸、15% 草酸盐、9% 甲酸盐和 22% 的碳酸盐[50];在 2M KOH + 2M 丙三醇的溶液中,0.1 A 和 0.6~0.7V 的条件下电解池工作 15h 可产生 35% 甘油酸、36% 丙醇二醇盐、3% 乙醇酸、14% 草酸、2.5% 甲酸盐和 12.5% 碳酸盐[97]。基于上述结果,人们提出了丙三醇氧化的反应机理,如图 4.4 所示:大部分的 OH 首先被氧化产生乙醇酸,其余的 OH 被氧化生成丙醇二酸,随着 C—C 键的断裂进一步产生更多的乙醇酸;乙醇酸和甲酸进一步氧化分别生成草酸和 $CO_2$。

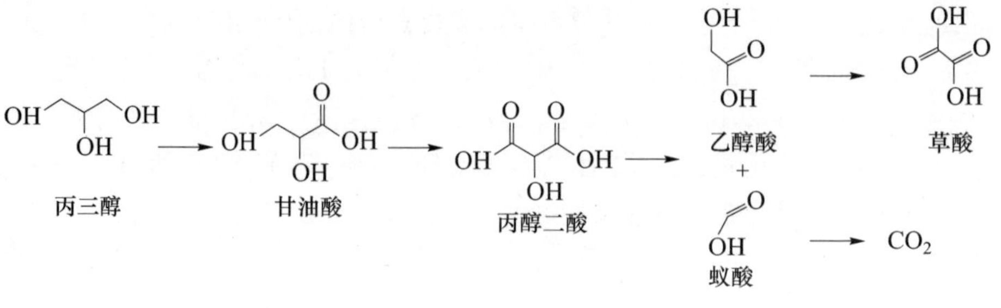

图 4.4 Pd 催化剂上丙三醇的氧化机理[50]

近期,Kwon 和 Koper 使用自己研发的在线采样和线下高效液相色谱分析系统，研究了在 Pt 和 Au 电极上丙三醇的电化学氧化机理[86]，阐述了外加电位、催化剂（Pt 和 Au）以及氧化产物分布之间的内在联系：使用 Pt 电极时，在较低的电势下只检测到了甘油酸（温度 25℃，电压小于 0.4V（相对于标准氢电极（RHE）），溶液为 0.1M NaOH + 0.1M 丙三醇）；电压升高时，由于 C—C 键的断裂，产生了乙醇酸和甲酸；当电极电势继续增长到约 0.5V 时，则能检测到丙醇二酸和草酸的生成。在 Au 电极上，丙三醇氧化的起始电位（在 0.65V）要比 Pt 高一些；在电压低于 0.8V 时产物只有乙醇酸，而电压高于 0.8V 时则会生成乙醇酸和甲酸；在 1 ~ 1.8V 的电压范围内都没有检测到丙醇二酸和草酸。该方法通过利用液相色谱，使循环伏安过程中对可溶性反应产物的检测得以实现，并对复杂多步电极反应的机理有了新的认识。

与当前的化学计量氧化法相比较时，在适当的条件下（如 30 ~ 80℃，3 ~ 10bar①），在水相多相催化体系中使用分子氧来控制丙三醇的部分（选择性）氧化是一种非常有吸引力的方法，同时由于其环境友好的优点，近年来人们对其进行了广泛的研究[89,98-105]。研究发现，Pt、Pd、Rh、和 Au 等贵金属催化剂都拥有优异的催化活性、选择性和稳定性。实验结果表明，在酸性介质中，PtBi 催化剂上生成 DHA 的选择性可高达 35%，而在碱性介质中多数 OH 会优先被氧化生成多种不同的产物，如 $C_3$ 酸（甘油酸、丙醇二酸）、$C_2$ 酸（乙醇酸、草酸）和 $C_1$ 酸（甲酸）。在多相催化中，催化剂的尺寸、结构、载体（C 或氧化物）、反应条件（如温度、氧分压以及催化剂和丙三醇的比例）和氧化剂（$O_2$ 或 $H_2O_2$）都会影响催化剂的选择性。在多相催化氧化丙三醇过程中，Au 是一种丙三醇氧化非常独特的催化剂：没有基质存在时，飞行时间接近于零（没有反应），在合适的 pH 值下，丙三醇的转化率可高达 56% 并全部转化为甘油酸[106,107]。使用 $H_2O_2$ 做氧化剂可得到高选择性的乙醇酸。图 4.5 所示为电化学氧化路径与多相催化氧化的路径。在多元醇的多相催化氧化和电催化氧化方面需要更多的实验和理论研究工作，以推动新型催化剂的发展得到更高附加值的化合物和电流。

阴离子交换膜直接丙三醇燃料电池在研究中也表现出了优异的性能，例如，以 Pd - Ni - Zn/C 催化剂为阳极、以 Fe - Co - N/C 为阴极的 AEM-DGFC 最大功率密度达到了 120mW/cm$^2$，这与质子交换膜基直接丙三醇燃料电池的性能相当，其功率密度是当前使用的以丙三醇为燃料的生物燃料电池的 2 ~ 3 个数量级（通常小于 1mW/cm$^2$）[5]。赤藓糖醇和木糖醇等比乙二醇和丙三醇更高级的多元醇已经被用于以 PtRu/C 为阳极催化剂的阴离子交换膜直接醇类燃料电池[108]，在 AEM-DGFC 中的性能比以丙三醇为燃料时要低一些，反应产物和详细的机理还需要进一步探究。

---

① 1bar = $10^5$ Pa。

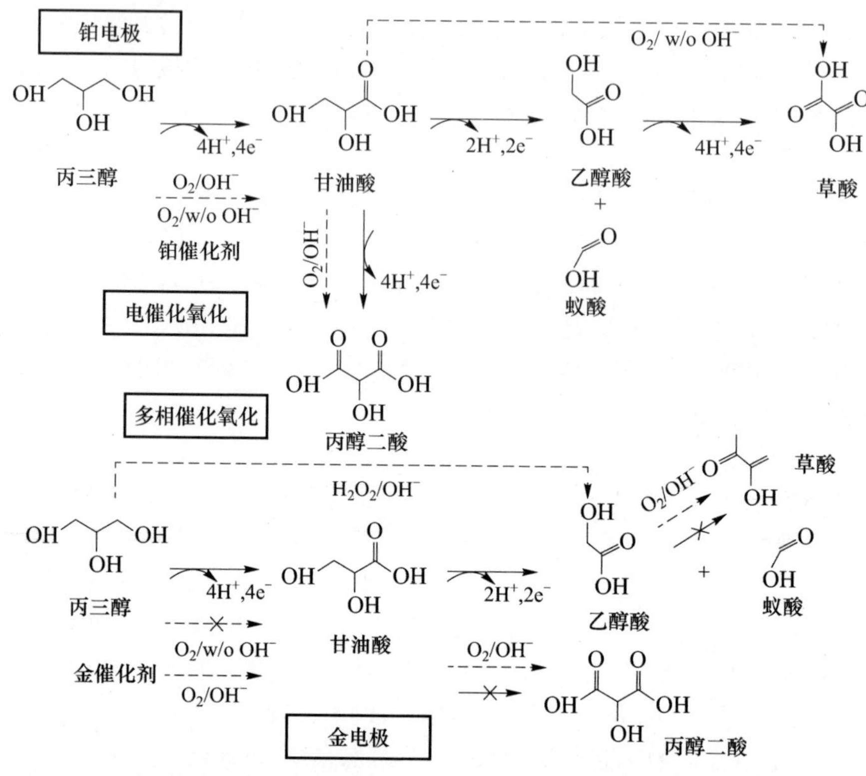


图 4.5 利用在线收集和线下高效液相色谱分析法提出的丙三醇电化学氧化路径(标记为实线箭头)[87],文献报道的丙三醇非均相催化氧化路径(标记为虚线箭头)[89]

## 4.5 金属电催化剂的合成方法

为了提高直接醇类燃料电池中醇类的电化学氧化性能,开发高活性且实用性强的电催化剂是十分必要的。催化剂的总体催化性能是由以下几个因素共同决定的:表面金属的局部电子性质(d 带偏移和电子效应)[109]、有效晶面的暴露(几何效应)[110,111]和不同金属的表面排布(整体效应)[112]。在过去的几十年间,研究者广泛研究了与理论计算相结合的单晶催化剂,并对催化剂结构与功能之间关系提出了一些有价值的见解[110,113]。例如:研究过程发现 CO 氧化的过电位顺序为 Pt (111) < Pt (554) < Pt (553),Pt (553)与 Pt (111)之间的峰值电位差达到了 0.17 V[114];Pt(111) - Ni(111)催化剂能通过电子结构的修饰(充分利用 d 带中心偏移)将表面覆盖的 $OH^-$ 脱去,这样可以使氧还原反应活性提高 90 倍[109,115]。然而,仿照单晶结构来准确合成实用催化剂仍是一个巨大的挑战:实用催化剂合成过程中的关键因素包括控制颗粒尺寸、粒径分布、颗粒形状、电子结构、化学组分、表面组成、整体分布、合金度、催化剂材料的氧含量等[116-118],

都会影响其性能;同时,催化剂和载体在苛刻的电化学环境下的长期稳定性是另外一个需要关心的问题[119]。因此,催化剂制备方法成为了决定其电化学催化剂活性、选择性和稳定性的一个关键因素。

制备实用催化剂总体上可分为"自上而下"的物理法(由宏观到纳米尺度)和"自下而上"的化学法(由分子、原子水平到纳米级别)两类[120]。物理法利用热蒸发或溅射等方法,在真空下将金属进行雾化[121-123];通过物理法制备的金属催化剂具有低污染的特点,可被用于机理研究;然而,采用物理法制备的金属催化剂其尺寸大小、分布和形貌都无法控制。和物理法相比较,化学法更易精确地控制颗粒的尺寸、形貌和结构[124]。近期出现的湿化学法合成技术最有希望能够准确控制纳米结构金属催化剂的大小、形状、结构和表面晶面[118,124-140],在制备高性能催化剂方面显示了巨大的潜力。典型的湿化学合成过程:化学还原溶剂中的金属前驱体使其成核,进而添加或不添加稳定剂,在合适的碳载体上控制核的生长,以得到最终想要的金属纳米粒子。尽管利用电化学合成法(如欠电位沉积法)准确制备核壳结构金属纳米粒子已经取得了很大的进步,但以下部分还是主要介绍利用化学还原法合成碳载体电化学催化剂的一些最新进展。

## 4.5.1 浸渍法

浸渍法包括两步:首先将多孔碳载体浸泡在金属前驱体溶液中,然后使用HCHO、HCOONa、$NaBH_4$、$NH_2NH_2$ 或 $H_2$ 等还原剂在适当的条件下将金属前驱体还原为金属纳米粒子[143-158]。当核形成以后,金属颗粒的生长主要会受到碳载体孔的限制,根据前驱体的渗透和浸润情况,多孔基底的形貌和孔径分布也会在影响纳米粒子的生长方面起到重要的作用;此外,还原剂的还原动力学和质量传递也会影响成核的数量和速率,进而控制颗粒的尺寸和粒径分布状况。为了在碳载体上得到均匀分散的金属粒子,使用乙醇或丙醇作溶剂,对碳载体表面进行氧化处理,对金属纳米粒子的负载会有所帮助。在适当的合成条件下,通过浸渍法制备的颗粒尺寸可以控制在10nm左右。由于浸渍法操作简单且容易进行扩大化生产,近年来成为了电化学催化剂的制备过程中最常用的方法。它主要的缺点是不能准确控制颗粒的尺寸,除非多孔基底的孔径分布很窄;换言之,只有在高度有序的介孔碳中才能准确控制金属颗粒的尺寸[147]。使用浸渍法制备的催化剂,其形状和结构也很难控制。研究使用包含两种不同金属(如Pt和Ru)的有机分子作为前驱体,代表了浸渍技术的一项重大进展。Lukehart等合成了($\eta$-$C_2H_4$)(Cl)Pt($\mu$-Cl)(2)Ru(Cl)($\eta$(3):$\eta$(3)-2,7-二甲基辛二烯)分子,用XC-72R炭黑作为载体,使用浸渍法在适当的气氛下处理后得到了质量分数分别为16%和50%两种PtRu/C合金催化剂,其粒径分别为3.4mm和5.0nm[146]。虽然该法制备的PtRu/C催化剂与商业PtRu/C催化剂相比具有更高的甲醇氧化活性,但是其复杂的合成过程限制了其在大规模生产方面的应用。

### 4.5.2 胶体法

由于可精确控制材料的尺寸、形状和结构等优点，胶体法广泛用于金属纳米颗粒的制备[125-135,137-141]。胶体法主要包括两个步骤：首先是制备出金属胶团，然后再将其负载在碳载体上。该方法的一项关键步骤为通过添加稳定剂来阻止胶体的团聚，常用的稳定剂包括聚合物、共聚物、表面活性剂、配体、溶剂、长链醇类和金属有机化合物等。成核和粒子生长过程的有效控制和分离是调节胶体尺寸的关键因素，通常利用有机分子在金属表面的空间位阻或纳米粒子之间的静电排斥作用来得到均匀的金属粒径分布。Watanabe 等发明了一种简洁的氧化物胶体法来制备 PtRu/C 催化剂：首先在加入反应物时，通过控制溶液的 pH 值，制备出了 PtRu 的胶体氧化物；然后向胶体氧化物中鼓入 $H_2$ 将其还原，得到高分散小尺寸(2~3nm)的 PtRu/C 催化剂[159]。Bönnemann 和 Richards 设计了一种巧妙的有机相还原法来准确控制催化剂的尺寸和粒径分布：在四氢呋喃溶液中使用 $NR_4BR_3H$ 还原有机金属前驱体；使用该方法，已经成功制备出 PtRu/C、PtRuSn/C、PtRuW/C 和 PtRuMo/C 等二元和三元的催化剂，并且具有比商业 PtRu/C 催化剂更优异的催化性能[48,160]。

#### 4.5.2.1 多元醇法

多元醇法已经在多元醇或二醇溶液中的单金属或多金属胶体的制备过程被广泛应用。在该方法中，多元醇作为溶剂的同时还被用作随温度变化的还原剂(图 4.6)[126,133,139,161-164]；PVP 可用作稳定剂，可以控制颗粒的尺寸、形状和结构。该方法的关键步骤为控制无机前驱体在升温过程中还原时的温度。研究过程发现，反应速率与形貌有密切关系：在较高反应速率下，得到的颗粒形貌较好，而在较低的还原速率下，核的形成和生长则属于动力学控制过程，最终的形貌也会偏离热力学有利的形状。因此，合理的动力学调控可以准确地控制颗粒的尺寸和形状，尤其是在晶种阶段。Pt 的一些纳米结构可以通过这种方法来准确合成，如纳米枝晶[139]、纳米棒和纳米条[152]等。该法的缺点一是稳定剂可能会导致部分催化活性的损失，若对制备出的催化剂进行后处理来除去稳定剂，又可能会导致催化剂颗粒的团聚或形貌结构发生变化。

Wang 等报道了在不添加 PVP 的条件下，利用多元醇法制备出了均匀的 Pt、Rh 和 Ru 胶体，其平均粒径达到了 2~4nm[165]；再其后续研究中，负载了贵金属或贵金属/过渡金属的纳米颗粒也陆续被制备出来[60-65,166-179]。Xin 等成功合成了碳负载的 Pt、PtRu、PtPd、PtIr、PtSn、PtW 和 PtFe 催化剂，将其粒径控制在了 2~5nm[60-65,166,168,170-173,175,177]，这种制备方法的优点是简单且易于进行大规模生产。在该合成体系中，$H_2PtCl_6$、$RuCl_3$、$PdCl_2$、$SnCl_2$、$FeCl_2$ 等无机化合物均可作为金属前驱体，同时体系中水的含量可以控制粒子的尺寸和分布；在碱性、

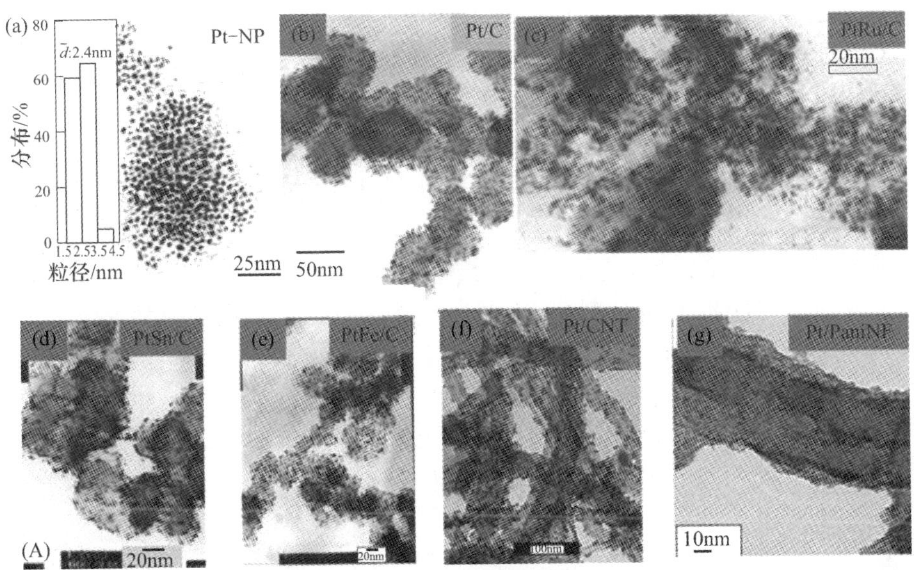


图4.6　使用多元醇法合成的Pt和Pt基电催化剂(不含PVP)。
(a) Pt纳米粒子($D=2.4$nm)[165]；(b) Pt/C(40%(质量分数)，$D=2.9$nm)[168]；(c) PtRu/C (20%(质量分数)Pt，$D=1.9$nm)[60]；(d) PtSn/C(20%(质量分数)Pt，$D=1.9$nm)[62]；(e) PtFe/C(20%(质量分数)Pt，$D=3.4$nm)[173]；(f) Pt/CNT (30%(质量分数)，$D=4.46$nm)[174]；(g) Pt/聚苯胺纳米纤维(PaniNF，30%(质量分数)，$D=2.1$nm)[176]

135～150℃的条件下进行还原反应3～4h后，调整溶液的pH值为2～3，将金属胶体破胶后再负载到碳载体上，通过此种方法可得到高性能的合金结构催化剂。由于此方法的步骤较为简单且原料成本低廉，在大规模生产过程中很有吸引力。然而，该法并不能成功应用在由贵金属/过渡金属构成的合金催化剂上，由于乙二醇是一种弱还原剂，不能完全还原过渡金属，因此在最终的双金属催化剂中过渡金属的含量往往很低，这也是在PtFe/C催化剂中Fe的含量只达到设定值的1/3～1/4的原因[173]。

#### 4.5.2.2　有机相合成法

2000年，IBM公司的Sun等利用有机相合成法，成功制备出了直径为4～5nm的Fe-Pt磁性纳米粒子[129,180]。这种方法随后被迅速推广到金属催化剂的制备过程中[134,135,137,181-188]。在非极性有机溶剂中，贵金属和过渡金属的前驱体可以更加紧密地接触并且氧化还原电位也更加接近，更有利于形成均匀的二元金属晶核，从而可以控制二元纳米粒子的生成。在合成过程中，过渡金属前驱体可以被有机强还原剂完全还原，如$LiBetH_3$；使用不同的C18、C16表面活性剂作稳定剂时，可以选择性吸附在特定的金属晶面上，这样不仅可以阻止纳米粒子的团聚，同时还能控制金属核的生长，得到理想形貌（如纳米线、纳米棒和纳米叶）[134,184,186,188]和结构（如核壳结构）的催化剂[135]。

图4.7(a)简要地展示了该制备方案。如图4.7(b)所示，Sun等第一次使用了一套简洁的方案来制备直径为4～5nm的Fe－Pt磁性材料。同时通过此种方法制备的PtCr/C、PtCo/CNT和PdNi/C催化剂，粒径为2～5nm，并且都具有优异的合金结构（图4.7(c)～(e)）。如图4.7(f)所示，作者通过调整油酰胺和十八稀的比例成功合成了直径为2～3nm、长度为10～100nm的PdFe纳米线。利用该方法制备的PdNi/C催化剂在碱性电解液中表现出了很高的甲醇氧化反应活性，这可能与表面Pd和Ni特殊的相互作用有关。

使用该法制备的催化剂在电催化应用之前应该除去反应过程中添加的表面活性剂。据文献报道，表面活性剂在250℃的温度下处理4h即可除去[185]，研究还发现，使用有机酸处理和电化学测试有利于表面活性剂的去除，并能获得更高性能的催化剂[187]。

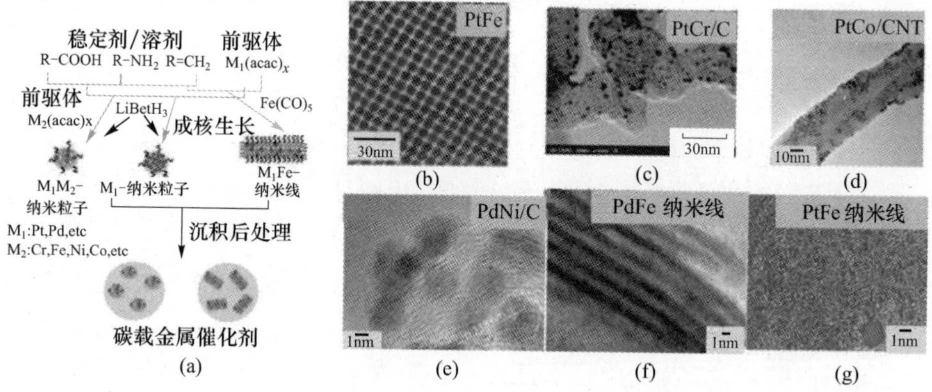


图4.7 (a)有机相合成法的示意图，通过此种方法制备催化剂；(b)PtFe纳米颗粒[129]；(c) PtCr/C (28%（质量分数），$D=2.3$nm)[187]；(d) PtCo/CNT (20%（质量分数）Pt，$D=2.0$nm)[180]；(e) PdNi/C (20%（质量分数）Pt，$D=2.4$nm)[187]；(f) PdFe纳米叶($D=1.8$nm，$L=100$nm)[190]；(g) PtFe纳米线($D=2.7$nm)[188]

### 4.5.3 微乳液法

微乳液法是将两个分别包含了还原剂和金属前驱体的溶液混合在一起，金属前驱体的还原反应被限定在微乳液中进行，其优点在于能够同时控制合金纳米颗粒的尺寸和组成[189-196]。微乳液就是极小的一滴由表面活性剂包裹的前驱体溶液。微乳液是连续分散的均匀液相，它与包含了液相的金属前驱体是不互溶的，其尺寸为几纳米到几百纳米，这由表面活性剂分子和两个不相溶液相之间自由能的差异引起的平衡表面自由能决定；水包油微乳液体系中，分散相为油相，水则形成了连续介质；反相微乳液则是油包水的微乳液体系。由于化学反应在微乳液中进行，微乳液充当纳米尺度的反应器，因此得到产物的粒径分布非常窄（如 2 ~ 5nm）。反应速率快，将肼和 $NaBH_4$ 等还原剂加到微乳液中，反应过程只需要几分钟。金属纳米颗粒的尺寸和粒径分布可以通过将还原剂也限定在单独乳液中的双微乳液法来进行控制。通过调节水和表面活性剂的物质的量比可以控制纳米颗粒的尺寸。此方法的另外一个优势是可以用来在碳载体上合成二元金属电催化剂，并且能得到优异的合金结构。微乳液法的缺点是不能控制形貌，同时在制备过程中使用了昂贵的表面活性剂分子，还需要额外的清洗步骤，大大限制了该法在大规模生产中的应用。

### 4.5.4 其他方法

近年来，一些非传统的合成技术也已经应用在催化剂的工业化制备过程中，例如：3M 公司的研究人员开发了一种简洁的 PVD 法来制备薄片状的催化剂（NTFC），并表现出了非凡的活性和稳定性[123]；卡博特公司成功开发出了喷雾转换反应法[197]，首先生成包含了金属前驱体和碳载体的液滴，然后在设定的温度和压力下使其热分解，最终在碳载体上得到分布均匀的催化剂纳米颗粒，制备出的 PtRu/C 催化剂粒径为 2 ~ 4nm，并表现出了很高的催化剂活性和出色的稳定性，同时也降低了成本。

## 4.6 阳极催化剂载体 – 碳纳米材料

催化剂载体是电化学催化剂中必不可少的一部分，其主要功能是使 Pt 纳米颗粒能够高度分散以防止团聚，并为电极的三相反应界面提供持续电子传递路径。在燃料电池的发展史上，利用高度分散的 Pt/C 催化剂替代传统的 Pt 黑具有里程碑式的意义，它使 Pt 在电极上的负载量减少了一个数量级。合适的碳载体应具有出色的导电能力、大的比表面积、合理的孔结构以及良好的电化学稳定性[116]。炭黑作为载体，被广泛用于低温燃料电池催化剂中，可由油炉和乙炔制备而成。炭黑的优点在于具有较大的比表面积和很好的导电性。油炉炭黑如 Vulcan XC-72，由于它的低成本和高的可用性，已经被广泛用作电催化的载体，其比表面积达到了 200 ~ 300$m^2/g$；其缺点是含有大量孔径小于 2nm 的微孔，而反应气体在小于 2nm 的孔径中无法顺利供应；另外，有研究报道炭黑的电化学稳定性在燃料电池的实际应用中仍存在问题，在高电势下操作时碳载体被腐蚀会导致 Pt 活性组分的团聚和侵蚀[119,198]。因此，寻找低温燃料电池中更加稳定的催化剂载体迫在眉睫。近几十年来，人们发现并合成了各种不同类型的碳纳米材料，如富勒烯、碳纳米管、石墨烯和介孔碳等，大大促进了纳米技术和纳米材料中新研究领域的发展。这些碳纳米材料作为下一代电催化载体材料已经被广泛的研究，并取得了许多令人振奋的结果。

### 4.6.1 碳纳米管

碳纳米管作为一种新颖的催化材料，由于其高纵横比和独特的电子特性[166,167,172,174,175,177,178,199-206]，自发现以来就受到了人们的广泛关注[199]。碳纳米管是碳的一种同素异形体，拥有圆柱形的纳米结构，可由石墨片的弯曲制得。碳纳米管可分为单壁纳米碳管（SWNT）、双壁纳米碳管（DWNT）和多壁纳米碳管（MWNT）。

碳纳米管可以通过电弧放电、化学气相沉积、激光辐射等方法制备，其管径为0.7～100nm，管的长度也从亚微米到厘米不等。碳纳米管是一种独特的一维碳材料，拥有出色的导电性能，其中多壁纳米碳管的电导率为1000～2000S/cm，这与炭黑XC-72的电导率4～10S/cm相比要高得多。此外，与拥有50%微孔的XC-72相比，碳纳米管没有微孔结构。

Che等首次利用多孔氧化铝为模板来制备直径为200nm、壁厚为20nm的碳纳米管（图4.8）。在将PtRu前驱体浸渍到碳纳米管上之后，利用HF将$Al_2O_3$框架除去；在580℃、$H_2$气氛下还原3h后得到了细小均匀的PtRu纳米颗粒（(1.59±0.3)nm）。研究发现，碳纳米管负载的PtRu纳米颗粒在酸性电解液中拥有很大的甲醇氧化特征峰[200]。利用相同的方法，Rajesh等考察了一系列直径为200nm的MWNT的甲醇氧化性能，并发现它们的活性顺序为PtRu/MWNT > Pt－$WO_3$/MWNT > PtRu/XC-72[203]。

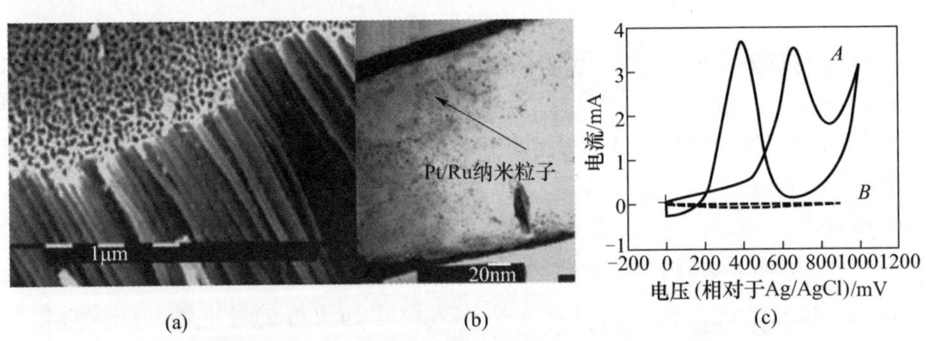


图4.8 (a)碳纳米管和(b)PtRu/CNT的TEM图以及(c) 2M甲醇+1M $H_2SO_4$条件下甲醇氧化循环伏安曲线。(c)中*A*为沉积了PtRu纳米颗粒后的曲线，*B*为沉积了PtRu纳米颗粒前曲线[200]

在使用碳纳米管之前，需要使用含氧的官能团对碳纳米管的外壁表面进行修饰，使其可以固定金属纳米粒子：Li等使用$H_2SO_4$－$HNO_3$混合溶液对多壁纳米碳管的表面进行处理，各种不同的官能团，如羟基(—OH)、羰基(—CO)和羧基(—COOH)可以被接枝于碳纳米管的表面。Li等利用多元醇法制备了10%（质量分数）Pt/MWNT，作为直接甲醇燃料电池阴极催化剂，发现其峰值功率密度达到了 43%[186,202]；Liu 等利用微波辅助多元醇法制备了直径为 2～6nm 的 PtRu/CNT，该催化剂在单电池测试中展示出了与商业 PtRu/C（E-TEK）催化剂相当的甲醇氧化活性[167]；由 Iijima 制备的单壁纳米碳管是一种特殊类型的碳纳米管，将 Pt 和 PtRu 纳米颗粒沉积在 SWNH 的外壁上，在直接醇燃料电池中显示了很高的催化活性，可以归因于高的电导率（低内阻）、高纯度（低硫含量）和薄的催化层厚度。近期，Li 等对比了 3 种不同类型的 PtRu（2～5nm）催化剂，分别以碳纳米管（MWNT、DWNT 和 SWNT）为载体，该系列催化剂显示出了截然不同的甲醇氧化活性（图 4.9），其中 PtRu/DWNT 表现出最优的氧化活性：它可减少贵金属 75% 的用量，且仍保留 68% 的功率密度[177]，需要进一步细致的研究来解释由小直径碳纳米管所带来的优异性能。

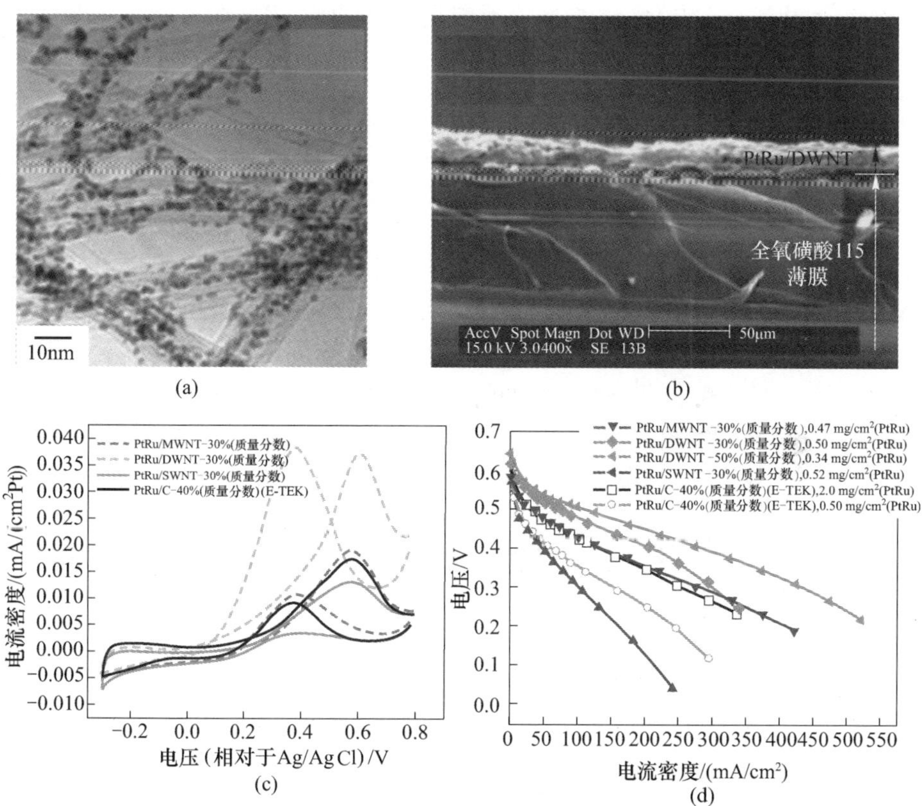


图 4.9 （a）PtRu/DWNT（50%（质量分数））的 TEM 图；（b）PtRu/DWNT 薄膜的 SEM 图；（c）在 0.5M $H_2SO_4$ + 0.5M 甲醇中 PtRu/CNT 催化剂的甲醇氧化曲线；（d）PtRu/CNT 为阳极催化剂的单电池极化曲线[177]

为了进一步提高催化剂的利用率和提高反应物的质量传递速率，人们开发了一种有序的电极结构的技术，可以使碳纳米管直接生长在气体扩散层上。Sun 等直接将碳纳米管长在了 Co-Ni 催化剂上面，然后通过离子交换技术负载上尺寸很小的 Pt 纳米颗粒(1.2 nm),Yan 等使用欠电位沉积法将 Pt 纳米颗负载到生长在碳纸上的碳纳米管上面[205]。这两种方法制备的电极性能都高于 Pt/C 电极。Yan 等进一步利用过滤法制备了有序的碳纳米管电极:使用亲水性尼龙过滤器,以形成一种疏水的碳纳米管催化剂薄膜,由 Pt/MWNT 薄膜组成的膜电极在质量传递控制区域表现出了很好的单电池性能[175]。

碳纳米管具有很高的石墨化结构,与炭黑相比表现出了更稳定的电化学性能。炭黑的腐蚀电流为 0.5mA/mg,MWNT 为 0.2mA/mg;在 0.9V 的电压下,Pt/MWNT 在 60℃,0.5M 的 $H_2SO_4$ 溶液中测试 170h 后,其电化学活性面积衰减了 25%,而 Pt/C 衰减了 80%;Pt/MWNT 的平均粒径从 2.8 nm 增长到了 3.0nm,而 Pt/C 的粒径则从 2.5nm 增长到了 5.0nm[207]。因此,碳纳米管优异的电化学稳定性,使其很有希望成为新的燃料电池催化剂的载体。

### 4.6.2 碳纳米纤维

碳纳米纤维(CNF)是一种重要的一维碳纳米材料,人们已经广泛地探讨了其作为电催化剂载体的可能性[146,151,203,208-212]。碳纳米纤维可以通过化学气相沉积法制备,与碳纳米管相比,价格更低廉。碳纳米纤维可被制成不同的形状,例如薄片状、丝带状和人字形。Bessel 等采用浸渍法成功制备出了小颗粒的 Pt/CNF,有趣的是该催化剂中 Pt 纳米颗粒看起来像是盘状的,这表明金属和载体之间有很强的相互作用;5%(质量分数)Pt/CNF 表现出了与 30%(质量分数)Pt/CB 相媲美的甲醇氧化活性。Lukehart 等制备出了粒径为 5~9nm,质量分数为 40% 的 PtRu/CNF 催化剂,与未负载的 PtRu 催化剂相比其甲醇氧化活性提高了 50%[146,151]。近期,Li 等报道碳纳米纤维基催化剂可以有效增加电极中全氟磺酸的含量(30%~50%),这是由于一维碳纳米纤维可以提供持续的电子传输路径所致,该研究为利用一维碳材料在制备高性能燃料电池电极中的应用提供了新的思路。

### 4.6.3 有序介孔碳

有序的介孔碳材料拥有极小孔径分布的有序介孔结构。均匀的小金属催化剂即使用浸渍法也很容易制备。在直接甲醇燃料电池中,有序介孔碳作为一种碳载体材料已经被广泛地研究[147-213]。有序的介孔碳可利用有序介孔二氧化硅为硬模板剂通过纳米铸造法制备而成[147,213-216]:将碳的前驱体嵌入到有序介孔二氧化硅模版中,然后在惰性气体保护下进行高温碳化,最后去除硅模板就得到了有序介孔碳材料。介孔碳的物理化学性质,如孔径、连通性、形貌、表面官能团、导电性和热稳定性都可以通过调整介孔二氧化硅模板、碳前驱体、碳化温度、加热环境以及后处理等多种手段来进行调节。也可以利用软模板法(自组装),通过碳前驱体和有机模板剂之间的自组装来直接合成有序介孔碳[217-219]。所制备的有序介孔碳材料拥有规整的介孔、大的比表面积（700～2000$m^2$/g）和高的孔容（1～2$cm^3$/g）。由于有序介孔碳大的比表面积和独特的孔隙结构，小于3nm的Pt、PtRu纳米颗粒很容易通过浸渍法控制合成。以Pt/OMC为阴极催化剂的直接甲醇燃料电池在50℃、0.45V的电压下功率密度达到了104mW/$cm^2$，可以将Pt的负载量减少3倍（从6mg/$cm^2$到2mg/$cm^2$）[220]。Samsung发现一个25$cm^2$大小的直接甲醇燃料电池在室温条件下使用0.75mol/L的甲醇溶液，在8V时的功率密度达到了80mW/$cm^2$，消耗100mL甲醇燃料可供一台笔记本电脑工作10h。

### 4.6.4 石墨烯片（GNS）

石墨烯作为一种单原子层的二维碳材料，在基础科学和实际应用领域都引起了人们的广泛关注[221,222]。石墨烯具有大的比表面积、高的导电性、独特的石墨化平面结构和潜在的低成本等特点使其有望成为燃料电池的催化剂载体[223-227]。另外，石墨烯纳米片还能改变它上面所负载Pt的性质：Pt/GNS的电流密度（0.12mA/$cm^2$）与Pt/CB（0.03mA/$cm^2$）相比提高了4倍；CO氧化研究表明，Pt/GNS对CO的吸附率与Pt/C相比要低40倍，这是由于在GNS上负载上的Pt纳米粒子小于0.5nm，这将使铂获得特殊的电子性能，进而改变它的催化活性[224]；在进行完电化学耐久性测试后，Pt-功能化石墨烯保留了49.8%的原始电化学表面积，然而商业催化剂则只能保留33.6%，因此，功能化石墨烯上的Pt在电化学耐久性测试中要比商业催化剂稳定，这是由于它拥有更多的石墨化结构[223]。

## 4.7 未来的挑战和机遇

低温直接乙醇燃料电池作为未来的电化学能源已经引起人们广泛的关注。阳极催化剂是决定电池整体性能和寿命的最关键部件之一。为了弄清楚催化剂结构和功能之间的关系、原位表征反应中间产物、阐明电催化反应机理以开发新型的催化材料，人们进行了大量而广泛的探索。虽然在制备高效燃料电池催化剂方面已经取得了很大的进步，然而在纳米或者原子尺度大规模、准确地制备出高活性、持久可靠、稳定的燃料电池催化剂仍未实现。新型的合成方法应该具有更廉价、简单并容易实现扩大化生产等优点；催化剂应该具有的异质纳米结构或是直接沉积在电极上的；在实际的电化学反应过程中常会发生纳米结构催化剂的重组（改变表面的组分、晶面、整体构型和电子结构），这些都会显著影响它们的活性稳定性。在催化剂实际应用之前，了解它们在实际电催化过程中的重组，以对其结构改变产生更深刻的认识是很有必要的。随着纳米技术的快速发展，原位表征技术和先进的计算能力都将会对先进催化材料的开发产生巨大的帮助。

理想的乙醇氧化产物是 $CO_2$。然而，直接 $C_{2+}$ 醇类燃料电池中 C—C 键的断裂仍面临巨大的挑战，尤其是在低温和低的阳极过电势条件下（如温度低于 90℃）。对于乙醇的初步氧化，纳米结构的 PtRhSn/C 在提升反应动力学和破坏 C—C 方面都有很好的效果。若将组合化学法与理论计算结合起来，可有效促进乙醇完全氧化领域高效的三元或四元 PtSn 基催化剂的发展。

另外，高的反应动力学有时与乙醇的局部氧化有关。利用可再生生物多元醇联产得到电和更高价值化合物，将会使生物燃料的生产更为有用和吸引人。有效的联产是以高选择性电催化剂的发展为基础的。催化剂对多元醇的电化学氧化与催化剂的种类、结构、操作电势以及反应条件有关[96,228-233]。它们之间的关系可以通过实验、分析和理论研究工具等来分析清楚。近些年，利用多相催化生产生物可再生化合物已经取得了显著的进步。学习由选择性多相催化氧化发展出来的知识，可能会引导低温乙醇燃料电池高选择性电催化氧化催化剂的发展。

研究直接醇类燃料电池阴极上的氧还原反应是一项长期的科学挑战。即使是活性最强的 Pt 催化剂，在开路电压下它的过电势也超过了 200mV；醇类的交叉反应导致阴极上产生了短路电压（额外损失 150～200mV），这样燃料电池的效率和燃料利用率都会显著降低。因此，开发高活性、低贵金属载量的阳极催化剂和乙醇容忍性好的非贵金属阴极催化剂是制备高活性的直接醇类燃料电池的迫切需求。因为阴离子可以从阴极扩散至阳极，醇类的交叉问题将会导致阴离子交换膜电池性能大幅降低。然而，现在使用的阴离子交换膜的一个很大的弊端在于它们低的阴离子电导率和低的化学稳定性（短的寿命）：NaOH 和 KOH 液体基底必须要和乙醇燃料混合在一起以提高催化剂活性位点周围的 pH 值，才能最终提高直接醇类燃料电池的性能；碳酸盐往电极孔里渗透也会减少反应气体的质量传递进而降低燃料电池的性能。因此，发展新型聚合物以制备高性能阴离子交换膜应该被看作是与研发下一代低液基或零液基低温醇类燃料电池催化剂同等重要的事情。

## 致　谢

衷心感谢李博士课题组成员的贡献和密歇根理工大学的美国国家科学基金会（cbet－1032547 和 cbet－1159448）、美国化学学会石油研究基金，以及周博士的大力支持！

## 参考文献

[1] Wasmus, S. and Küver, A. (1999) Methanol oxidation and direct methanol fuel cells: a selective review.

*Journal of Electroanalytical Chemistry*, **461** (1 -2), 14 -31.

[2] Lamy, C., Belgsir, E. M., and Léger, J. M. (2001) Electrocatalytic oxidation of aliphatic alcohols: application to the direct alcohol fuel cell (DAFC). *Journal of Applied Electrochemistry*, **31** (7), 799 -809.

[3] Gasteiger, H. A., Kocha, S. S., Sompalli, B., and Wagner, F. T. (2005) Activity benchmarks and requirements for Pt, Pt-alloy, and non-Pt oxygen reduction catalysts for PEMFCs. *Applied Catalysis B: Environmental*, **56** (1 -2), 9 -35.

[4] Aricò, A. S., Baglio, V., and Antonucci, V. (2009) Direct methanol fuel cells: history, status and perspectives, in *Electrocatalysis of Direct Methanol Fuel Cells: From Fundamentals to Applications* (eds. H. Liu and J. Zhang), Wiley-VCH Verlag GmbH, Weinheim, pp. 1 -78.

[5] Bianchini, C. and Shen, P. K. (2009) Palladium-based electrocatalysts for alcohol oxidation in half cells and in direct alcohol fuel cells. *Chemical Reviews*, **109** (9), 4183 -4206.

[6] Lamy, C., Coutanceau, C., and Leger, J. -M. (2009) The direct ethanol fuel cell: a challenge to convert bioethanol cleanly into electric energy, in *Catalysis for Sustainable Energy Production* (eds P. Barbaro and C. Bianchini), Wiley-VCH Verlag GmbH, Weinheim, pp. 1 -46.

[7] Antolini, E. and Gonzalez, E. R. (2010) Alkaline direct alcohol fuel cells. *Journal of Power Sources*, **195** (11), 3431 -3450.

[8] Yu, E. H., Krewer, U., and Scott, K. (2010) Principles and materials aspects of direct alkaline alcohol fuel cells. *Energies*, **3** (8), 1499 -1528.

[9] U. S. DOE (2011) 2010 Annual Progress Report: DOE Hydrogen Program, in Related Information, U. S. Department of Energy.

[10] Huber, G. W., Iborra, S., and Corma, A. (2006) Synthesis of transportation fuels from biomass: chemistry, catalysts, and engineering. *Chem Rev.*, **37** (52), 4044 -4098.

[11] van Santen, R. A. (2007) Renewable catalytic technologies: a perspective, in *Catalysis for Renewables: From Feedstock to Energy Production* (eds G. Centi and R. A. van Santen), Wiley-VCH Verlag GmbH, Weinheim, pp. 1 -19.

[12] Huber, G. W. (2008) *Breaking the Chemical and Engineering Barriers to Lignocellulosic Biofuels: Next Generation Hydrocarbon Biorefineries*, National Science Foundation, Chemical, Bioengineering, Environmental and Transport Systems Division, Washington, DC.

[13] Wang, Q., Sun, G. Q., Jiang, L. H., Xin, Q., Sun, S. G., Jiang, Y. X., Chen, S. P., Jusys, Z., and Behm, R. J. (2007) Adsorption and oxidation of ethanol on colloid-based Pt/C, PtRu/C and Pt3Sn/C catalysts: *in situ* FTIR spectroscopy and on-line DEMS studies. *Physical Chemistry Chemical Physics*, **9** (21), 2686 -2696.

[14] Justi, E. W. and Winsel, A. W. (1955) 821688.

[15] Bagotzky, V. S. and Vassilyev, Y. B. (1967) Mechanism of electro-oxidation of methanol on the platinum electrode. *Electrochimica Acta*, **12** (9), 1323 -1343.

[16] Cathro, K. (1969) The oxidation of watersoluble organic fuels using platinum-tin catalysts. *Journal of Electroanalytical Chemistry and Interfacial Electrochemistry*, **16**, 1608 -1611.

[17] Watanabe, M. and Motoo, S. (1975) Electrocatalysis by ad-atoms: Part II. Enhancement of the oxidation of methanol on platinum by ruthenium ad-atoms. *Journal of Electroanalytical Chemistry and Interfacial Electrochemistry*, **60** (3), 267 -273.

[18] Janssen, M. M. P. and Moolhuysen, J. (1976) Binary systems of platinum and a second metal as oxidation catalysts for methanol fuel cells. *Electrochimica Acta*, **21** (11), 869 -878.

[19] Andrew, M. R., McNicol, B. D., Short, R. T., and Drury, J. S. (1977) Electrolytes for methanol-air

fuel cells. I. The performance of methanol electrooxidation catalysts in sulphuric acid and phosphoric acid electrolytes. *Journal of Applied Electrochemistry*, **7** (2), 153 – 160.

[20] Markovic, N. M. , Widelov, A. , Ross, P. N. , Monterio, O. R. , and Brown, I. G. (1977) Electrooxidation of CO and $CO/H_2$ mixtures on a Pt – Sn catalyst prepared by an implantation method. *Catalysis Letters*, **43**, 161 – 166.

[21] Beden, B. , Kadirgan, F. , Lamy, C. , and Leger, J. M. (1981) Electrocatalytic oxidation of methanol on platinum-based binary electrodes. *Journal of Electroanalytical Chemistry and Interfacial Electrochemistry*, **127** (1 – 3), 75 – 85.

[22] Shibata, M. and Motoo, S. (1985) Electrocatalysis by ad-atoms: Part XI. Enhancement of acetaldehyde oxidation by Shole control and oxygen adsorbing ad-atoms. *Journal of Electroanalytical Chemistry and Interfacial Electrochemistry*, **187** (1), 151 – 159.

[23] Cameron, D. S. , Hards, G. A. , Harrison, B. , and Potter, R. J. (1987) Direct methanol fuel cells. *Platinum Metals Reviews*, **31**, 173 – 181.

[24] Aramata, A. , Kodera, T. , and Masuda, M. (1988) Electrooxidation of methanol on platinum bonded to the solid polymer electrolyte, Nafion. *Journal of Applied Electrochemistry*, **18** (4), 577 – 582.

[25] Goodenough, J. B. , Hamnett, A. , Kennedy, B. J. , Manoharan, R. , and Weeks, S. A. (1988) Methanol oxidation on unsupported and carbon supported Pt + Ru anodes. *Journal of Electroanalytical Chemistry and Interfacial Electrochemistry*, **240**(1 – 2), 133 – 145.

[26] Parsons, R. and VanderNoot, T. (1988) The oxidation of small organic molecules: a survey of recent fuel cell related research. *Journal of Electroanalytical Chemistry and Interfacial Electrochemistry*, **257** (1 – 2), 9 – 45.

[27] Ticanelli, E. , Beery, J. G. , Paffett, M. T. , and Gottesfeld, S. (1989) An electrochemical, ellipsometric, and surface science investigation of the PtRu bulk alloy surface. *Journal of Electroanalytical Chemistry and Interfacial Electrochemistry*, **258** (1), 61 – 77.

[28] Christensen, P. A. , Hamnett, A. , and Troughton, G. L. (1993) The role of morphology in the methanol electrooxidation reaction. *Journal of Electroanalytical Chemistry*, **362** (1 – 2), 207 – 218.

[29] Marković, N. M. , Gasteiger, H. A. , Ross, P. N. , Jr. , Jiang, X. , Villegas, I. , and Weaver, M. J. (1995) Electro-oxidation mechanisms of methanol and formic acid on Pt-Ru alloy surfaces. *Electrochimica Acta*, **40** (1), 91 – 98.

[30] McBreen, J. and Mukerjee, S. (1995) *In situ* X-ray absorption studies of a Pt – Ru electrocatalyst. *Journal of the Electrochemical Society*, **142** (10), 3399 – 3404.

[31] Ravikumar, M. K. and Shukla, A. K. (1996) Effect of methanol crossover in a liquidfeed polymer-electrolyte direct methanol fuel cell. *Journal of the Electrochemical Society*, **143** (8), 2601 – 2606.

[32] Gurau, B. , Viswanathan, R. , Liu, R. , Lafrenz, T. J. , Ley, K. L. , Smotkin, E. S. , Reddington, E. , Sapienza, A. , Chan, B. C. , Mallouk, T. E. , and Sarangapani, S. (1998) Structural and electrochemical characterization of binary, ternary, and quaternary platinum alloy catalysts for methanol electro-oxidation1. *The Journal of Physical Chemistry B*, **102** (49), 9997 – 10003.

[33] Reddington, E. , Sapienza, A. , Gurau, B. , Viswanathan, R. , Sarangapani, S. , Smotkin, E. S. , and Mallouk, T. E. (1998) Combinatorial electrochemistry: a highly parallel, optical screening method for discovery of better electrocatalysts. *Science*, **280** (5370), 1735 – 1737.

[34] Sun, S. G. (1998) Studying electrocatalytic oxidation of small organic molecules with in-situ infra spectroscopy, in *Electrocatalysis* (*Frontiers in Electrochemistry*) (eds J. Lipkowski and P. N. Ross), Wiley-VCH Verlag GmbH, Weinheim, pp. 243 – 290.

[35] Liu, R., Iddir, H., Fan, Q., Hou, G., Bo, A., Ley, K. L., Smotkin, E. S., Sung, Y. E., Kim, H., Thomas, S., and Wieckowski, A. (2000) Potential-dependent infrared absorption spectroscopy of adsorbed CO and X-ray photoelectron spectroscopy of arc-melted single-phase Pt, PtRu, PtOs, PtRuOs, and Ru electrodes. *The Journal of Physical Chemistry B*, **104** (15), 3518 – 3531.

[36] Iwasita, T. (2002) Electrocatalysis of methanol oxidation. *Electrochimica Acta*, **47** (22 – 23), 3663 – 3674.

[37] Piela, P., Eickes, C., Brosha, E., Garzon, F., and Zelenay, P. (2004) Ruthenium crossover in direct methanol fuel cell with Pt – Ru black anode. *Journal of the Electrochemical Society*, **151** (12), A2053 – A2059.

[38] Cao, D., Lu, G. Q., Wieckowski, A., Wasileski, S. A., and Neurock, M. (2005) Mechanisms of methanol decomposition on platinum: a combined experimental and *ab initio* approach. *The Journal of Physical Chemistry B*, **109** (23), 11622 – 11633.

[39] Brockris, J. O. M. and Wroblowa, H. (1964) Activity of electrolytically deposited platinum and ruthenium by the electrooxidation of methanol. *Journal of Electroanalytical Chemistry and Interfacial Electrochemistry*, **7**, 428.

[40] Petry, G. A., Podlovchenko, B. I., Frukmin, A. N., and Lal, H. (1965) The behavior of platinized-platinum and platinumruthenium electrodes in methanol solutions. *Journal of Electroanalytical Chemistry and Interfacial Electrochemistry*, **10**, 253.

[41] Méli, G., Léger, J. M., Lamy, C., and Durand, R. (1993) Direct electrooxidation of methanol on highly dispersed platinum-based catalyst electrodes: temperature effect. *Journal of Applied Electrochemistry*, **23** (3), 197 – 202.

[42] Rolison, D. R., Hagans, P. L., Swider, K. E., and Long, J. W. (1999) Role of hydrous ruthenium oxide in Pt – Ru direct methanol fuel cell anode electrocatalysts: the importance of mixed electron/proton conductivity. *Langmuir*, **15** (3), 774 – 779.

[43] Long, J. W., Stroud, R. M., Swider-Lyons, K. E., and Rolison, D. R. (2000) How to make electrocatalysts more active for direct methanol oxidation avoid PtRu bimetallic alloys! . *The Journal of Physical Chemistry B*, **104** (42), 9772 – 9776.

[44] Rolison, D. R. (2003) Catalytic nanoarchitectures: the importance of nothing and the unimportance of periodicity. *Science*, **299** (5613), 1698 – 1701.

[45] Ren, X., Wilson, M. S., and Gottesfeld, S. (1996) High performance direct methanol polymer electrolyte fuel cells. *Journal of the Electrochemical Society*, **143** (1), L12 – L15.

[46] Kim, T., Kobayashi, K., Takahashi, M., and Nagai, M. (2005) Effect of Sn in Pt – Ru – Sn ternary catalysts for $CO/H_2$ and methanol electrooxidation. *Chemistry Letters*, **34** (6), 798 – 799.

[47] Page, T., Johnson, R., Hormes, J., Noding, S., and Rambabu, B. (2000) A study of methanol electro-oxidation reactions in carbon membrane electrodes and structural properties of Pt alloy electrocatalysts by EXAFS. *Journal of Electroanalytical Chemistry*, **485** (1), 34 – 41.

[48] Park, K.-W., Choi, J.-H., Kwon, B.-K., Lee, S.-A., Sung, Y.-E., Ha, H.-Y., Hong, S.-A., Kim, H., and Wieckowski, A. (2002) Chemical and electronic effects of Ni in Pt/Ni and Pt/Ru/Ni alloy nanoparticles in methanol electrooxidation. *The Journal of Physical Chemistry B*, **106** (8), 1869 – 1877.

[49] Antolini, E., Salgado, J. R. C., and Gonzalez, E. R. (2006) The methanol oxidation reaction on platinum alloys with the first row transition metals: the case of Pt – Co and – Ni alloy electrocatalysts for DMFCs: a short review. *Applied Catalysis B: Environmental*, **63** (1 – 2), 137 – 149.

[50] Bambagioni, V., Bianchini, C., Marchionni, A., Filippi, J., Vizza, F., Teddy, J., Serp, P., and

Zhiani, M. (2009) Pd and Pt – Ru anode electrocatalysts supported on multi-walled carbon nanotubes and their use in passive and active direct alcohol fuel cells with an anion-exchange membrane (alcohol = methanol, ethanol, glycerol). *Journal of Power Sources*, **190** (2), 241 – 251.

[51] Farrell, A. E., Plevin, R. J., Turner, B. T., Jones, A. D., O'Hare, M., and Kammen, D. M. (2006) Ethanol can contribute to energy and environmental goals. *Science*, **311** (5760), 506 – 508.

[52] Simões, F. C., dos Anjos, D. M., Vigier, F., Léger, J. M., Hahn, F., Coutanceau, C., Gonzalez, E. R., Tremiliosi-Filho, G., de Andrade, A. R., Olivi, P., and Kokoh, K. B. (2007) Electroactivity of tin modified platinum electrodes for ethanol electrooxidation. *Journal of Power Sources*, **167** (1), 1 – 10.

[53] Iwasita, T. and Pastor, E. (1994) A DEMS and FTIR spectroscopic investigation of adsorbed ethanol on polycrystalline platinum. *Electrochimica Acta*, **39** (4), 531 – 537.

[54] Pastor, E. and Iwasita, T. (1994) D/H exchange of ethanol at platinum electrodes. *Electrochimica Acta*, **39** (4), 547 – 551.

[55] Perez, J. M., Beden, B., Hahn, F., Aldaz, A., and Lamy, C. (1989) "*In situ*" infrared reflectance spectroscopic study of the early stages of ethanol adsorption at a platinum electrode in acid medium. *Journal of Electroanalytical Chemistry and Interfacial Electrochemistry*, **262** (1 – 2), 251 – 261.

[56] Rousseau, S., Coutanceau, C., Lamy, C., and Léger, J. M. (2006) Direct ethanol fuel cell (DEFC): electrical performances and reaction products distribution under operating conditions with different platinum-based anodes. *Journal of Power Sources*, **158** (1), 18 – 24.

[57] Lamy, C., Lima, A., LeRhun, V., Delime, F., Coutanceau, C., and Léger, J.-M. (2002) Recent advances in the development of direct alcohol fuel cells (DAFC). *Journal of Power Sources*, **105** (2), 283 – 296.

[58] Lamy, C., Rousseau, S., Belgsir, E. M., Coutanceau, C., and Léger, J. M. (2004) Recent progress in the direct ethanol fuel cell: development of new platinum-tin electrocatalysts. *Electrochimica Acta*, **49** (22 – 23), 3901 – 3908.

[59] Ermete, A. (2007) Catalysts for direct ethanol fuel cells. Journal of Power Sources, **170** (1), 1 – 12.

[60] Zhou, W. J., Zhou, Z. H., Song, S. Q., Li, W. Z., Sun, G. Q., Tsiakaras, P., and Xin, Q. (2003) Pt based anode catalysts for direct ethanol fuel cells. *Applied Catalysis B: Environmental*, **46** (2), 273 – 285.

[61] Zhao, X., Li, W., Jiang, L., Zhou, W., Xin, Q., Yi, B., and Sun, G. (2004) Multi-wall carbon nanotube supported Pt – Sn nanoparticles as an anode catalyst for the direct ethanol fuel cell. *Carbon*, **42** (15), 3263 – 3265.

[62] Zhou, W. J., Li, W. Z., Song, S. Q., Zhou, Z. H., Jiang, L. H., Sun, G. Q., Xin, Q., Poulianitis, K., Kontou, S., and Tsiakaras, P. (2004) Bi- and tri-metallic Pt-based anode catalysts for direct ethanol fuel cells. *Journal of Power Sources*, **131** (1 – 2), 217 – 223.

[63] Zhou, W. J., Song, S. Q., Li, W. Z., Sun, G. Q., Xin, Q., Kontou, S., Poulianitis, K., and Tsiakaras, P. (2004) Pt-based anode catalysts for direct ethanol fuel cells. *Solid State Ionics*, **175** (1 – 4), 797 – 803.

[64] Zhou, W. J., Zhou, B., Li, W. Z., Zhou, Z. H., Song, S. Q., Sun, G. Q., Xin, Q., Douvartzides, S., Goula, M., and Tsiakaras, P. (2004) Performance comparison of low-temperature direct alcohol fuel cells with different anode catalysts. *Journal of Power Sources*, **126** (1 – 2), 16 – 22.

[65] Zhou, W. J., Song, S. Q., Li, W. Z., Zhou, Z. H., Sun, G. Q., Xin, Q., Douvartzides, S., and Tsiakaras, P. (2005) Direct ethanol fuel cells based on PtSn anodes: the effect of Sn content on the fuel cell performance. *Journal of Power Sources*, **140** (1), 50 – 58.

[66] Vigier, F., Coutanceau, C., Hahn, F., Belgsir, E. M., and Lamy, C. (2004) On the mechanism of ethanol electro-oxidation on Pt and PtSn catalysts: electrochemical and in situ IR reflectance spectroscopy studies. *Journal of Electroanalytical Chemistry*, **563** (1), 81-89.

[67] Vigier, F., Rousseau, S., Coutanceau, C., Leger, J.-M., and Lamy, C. (2006) Electrocatalysis for the direct alcohol fuel cell. *Topics in Catalysis*, **40** (1), 111-121.

[68] Liu, P., Logadottir, A., and Nørskov, J. K. (2003) Modeling the electro-oxidation of CO and $H_2/CO$ on Pt, Ru, PtRu and $Pt_3Sn$. *Electrochimica Acta*, **48** (25-26), 3731-3742.

[69] de Souza, J. P. I., Queiroz, S. L., Bergamaski, K., Gonzalez, E. R., and Nart, F. C. (2002) Electro-oxidation of ethanol on Pt, Rh, and PtRh electrodes: a study using DEMS and *in-situ* FTIR techniques. *The Journal of Physical Chemistry B*, **106** (38), 9825-9830.

[70] Colmati, F., Antolini, E., and Gonzalez, E. R. (2008) Preparation, structural characterization and activity for ethanol oxidation of carbon supported ternary Pt-Sn-Rh catalysts. *Journal of Alloys and Compounds*, **456** (1-2), 264-270.

[71] Kowal, A., Li, M., Shao, M., Sasaki, K., Vukmirovic, M. B., Zhang, J., Marinkovic, N. S., Liu, P., Frenkel, A. I., and Adzic, R. R. (2009) Ternary $Pt/Rh/SnO_2$ electrocatalysts for oxidizing ethanol to $CO_2$. *Nature Materials*, **8** (4), 325-330.

[72] Spendelow, J. S. and Wieckowski, A. (2007) Electrocatalysis of oxygen reduction and small alcohol oxidation in alkaline media. *Physical Chemistry Chemical Physics*, **9** (21), 2654-2675.

[73] Varcoe, J. R. and Slade, R. C. T. (2005) Prospects for alkaline anion-exchange membranes in low temperature fuel cells. *Fuel Cells*, **5** (2), 187-200.

[74] Yanagi, H. and Fukuta, K. (2008) Anion exchange membrane and ionomer for alkaline membrane fuel cells (AMFCs). *ECS Transactions*, **16** (2), 257-262.

[75] Blizanac, B. B., Ross, P. N., and Markovic, N. M. (2007) Oxygen electroreduction on Ag(1 1 1): the pH effect. *Electrochimica Acta*, **52** (6), 2264-2271.

[76] Li, X., Popov, B. N., Kawahara, T., and Yanagi, H. (2011) Non-precious metal catalysts synthesized from precursors of carbon, nitrogen, and transition metal for oxygen reduction in alkaline fuel cells. *Journal of Power Sources*, **196** (4), 1717-1722.

[77] Xu, H. and Hou, X. (2007) Synergistic effect of modified Pt/C electrocatalysts on the performance of PEM fuel cells. *International Journal of Hydrogen Energy*, **32** (17), 4397-4401.

[78] Cui, G., Song, S., Shen, P. K., Kowal, A., and Bianchini, C. (2009) First-principles considerations on catalytic activity of Pd toward ethanol oxidation. *The Journal of Physical Chemistry C*, **113** (35), 15639-15642.

[79] Liang, Z. X., Zhao, T. S., Xu, J. B., and Zhu, L. D. (2009) Mechanism study of the ethanol oxidation reaction on palladium in alkaline media. *Electrochimica Acta*, **54** (8), 2203-2208.

[80] Chen, Y., Zhuang, L., and Lu, J. (2007) Non-Pt anode catalysts for alkaline direct alcohol fuel cells. *Chinese Journal of Catalysis*, **28** (10), 870-874.

[81] He, Q., Chen, W., Mukerjee, S., Chen, S., and Laufek, F. (2009) Carbon-supported PdM (M = Au and Sn) nanocatalysts for the electrooxidation of ethanol in high pH media. *Journal of Power Sources*, **187** (2), 298-304.

[82] Shen, P. K. and Xu, C. (2006) Alcohol oxidation on nanocrystalline oxide Pd/C promoted electrocatalysts. *Electrochemistry Communications*, **8** (1), 184-188.

[83] Bianchini, C., Bambagioni, V., Filippi, J., Marchionni, A., Vizza, F., Bert, P., and Tampucci, A. (2009) Selective oxidation of ethanol to acetic acid in highly efficient polymer electrolyte membranedirect

ethanol fuel cells. *Electrochemistry Communications*, **11** (5), 1077 – 1080.

[84] Bambagioni, V., Bevilacqua, M., Filippi, J., Marchionni, A., Moneti, S., Vizza, F., and Bianchini, C. (2010) Direct alcohol fuel cells as chemical reactors for the sustainable production of energy and chemicals: energy and chemicals from renewables by electrocatalysis. *Chemistry Today*, **28** (3), 518 – 528.

[85] Santasalo-Aarnio, A., Kwon, Y., Ahlberg, E., Kontturi, K., Kallio, T., and Koper, M. T. M. (2011) Comparison of methanol, ethanol and iso-propanol oxidation on Pt and Pd electrodes in alkaline media studied by HPLC. *Electrochemistry Communications*, **13** (5), 466 – 469.

[86] Kwon, Y. and Koper, M. T. M. (2010) Combining voltammetry with HPLC: application to electro-oxidation of glycerol. *Analytical Chemistry*, **82** (13), 5420 – 5424.

[87] Kwon, Y., Lai, S. C. S., Rodriguez, P., and Koper, M. T. M. (2011) Electrocatalytic oxidation of alcohols on gold in alkaline media: base or gold catalysis. *Journal of the American Chemical Society*, **133** (18), 6914 – 6917.

[88] Roquet, L., Belgsir, E. M., Léger, J. M., and Lamy, C. (1994) Kinetics and mechanisms of the electrocatalytic oxidation of glycerol as investigated by chromatographic analysis of the reaction products: potential and pH effects. *Electrochimica Acta*, **39** (16), 2387 – 2394.

[89] Zope, B. N., Hibbitts, D. D., Neurock, M., and Davis, R. J. (2010) Reactivity of the gold/water interface during selective oxidation catalysis. *Science*, **330** (6000), 74 – 78.

[90] Demarconnay, L., Brimaud, S., Coutanceau, C., and Léger, J. M. (2007) Ethylene glycol electrooxidation in alkaline medium at multi-metallic Pt based catalysts. *Journal of Electroanalytical Chemistry*, **601** (1 – 2), 169 – 180.

[91] Matsuoka, K., Iriyama, Y., Abe, T., Matsuoka, M., and Ogumi, Z. (2005) Electro-oxidation of methanol and ethylene glycol on platinum in alkaline solution: poisoning effects and product analysis. *Electrochimica Acta*, **51** (6), 1085 – 1090.

[92] Chang, S. C., Ho, Y., and Weaver, M. J. (1991) Applications of real-time FTIR spectroscopy to the elucidation of complex electroorganic pathways: electrooxidation of ethylene glycol on gold, platinum, and nickel in alkaline solution. *Journal of the American Chemical Society*, **113** (25), 9506 – 9513.

[93] Kohlmüller, H. (1976) Anodic oxidation of ethylene glycol with noble metal alloy catalysts. *Journal of Power Sources*, **1** (3), 249 – 256.

[94] Kadirgan, F., Beden, B., and Lamy, C. (1983) Electrocatalytic oxidation of ethylene-glycol: Part II. Behaviour of platinum-ad-atom electrodes in alkaline medium. *Journal of Electroanalytical Chemistry and Interfacial Electrochemistry*, **143** (1 – 2), 135 – 152.

[95] Dalbay, N. and Kadirgan, F. (1991) Electrolytically co-deposited platinum: palladium electrodes and their electrocatalytic activity for ethylene glycol oxidation – a synergistic effect. *Electrochimica Acta*, **36** (2), 353 – 356.

[96] Simões, M., Baranton, S., and Coutanceau, C. (2010) Electro-oxidation of glycerol at Pd based nanocatalysts for an application in alkaline fuel cells for chemicals and energy cogeneration. *Applied Catalysis B: Environmental*, **93** (3 – 4), 354 – 362.

[97] Bambagioni, V., Bevilacqua, M., Bianchini, C., Filippi, J., Lavacchi, A., Marchionni, A., Vizza, F., and Shen, P. K. (2010) Self-sustainable production of hydrogen, chemicals, and energy from renewable alcohols by electrocatalysis. *ChemSusChem*, **3** (7), 851 – 855.

[98] Besson, M. and Gallezot, P. (2000) Selective oxidation of alcohols and aldehydes on metal catalysts. *Catalysis Today*, **57** (1 – 2), 127 – 141.

[99] Mallat, T. and Baiker, A. (2004) Oxidation of alcohols with molecular oxygen on solid catalysts. *Chemi-*

*cal Reviews*, **104** (6), 3037 – 3058.

[100] Ketchie, W. C., Fang, Y. -L., Wong, M. S., Murayama, M., and Davis, R. J. (2007) Influence of gold particle size on the aqueous-phase oxidation of carbon monoxide and glycerol. *Journal of Catalysis*, **250** (1), 94 – 101.

[101] Corma, A. and Garcia, H. (2008) Supported gold nanoparticles as catalysts for organic reactions. *Chemical Society Reviews*, **37** (9), 2096 – 2126.

[102] Della Pina, C., Falletta, E., Prati, L., and Rossi, M. (2008) Selective oxidation using gold. *Chemical Society Reviews*, **37** (9), 2077 – 2095.

[103] Hutchings, G. J. (2008) Nanocrystalline gold and gold palladium alloy catalysts for chemical synthesis. *Chemical Communications*, (10), 1148 – 1164.

[104] Zhou, C. -H., Beltramini, J. N., Fan, Y. -X., and Lu, G. Q. (2008) Chemoselective catalytic conversion of glycerol as a biorenewable source to valuable commodity chemicals. *Chemical Society Reviews*, **37** (3), 527 – 549.

[105] Dimitratos, N., Lopez-Sanchez, J., and Hutchings, G. (2009) Green catalysis with alternative feedstocks. *Topics in Catalysis*, **52** (3), 258 – 268.

[106] Carrettin, S., McMorn, P., Johnston, P., Griffin, K., and Hutchings, G. J. (2002) Selective oxidation of glycerol to glyceric acid using a gold catalyst in aqueous sodium hydroxide. *Chemical Communications*, (7), 696 – 697.

[107] Carrettin, S., McMorn, P., Johnston, P., Griffin, K., Kiely, C. J., and Hutchings, G. J. (2003) Oxidation of glycerol using supported Pt, Pd and Au catalysts. *Physical Chemistry Chemical Physics*, **5** (6), 1329 – 1336.

[108] Matsuoka, K., Iriyama, Y., Abe, T., Matsuoka, M., and Ogumi, Z. (2005) Alkaline direct alcohol fuel cells using an anion exchange membrane. *Journal of Power Sources*, **150** (0), 27 – 31.

[109] Stamenkovic, V. R., Fowler, B., Mun, B. S., Wang, G., Ross, P. N., Lucas, C. A., and Marković, N. M. (2007) Improved oxygen reduction activity on $Pt_3Ni(111)$ via increased surface site availability. *Science*, **315** (5811), 493 – 497.

[110] Marković, N. M. and Ross, P. N., Jr. (2002) Surface science studies of model fuel cell electrocatalysts. *Surface Science Reports*, **45** (4 – 6), 117 – 229.

[111] Tian, N., Zhou, Z. -Y., and Sun, S. -G. (2008) Platinum metal catalysts of highindex surfaces: from single-crystal planes to electrochemically shape-controlled nanoparticles. *The Journal of Physical Chemistry C*, **112** (50), 19801 – 19817.

[112] Arenz, M., Stamenkovic, V., Schmidt, T. J., Wandelt, K., Ross, P. N., and Markovic, N. M. (2003) The electrooxidation of formic acid on Pt – Pd single crystal bimetallic surfaces. *Physical Chemistry Chemical Physics*, **5** (19), 4242 – 4251.

[113] Somorjai, G. A. (1994) *Introduction to Surface Chemistry and Catalysis*, John Wiley & Sons, Inc., New York.

[114] Lebedeva, N. P., Koper, M. T. M., Herrero, E., Feliu, J. M., and van Santen, R. A. (2000) Cooxidation on stepped Pt[$n$(111) × (111)] electrodes. *Journal of Electroanalytical Chemistry*, **487** (1), 37 – 44.

[115] Stamenkovic, V. R., Mun, B. S., Arenz, M., Mayrhofer, K. J. J., Lucas, C. A., Wang, G., Ross, P. N., and Markovic, N. M. (2007) Trends in electrocatalysis on extended and nanoscale Pt-bimetallic alloy surfaces. *Nature Materials*, **6** (3), 241 – 247.

[116] Ralph, T. R. and Hogarth, M. P. (2002) Catalysis for low temperature fuel cells Part 1: the cathode

challenges. *Platinum Metals Review*, **46** (1), 3 – 14.

[117] Ermete, A. (2003) Formation of carbon-supported PtM alloys for low temperature fuel cells: a review. *Materials Chemistry and Physics*, **78** (3), 563 – 573.

[118] Somorjai, G., Tao, F., and Park, J. (2008) The nanoscience revolution: merging of colloid science, catalysis and nanoelectronics. *Topics in Catalysis*, **47** (1), 1 – 14.

[119] Borup, R., Meyers, J., Pivovar, B., Kim, Y. S., Mukundan, R., Garland, N., Myers, D., Wilson, M., Garzon, F., Wood, D., Zelenay, P., More, K., Stroh, K., Zawodzinski, T., Boncella, J., McGrath, J. E., Inaba, M., Miyatake, K., Hori, M., Ota, K., Ogumi, Z., Miyata, S., Nishikata, A., Siroma, Z., Uchimoto, Y., Yasuda, K., Kimijima, K.-I., and Iwashita, N. (2007) Scientific aspects of polymer electrolyte fuel cell durability and degradation. *Chemical Reviews*, **107** (10), 3904 – 3951.

[120] Chan, K.-Y., Ding, J., Ren, J., Cheng, S., and Tsang, K. Y. (2004) Supported mixed metal nanoparticles as electrocatalysts in low temperature fuel cells. *Journal of Materials Chemistry*, **14** (4), 505 – 516.

[121] Sasaki, T., Koshizaki, N., Terauchi, S., Umehara, H., Matsumoto, Y., and Koinuma, M. (1997) Preparation of Pt/ $TiO_2$ nanocomposite films using cosputtering method. *Nanostructured Materials*, **8** (8), 1077 – 1083.

[122] Yan, X. M., Ni, J., Robbins, M., Park, H. J., Zhao, W., and White, J. M. (2002) Silver nanoparticles synthesized by vapor deposition onto an ice matrix. *Journal of Nanoparticle Research*, **4** (6), 525 – 533.

[123] Debe, M. K., Schmoeckel, A. K., Vernstrom, G. D., and Atanasoski, R. (2006) High voltage stability of nanostructured thin film catalysts for PEM fuel cells. *Journal of Power Sources*, **161** (2), 1002 – 1011.

[124] Astruc, D. (2008) *Nanoparticles and Catalysis*, Wiley-VCH Verlag GmbH, Weinheim.

[125] Ahmadi, T. S., Wang, Z. L., Green, T. C., Henglein, A., and El-Sayed, M. A. (1996) Shape-controlled synthesis of colloidal platinum nanoparticles. *Science*, **272** (5270), 1924 – 1925.

[126] Toshima, N. and Yonezawa, T. (1998) Bimetallic nanoparticles-novel materials for chemical and physical applications. *New Journal of Chemistry*, **22** (11), 1179 – 1201.

[127] Crooks, R. M., Zhao, M., Sun, L., Chechik, V., and Yeung, L. K. (2000) Dendrimerencapsulated metal nanoparticles: synthesis, characterization, and applications to catalysis. *Accounts of Chemical Research*, **34** (3), 181 – 190.

[128] Murray, C. B., Kagan, C. R., and Bawendi, M. G. (2000) Synthesis and characterization of monodisperse nanocrystals and close-packed nanocrystal assemblies. *Annual Review of Materials Science*, **30**, 545 – 610.

[129] Sun, S., Murray, C. B., Weller, D., Folks, L., and Moser, A. (2000) Monodisperse FePt nanoparticles and ferromagnetic FePt nanocrystal superlattices. *Science*, **287** (5460), 1989 – 1992.

[130] Bönnemann, H. and Richards, R. M. (2001) Nanoscopic metal particles: synthetic methods and potential applications. *European Journal of Inorganic Chemistry*, **2001** (10), 2455 – 2480.

[131] Jana, N. R., Gearheart, L., and Murphy, C. J. (2001) Seed-mediated growth approach for shape-controlled synthesis of spheroidal and rod-like gold nanoparticles using a surfactant template. *Advanced Materials*, **13** (18), 1389 – 1393.

[132] Roucoux, A., Schulz, J., and Patin, H. (2002) Reduced transition metal colloids: a novel family of reusable catalysts? *Chemical Reviews*, **102** (10), 3757 – 3778.

[133] Sun, Y. and Xia, Y. (2002) Shapecontrolled synthesis of gold and silver nanoparticles. *Science*, **298** (5601), 2176–2179.

[134] Wang, C., Hou, Y., Kim, J., and Sun, S. (2007) A general strategy for synthesizing FePt nanowires and nanorods. *Angewandte Chemie: International Edition*, **46** (33), 6333–6335.

[135] Alayoglu, S., Nilekar, A. U., Mavrikakis, M., and Eichhorn, B. (2008) Ru–Pt core-shell nanoparticles for preferential oxidation of carbon monoxide in hydrogen. *Nature Materials*, **7** (4), 333–338.

[136] Somorjai, G. and Park, J. (2008) Colloid science of metal nanoparticle catalysts in 2D and 3D structures: challenges of nucleation, growth, composition, particle shape, size control and their influence on activity and selectivity. *Topics in Catalysis*, **49** (3), 126–135.

[137] Wang, C., Daimon, H., Onodera, T., Koda, T., and Sun, S. (2008) A general approach to the size- and shape-controlled synthesis of platinum nanoparticles and their catalytic reduction of oxygen. *Angewandte Chemie*, **120** (19), 3644–3647.

[138] Chen, J., Lim, B., Lee, E. P., and Xia, Y. (2009) Shape-controlled synthesis of platinum nanocrystals for catalytic and electrocatalytic applications. *Nano Today*, **4** (1), 81–95.

[139] Lim, B., Jiang, M., Camargo, P. H. C., Cho, E. C., Tao, J., Lu, X., Zhu, Y., and Xia, Y. (2009) Pd–Pt bimetallic nanodendrites with high activity for oxygen reduction. *Science*, **324** (5932), 1302–1305.

[140] Xie, X., Li, Y., Liu, Z.-Q., Haruta, M., and Shen, W. (2009) Low-temperature oxidation of CO catalysed by $Co_3O_4$ nanorods. *Nature*, **458** (7239), 746–749.

[141] Zhang, J., Vukmirovic, M. B., Xu, Y., Mavrikakis, M., and Adzic, R. R. (2005) Controlling the catalytic activity of platinum-monolayer electrocatalysts for oxygen reduction with different substrates. *Angewandte Chemie: International Edition*, **44** (14), 2132–2135.

[142] Zhang, J., Sasaki, K., Sutter, E., and Adzic, R. R. (2007) Stabilization of platinum oxygen-reduction electrocatalysts using gold clusters. *Science*, **315** (5809), 220–222.

[143] Goodenough, J. B., Hamnett, A., Kennedy, B. J., Manoharan, R., and Weeks, S. A. (1990) Porous carbon anodes for the direct methanol fuel cell: I. The role of the reduction method for carbon supported platinum electrodes. *Electrochimica Acta*, **35** (1), 199–207.

[144] Román-Martínez, M. C., Cazorla-Amorós, D., Yamashita, H., and de Miguel, S., and Scelza, O. A. (1999) XAFS study of dried and reduced PtSn/C catalysts: nature and structure of the catalytically active phase. *Langmuir*, **16** (3), 1123–1131.

[145] Takasu, Y., Fujiwara, T., Murakami, Y., Sasaki, K., Oguri, M., Asaki, T., and Sugimoto, W. (2000) Effect of structure of carbon-supported PtRu electrocatalysts on the electrochemical oxidation of methanol. *Journal of the Electrochemical Society*, **147** (12), 4421–4427.

[146] Boxall, D. L., Deluga, G. A., Kenik, E. A., King, W. D., and Lukehart, C. M. (2001) Rapid synthesis of a $Pt_1Ru_1$/carbon nanocomposite using microwave irradiation: a DMFC anode catalyst of high relative performance. *Chemistry of Materials*, **13** (3), 891–900.

[147] Joo, S. H., Choi, S. J., Oh, I., Kwak, J., Liu, Z., Terasaki, O., and Ryoo, R. (2001) Ordered nanoporous arrays of carbon supporting high dispersions of platinum nanoparticles. *Nature*, **412** (6843), 169–172.

[148] Dickinson, A. J., Carrette, L. P. L., Collins, J. A., Friedrich, K. A., and Stimming, U. (2002) Preparation of a Pt–Ru/C catalyst from carbonyl complexes for fuel cell applications. *Electrochimica Acta*, **47** (22–23), 3733–3739.

[149] Fujiwara, N., Yasuda, K., Ioroi, T., Siroma, Z., and Miyazaki, Y. (2002) Preparation of platinum-

ruthenium onto solid polymer electrolyte membrane and the application to a DMFC anode. *Electrochimica Acta*, **47** (25), 4079 – 4084.

[150] Pozio, A., Silva, R. F., De Francesco, M., Cardellini, F., and Giorgi, L. (2002) A novel route to prepare stable Pt – Ru/C electrocatalysts for polymer electrolyte fuel cell. *Electrochimica Acta*, **48** (3), 255 – 262.

[151] Steigerwalt, E. S., Deluga, G. A., and Lukehart, C. M. (2002) Pt-Ru/carbon fiber nanocomposites: synthesis, characterization, and performance as anode catalysts of direct methanol fuel cells. A search for exceptional performance. *The Journal of Physical Chemistry B*, **106** (4), 760 – 766.

[152] Xiong, L., Kannan, A. M., and Manthiram, A. (2002) Pt – M (M = Fe, Co, Ni and Cu) electrocatalysts synthesized by an aqueous route for proton exchange membrane fuel cells. *Electrochemistry Communications*, **4** (11), 898 – 903.

[153] Venkataraman, R., Kunz, H. R., and Fenton, J. M. (2003) Development of new CO tolerant ternary anode catalysts for proton exchange membrane fuel cells. *Journal of the Electrochemical Society*, **150** (3), A278 – A284.

[154] Kawaguchi, T., Sugimoto, W., Murakami, Y., and Takasu, Y. (2004) Temperature dependence of the oxidation of carbon monoxide on carbon supported Pt, Ru, and PtRu. *Electrochemistry Communications*, **6** (5), 480 – 483.

[155] Lizcano-Valbuena, W. H., de Azevedo, D. C., and Gonzalez, E. R. (2004) Supported metal nanoparticles as electrocatalysts for low-temperature fuel cells. *Electrochimica Acta*, **49** (8), 1289 – 1295.

[156] Vigier, F., Coutanceau, C., Perrard, A., Belgsir, E. M., and Lamy, C. (2004) Development of anode catalysts for a direct ethanol fuel cell. *Journal of Applied Electrochemistry*, **34** (4), 439 – 446.

[157] Li, W., Sun, G., Yan, Y., and Xin, Q. (2005) Supported noble metal electrocatalysts in low temperature fuel cells. *Progress in Chemistry*, **17** (5), 761 – 772.

[158] Cui, Z., Liu, C., Liao, J., and Xing, W. (2008) Highly active PtRu catalysts supported on carbon nanotubes prepared by modified impregnation method for methanol electro-oxidation. *Electrochimica Acta*, **53** (27), 7807 – 7811.

[159] Watanabe, M., Uchida, M., and Motoo, S. (1987) Preparation of highly dispersed Pt + Ru alloy clusters and the activity for the electrooxidation of methanol. *Journal of Electroanalytical Chemistry and Interfacial Electrochemistry*, **229** (1 – 2), 395 – 406.

[160] Götz, M. and Wendt, H. (1998) Binary and ternary anode catalyst formulations including the elements W, Sn and Mo for PEMFCs operated on methanol or reformate gas. *Electrochimica Acta*, **43** (24), 3637 – 3644.

[161] Hirai, H., Nakao, Y., and Toshima, N. (1979) Preparation of colloidal transition metals in polymers by reduction with alcohols or ethers. *Journal of Macromolecular Science: Part A - Chemistry*, **13** (6), 727 – 750.

[162] Toshima, N., Yonezawa, T., and Kushihashi, K. (1993) Polymer-protected palladium-platinum bimetallic clusters: preparation, catalytic properties and structural considerations. *Journal of the Chemical Society, Faraday Transactions*, **89** (14), 2537 – 2543.

[163] Camargo, P. H. C., Xiong, Y., Ji, L., Zuo, J. M., and Xia, Y. (2007) Facile synthesis of tadpole-like nanostructures consisting of Au heads and Pd tails. *Journal of the American Chemical Society*, **129** (50), 15452 – 15453.

[164] Xiong, Y., Cai, H., Wiley, B. J., Wang, J., Kim, M. J., and Xia, Y. (2007) Synthesis and mechanistic study of palladium nanobars and nanorods. *Journal of the American Chemical Society*, **129** (12),

3665 - 3675.

[165] Wang, Y. , Ren, J. , Deng, K. , Gui, L. , and Tang, Y. (2000) Preparation of tractable platinum, rhodium, and ruthenium nanoclusters with small particle size in organic media. *Chemistry of Materials*, **12** (6), 1622 - 1627.

[166] Li, W. , Liang, C. , Qiu, J. , Zhou, W. , Han, H. , Wei, Z. , Sun, G. , and Xin, Q. (2002) Carbon nanotubes as support for cathode catalyst of a direct methanol fuel cell. *Carbon*, **40** (5), 791 - 794.

[167] Liu, Z. , Lee, J. Y. , Chen, W. , Han, M. , and Gan, L. M. (2003) Physical and electrochemical characterizations of microwave-assisted polyol preparation of carbon-supported PtRu nanoparticles. *Langmuir*, **20** (1), 181 - 187.

[168] Zhou, Z. , Wang, S. , Zhou, W. ,Wang, G. , Jiang, L. , Li, W. ,Song,S. , Liu, J. , Sun, G. , and Xin, Q. (2003) Novel synthesis of highly active Pt/C cathode electrocatalyst for direct methanol fuel cell. *Chemical Communications*, (3), 394 - 395.

[169] Bock, C. , Paquet, C. , Couillard, M. , Botton, G. A. , and MacDougall, B. R. (2004) Size-selected synthesis of PtRu nano-catalysts: reaction and size control mechanism. *Journal of the American Chemical Society*, **126** (25), 8028 - 8037.

[170] Jiang, L. , Zhou, Z. , Li, W. , Zhou, W. , Song, S. , Li, H. , Sun, G. , and Xin, Q. (2004) Effects of treatment in different atmosphere on $Pt_3Sn/C$ electrocatalysts for ethanol electro-oxidation. *Energy & Fuels*, **18** (3), 866 - 871.

[171] Li, H. , Xin, Q. , Li, W. , Zhou, Z. , Jiang, L. , Yang, S. , and Sun, G. (2004) An improved palladium-based DMFCs cathode catalyst. *Chemical Communications*, (23), 2776 - 2777.

[172] Li, W. , Liang, C. , Qiu, J. , Li, H. , Zhou, W. , Sun, G. , and Xin, Q. (2004) Multi-walled carbon nanotubes supported Pt-Fe cathodic catalyst for direct methanol fuel cell. *Reaction Kinetics and Catalysis Letters*, **82** (2), 235 - 240.

[173] Li, W. , Zhou, W. , Li, H. , Zhou, Z. , Zhou, B. , Sun, G. , and Xin, Q. (2004) Nanostructured Pt-Fe/C as cathode catalyst in direct methanol fuel cell. *Electrochimica Acta*, **49** (7), 1045 - 1055.

[174] Xing, Y. (2004) Synthesis and electrochemical characterization of uniformly-dispersed high loading Pt nanoparticles on sonochemically-treated carbon nanotubes. *The Journal of Physical Chemistry B*, **108** (50), 19255 - 19259.

[175] Li, W. , Wang, X. , Chen, Z. , Waje, M. , and Yan, Y. (2005) Carbon nanotube film by filtration as cathode catalyst support for proton-exchange membrane fuel cell. *Langmuir*, **21** (21), 9386 - 9389.

[176] Chen, Z. , Xu, L. , Li, W. , Waje, M. , and Yan, Y. (2006) Polyaniline nanofibre supported platinum nanoelectrocatalysts for direct methanol fuel cells. *Nanotechnology*, **17** (20), 5254.

[177] Li, W. , Wang, X. , Chen, Z. , Waje, M. , and Yan, Y. (2006) Pt-Ru supported on double-walled carbon nanotubes as high-performance anode catalysts for direct methanol fuel cells. *The Journal of Physical Chemistry B*, **110** (31), 15353 - 15358.

[178] Knupp, S. L. , Li, W. , Paschos, O. , Murray, T. M. , Snyder, J. , and Haldar, P. (2008) The effect of experimental parameters on the synthesis of carbon nanotube/nanofiber supported platinum by polyol processing techniques. *Carbon*, **46** (10), 1276 - 1284.

[179] Avasarala, B. , Murray, T. , Li, W. , and Haldar, P. (2009) Titanium nitride nanoparticles based electrocatalysts for proton exchange membrane fuel cells. *Journal of Materials Chemistry*, **19** (13), 1803 - 1805.

[180] Sun, S. and Zeng, H. (2002) Size-controlled synthesis of magnetite nanoparticles. *Journal of the American Chemical Society*, **124** (28), 8204 - 8205.

[181] Liu, Z., Ada, E. T., Shamsuzzoha, M., Thompson, G. B., and Nikles, D. E. (2006) Synthesis and activation of PtRu alloyed nanoparticles with controlled size and composition. *Chemistry of Materials*, **18** (20), 4946–4951.

[182] Luo, J., Njoki, P. N., Lin, Y., Mott, D., and Wang, L. Y., and Zhong, C.-J. (2006) Characterization of carbon-supported AuPt nanoparticles for electrocatalytic methanol oxidation reaction. *Langmuir*, **22** (6), 2892–2898.

[183] Yano, H., Kataoka, M., Yamashita, H., Uchida, H., and Watanabe, M. (2007) Oxygen reduction activity of carbonsupported Pt–M (M = V, Ni, Cr, Co, and Fe) alloys prepared by nanocapsule method. *Langmuir*, **23** (11), 6438–6445.

[184] Li, W. and Haldar, P. (2009) Supportless PdFe nanorods as highly active electrocatalyst for proton exchange membrane fuel cell. *Electrochemistry Communications*, **11** (6), 1195–1198.

[185] Li, W., Chen, Z., Xu, L., and Yan, Y. (2010) A solution-phase synthesis method to highly active Pt–Co/C electrocatalysts for proton exchange membrane fuel cell. *Journal of Power Sources*, **195** (9), 2534–2540.

[186] Zhang, Z., More, K. L., Sun, K., Wu, Z., and Li, W. (2011) Preparation and characterization of PdFe nanoleaves as electrocatalysts for oxygen reduction reaction. *Chemistry of Materials*, **23** (6), 1570–1577.

[187] Zhang, Z., Xin, L., Sun, K., and Li, W. (2011) Pd–Ni electrocatalysts for efficient ethanol oxidation reaction in alkaline electrolyte. *International Journal of Hydrogen Energy*, **36** (20), 12686–12697.

[188] Zhang, Z., Li, M., Wu, Z., and Li, W. (2011) Ultra-thin PtFe-nanowires as durable electrocatalysts for fuel cells. *Nanotechnology*, **22** (1), 015602.

[189] Shinoda, K. and Friberg, S. (1975) Microemulsions: colloidal aspects. *Advances in Colloid and Interface Science*, **4** (4), 281–300.

[190] Bommarius, A. S., Holzwarth, J. F., Wang, D. I. C., and Hatton, T. A. (1990) Coalescence and solubilizate exchange in a cationic four-component reversed micellar system. *The Journal of Physical Chemistry*, **94** (18), 7232–7239.

[191] Wang, J., Ee, L. S., Ng, S. C., Chew, C. H., and Gan, L. M. (1997) Reduced crystallization temperature in a microemulsion-derived zirconia precursor. *Materials Letters*, **30** (1), 119–124.

[192] Wu, M.-L., Chen, D.-H., and Huang, T.-C. (2001) Preparation of Au/Pt bimetallic nanoparticles in water-in-oil microemulsions. *Chemistry of Materials*, **13** (2), 599–606.

[193] Liu, Z., Lee, J. Y., Han, M., Chen, W., and Gan, L. M. (2002) Synthesis and characterization of PtRu/C catalysts from microemulsions and emulsions. *Journal of Materials Chemistry*, **12** (8), 2453–2458.

[194] Zhang, X. and Chan, K.-Y. (2002) Water-in-oil microemulsion synthesis of platinum-ruthenium nanoparticles, their characterization and electrocatalytic properties. *Chemistry of Materials*, **15** (2), 451–459.

[195] Rojas, S., García-García, F. J., Järas, S., Martínez-Huerta, M. V., Fierro, J. L. G., and Boutonnet, M. (2005) Preparation of carbon supported Pt and PtRu nanoparticles from microemulsion: electrocatalysts for fuel cell applications. *Applied Catalysis A: General*, **285** (1–2), 24–35.

[196] Xiong, L. and Manthiram, A. (2005) Catalytic activity of Pt–Ru alloys synthesized by a microemulsion method in direct methanol fuel cells. *Solid State Ionics*, **176** (3–4), 385–392.

[197] Lei, H., Atanassova, P., Sun, Y., and Blizanac, B. (2009) State-of-the-art electrocatalysts for direct methanol fuel cells, in *Electrocatalysis of Direct Methanol Fuel Cells: From Fundamentals to Applications* (eds. H. Liu and J. Zhang), Wiley-VCH Verlag GmbH, Weinheim, pp. 197–226.

[198] Shao-Horn, Y., Ferreira, P., and la O, G. J. (2006) Instability of Pt/C electrocatalysts in proton exchange membrane fuel cells: a mechanistic investigation. *ECS Meeting Abstracts*, **503** (2), 200.

[199] Iijima, S. (1991) Helical microtubules of graphitic carbon. *Nature*, **354** (6348), 56-58.

[200] Che, G., Lakshmi, B. B., Fisher, E. R., and Martin, C. R. (1998) Carbon nanotubule membranes for electrochemical energy storage and production. *Nature*, **393** (6683), 346-349.

[201] Britto, P. J., Santhanam, K. S. V., Rubio, A., Alonso, J. A., and Ajayan, P. M. (1999) Improved charge transfer at carbon nanotube electrodes. *Advanced Materials*, **11** (2), 154-157.

[202] Li, W., Liang, C., Zhou, W., Qiu, J., Zhenhua, Sun, G., and Xin, Q. (2003) Preparation and characterization of multiwalled carbon nanotube-supported platinum for cathode catalysts of direct methanol fuel cells. *The Journal of Physical Chemistry B*, **107** (26), 6292-6299.

[203] Rajesh, B., Ravindranathan Thampi, K., Bonard, J. M., Xanthopoulos, N., Mathieu, H. J., and Viswanathan, B. (2003) Carbon nanotubes generated from template carbonization of polyphenyl acetylene as the support for electrooxidation of methanol. *The Journal of Physical Chemistry B*, **107** (12), 2701-2708.

[204] Sun, X., Li, R., Villers, D., Dodelet, J. P., and Désilets, S. (2003) Composite electrodes made of Pt nanoparticles deposited on carbon nanotubes grown on fuel cell backings. *Chemical Physics Letters*, **379** (1-2), 99-104.

[205] Wang, C., Waje, M., Wang, X., Tang, J. M., Haddon, R. C., and Yushan (2003) Proton exchange membrane fuel cells with carbon nanotube based electrodes. *Nano Letters*, **4** (2), 345-348.

[206] Li, L. and Xing, Y. (2007) Pt-Ru nanoparticles supported on carbon nanotubes as methanol fuel cell catalysts. *The Journal of Physical Chemistry C*, **111** (6), 2803-2808.

[207] Wang, X., Li, W., Chen, Z., Waje, M., and Yan, Y. (2006) Durability investigation of carbon nanotube as catalyst support for proton exchange membrane fuel cell. *Journal of Power Sources*, **158** (1), 154-159.

[208] Bessel, C. A., Laubernds, K., Rodriguez, N. M., and Baker, R. T. K. (2001) Graphite nanofibers as an electrode for fuel cell applications. *Journal of Physical Chemistry B*, **105** (6), 1115-1118.

[209] Serp, P., Corrias, M., and Kalck, P. (2003) Carbon nanotubes and nanofibers in catalysis. *Applied Catalysis A: General*, **253** (2), 337-358.

[210] Guo, J., Sun, G., Wang, Q., Wang, G., Zhou, Z., Tang, S., Jiang, L., Zhou, B., and Xin, Q. (2006) Carbon nanofibers supported Pt-Ru electrocatalysts for direct methanol fuel cells. *Carbon*, **44** (1), 152-157.

[211] Lee, K., Zhang, J., Wang, H., and Wilkinson, D. P. (2006) Progress in the synthesis of carbon nanotube- and nanofiber-supported Pt electrocatalysts for PEM fuel cell catalysis. *Journal of Applied Electrochemistry*, **36** (5), 507-522.

[212] Hsin, Y. L., Hwang, K. C., and Yeh, C.-T. (2007) Poly(vinylpyrrolidone)-modified graphite carbon nanofibers as promising supports for PtRu catalysts in direct methanol fuel cells. *Journal of the American Chemical Society*, **129** (32), 9999-10010.

[213] Chai, G. S., Yoon, S. B., Yu, J.-S., Choi, J.-H., and Sung, Y.-E. (2004) Ordered porous carbons with tunable pore sizes as catalyst supports in direct methanol fuel cell. *The Journal of Physical Chemistry B*, **108** (22), 7074-7079.

[214] Lee, J., Yoon, S., Hyeon, T., Oh, S. M., and Bum Kim, K. (1999) Synthesis of a new mesoporous carbon and its application to electrochemical double-layer capacitors. *Chemical Communications*, (21), 2177-2178.

[215] Ryoo, R. , Joo, S. H. , and Jun, S. (1999) Synthesis of highly ordered carbon molecular sieves via template-mediated structural transformation. *The Journal of Physical Chemistry B*, **103** (37), 7743 – 7746.

[216] Lu, A. H. and Schüth, F. (2006) Nanocasting: a versatile strategy for creating nanostructured porous materials. *Advanced Materials*, **18** (14), 1793 – 1805.

[217] Liang, C. , Hong, K. , Guiochon, G. A. , Mays, J. W. , and Dai, S. (2004) Synthesis of a large-scale highly ordered porous carbon film by self-assembly of block copolymers. *Angewandte Chemie: International Edition*, **43** (43), 5785 – 5789.

[218] Meng, Y. , Gu, D. , Zhang, F. , Shi, Y. , Yang, H. , Li, Z. , Yu, C. , Tu, B. , and Zhao, D. (2005) Ordered mesoporous polymers and homologous carbon frameworks: amphiphilic surfactant templating and direct transformation. *Angewandte Chemie*, **117** (43), 7215 – 7221.

[219] Tanaka, S. , Nishiyama, N. , Egashira, Y. , and Ueyama, K. (2005) Synthesis of ordered mesoporous carbons with channel structure from an organic-organic nanocomposite. *Chemical Communications*, (16), 2125 – 2127.

[220] Kim, H. -T. , You, D. J. , Yoon, H. -K. , Joo, S. H. , Pak, C. , Chang, H. , and Song, I. -S. (2008) Cathode catalyst layer using supported Pt catalyst on ordered mesoporous carbon for direct methanol fuel cell. *Journal of Power Sources*, **180** (2), 724 – 732.

[221] Novoselov, K. S. , Geim, A. K. , Morozov, S. V. , Jiang, D. , Zhang, Y. , Dubonos, S. V. , Grigorieva, I. V. , and Firsov, A. A. (2004) Electric field effect in atomically thin carbon films. *Science*, **306** (5696), 666 – 669.

[222] Geim, A. K. and Novoselov, K. S. (2007) The rise of graphene. *Nature Materials*, **6** (3), 183 – 191.

[223] Guo, S. , Dong, S. , and Wang, E. (2009) Three-dimensional Pt-on-Pd bimetallic nanodendrites supported on grapheme nanosheet: facile synthesis and used as an advanced nanoelectrocatalyst for methanol oxidation. *ACS Nano*, **4** (1), 547 – 555.

[224] Kou, R. , Shao, Y. , Wang, D. , Engelhard, M. H. , Kwak, J. H. , Wang, J. , Viswanathan, V. V. , Wang, C. , Lin, Y. , Wang, Y. , Aksay, I. A. , and Liu, J. (2009) Enhanced activity and stability of Pt catalysts on functionalized graphene sheets for electrocatalytic oxygen reduction. *Electrochemistry Communications*, **11** (5), 954 – 957.

[225] Li, Y. , Tang, L. , and Li, J. (2009) Preparation and electrochemical performance for methanol oxidation of Pt/grapheme nanocomposites. *Electrochemistry Communications*, **11** (4), 846 – 849.

[226] Seger, B. and Kamat, P. V. (2009) Electrocatalytically active graphene-platinum nanocomposites: role of 2-D carbon support in PEM fuel cells. *The Journal of Physical Chemistry C*, **113** (19), 7990 – 7995.

[227] Yoo, E. , Okata, T. , Akita, T. , Kohyama, M. , Nakamura, J. , and Honma, I. (2009) Enhanced electrocatalytic activity of Pt subnanoclusters on graphene nanosheet surface. *Nano Letters*, **9** (6), 2255 – 2259.

[228] Zhang, Z. Y. , Xin, L. , and Li, W. (2012) Electrocatalytic oxidation of glycerol on Pt/C in anion-exchange membrane fuel cell: cogeneration of electricity and valuable chemicals. *Applied Catalysis B: Environmental*, **119**, 40 – 48.

[229] Xin, L. , Zhang, Z. Y. , Qi, J. , Chadderdon, D. , and Li, W. Z. (2012) Electrocatalytic oxidation of ethylene glycol (EG) on supported Pt and Au catalysts in alkaline media: reaction pathway investigation in three-electrode cell and fuel cell reactors. *Applied Catalysis B: Environmental*, **125**, 85 – 94.

[230] Zhang, Z. Y. , Xin, L. , and Li, W. (2012) Supported gold nanoparticles as anode catalyst for anion-exchange membranedirect glycerol fuel cell (AEM – DGFC). *International Journal of Hydrogen Energy*, **37** (11), 9393 – 9401.

[231] Xin, L. , Zhang, Z. Y. , Wang, Z. C. , and Li, W. Z. (2012) Simultaneous generation of mesoxalic acid and electricity from glycerol on a gold anode catalyst in anion-exchange membrane fuel cells. *ChemCatChem*, **4** (8), 1105 - 1114.

[232] Zhang, Z. Y. , Xin, L. , Qi, J. , Chadderdon, D. J. , Sun, K. , Warsko, K. M. , and Li, W. Z. (2014) Selective electro-oxidation of glycerol to tartronate or mesoxalate on Au nanoparticle catalyst via electrode potential tuning in anion-exchange membrane electro-catalytic flow reactor. *Applied Catalysis B: Environmental*, **147**, 871 - 878.

[233] Qi, J. , Xin, L. , Chadderdon, D. J. , Qiu, Y. , Jiang, Y. , Benipal, N. , Liang, C. H. , andLi, W. Z. (2014) Electrocatalytic selective oxidation of glycerol to tartronate on Au/C anode catalysts in anion exchange membrane fuel cells with electricity cogeneration. *Applied Catalysis B:Environmental*, **154**, 360 - 368.

# 第5章

# 直接甲醇燃料电池膜材料

Bradley P. Ladewig, Benjamin M. Asquith, Jochen Meier-Haack

## 5.1 引　　言

本章综述了直接甲醇燃料电池(Direct Methanol Fuel Cells,DMFC)膜材料技术的现有状态,侧重于膜材料的研究进展(较少涉及商业产品)。本章的重点仅针对膜材料;然而考虑到相邻的燃料/氧化剂分散层、催化剂和载体与膜结合紧密,这些材料与膜材料必须兼容,因此本章也涉及除膜材料以外的这部分材料。尽管可以通过合成类似的阳离子交换膜时所使用的前驱体和溶剂估算膜的成本、尝试得到膜大致的成本效益,但是鉴于较难得到大部分膜材料的可靠成本数据,本章内容未对膜材料进行成本效益评价[1]。

## 5.2 直接甲醇燃料电池工作的基本原理

直接甲醇燃料电池是一类聚合物电解质膜(Polymer Electrolyte Membrane,PEM)燃料电池,通常采用阳离子交换膜来分隔阳极室和阴极室(碱性交换膜也是一种选择,这些将在本章后面讨论)。下面以采用阳离子交换膜的典型液相进料电池来说明 DMFC 工作的基本原理,如图 5.1 所示。

电池的燃料是甲醇的水溶液,溶液的浓度可以在很大范围内变化。选用高浓度的甲醇溶液可以减少电池体系内携带的水量(DMFC 的应用大部分是在可移动体系中,所以希望电池体系更小更轻),并为阳极催化剂提供较高的反应物浓度。但是,采用高浓度的甲醇溶液做为阳极进料存在着显著的缺点,最值得注意的就是高浓度梯度产生的驱动力导致甲醇在膜内传输的增加。这就是"甲醇渗透",也是 DMFC 研究中需要解决的顶级挑战之一。

甲醇在阳极反应,生成二氧化碳、质子和电子。质子通过膜传递到阴极,而

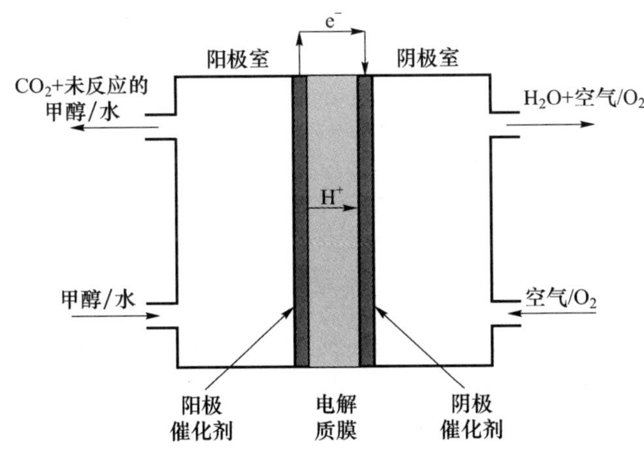


图 5.1　DMFC 单电池示意图

电子则通过集流体收集并通过外部电路做功。在阴极,电子再与质子和氧气结合生成水,产生的水必须从阴极除去。

## 5.3　直接甲醇燃料电池膜材料

正如 5.2 节所述,对于 DMFC 膜性能的要求非常简单。该膜应具有高质子传导性而不导电子。与此同时,它应该具有最小的甲醇渗透性;由于甲醇和水分子之间的相似性,这常常意味着该膜还应具有低的水渗透性。以上两点应该在燃料电池所需的工作温度下实现。考虑到低温下较差的电极过程动力学性能,通常希望 DMFC 在 100℃左右这样较高的温度下工作。

仅这两个特性已足够确定出适用于 DMFC 的有前途的候选材料;然而,如许多作者[2-4]强调的,还有更多的特性,如化学和热稳定性、加工性、成本、环境影响、是否易于制成膜电极等,在选择合适的膜材料时需考虑这些特性。

### 5.3.1　全氟磺酸膜

DMFC 中最广为人知、研究最多的膜材料是全氟磺酸膜,如 Nafion(图 5.2)。这类大分子在单一分子中具有两个不同的功能:①疏水性,这会影响到材料的化学稳定性和热稳定性;②亲水磺酸区,这部分负责水的更新和离子交换。在水存在时,膜相分离成亲水区和疏水区[5]。表征全氟磺酸膜相分离微结构的工作已经有很多,详细讨论已经超出了本章的范围,但有兴趣的读者可以通过 Mauritz 和 Moore 的综述[6]做进一步的了解。

全氟磺酸膜具有的几个关键特征使得它非常适合于 DMFC,包括:①优良的化学稳定性(归功于全氟化的分子骨架),这使得它在氧化和还原环境下具有抗

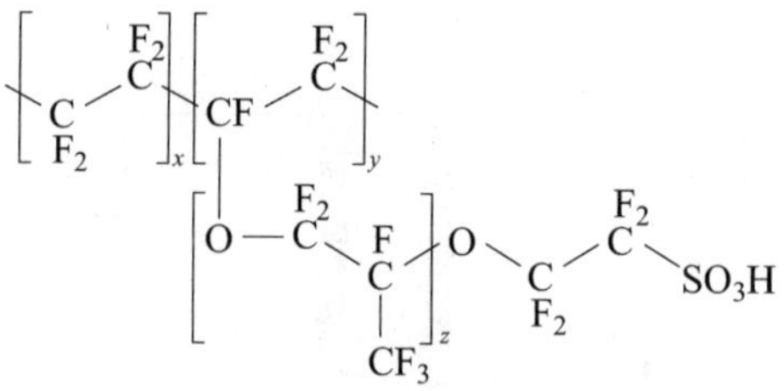

图 5.2　Nafion 的化学结构

降解的能力;②高的质子传导性,最高可达 0.2S/cm[7]。然而在高温下、特别是超过 90℃时,全氟磺酸膜失效,这主要是由于:①膜脱水降低了可用的电荷载流子,进而降低了质子传导性;②聚合物软化导致膜机械强度的降低;③气态物质的交叉渗透,这主要是氢质子交换膜燃料电池中的问题,但在 DMFC 中也不希望阴极的氧气和氮气渗透到阳极。

许多公司已经生产商业化的全氟磺酸膜,商品名称包括 Nafion、Flemion、Aciplex、Aquivion 和 Fumion[8]。然而当甲醇在电极上反应生成二氧化碳和水时,由于甲醇容易穿过该膜,基于此膜的直接甲醇燃料电池的效率是相当低的(减少电池的库仑效率)。因此,膜的修饰、尽可能地抑制甲醇渗透就变得尤为重要。

科学家们在膜的修饰上做了大量的探索工作,一些比较成功的例子包括在膜中掺杂功能化的无机纳米粒子。对这些工作的全面综述超出了本章的范围[2,3]。

### 5.3.2　聚苯乙烯基电解质

在苯乙烯基电解质中,有几种不同的类型被认为是有潜力的 DMFC 膜。这几类膜都含有磺酸基因(图 5.3)如全氟磺酸膜一样,但此类膜中磺酸基团的 p$Ka$ 不如全氟磺酸膜中磺酸基团的酸性高(高电负性的氟化物分子导致高电荷的 $SO_3^-$ 基团)。科学家们还研究了随机共聚物(磺酸基团随机分布在整个聚合物链中并且数量依赖于其在反应物中的占比),包括磺化聚苯乙烯[9-12]和苯乙烯与苯乙烯磺酸的随机共聚物[13,14]。磺酸化聚苯乙烯可以通过苯乙烯类单体与磺化苯乙烯类单体的共聚合或由聚苯乙烯进行直接的磺化反应处理(后磺化)而容易地合成。所述后磺化路线是最常见的,而且在这种方法中,膜的离子

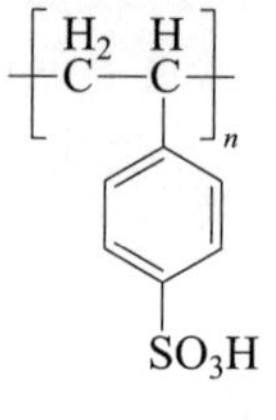

图 5.3　磺化聚苯乙烯的重复单元

交换容量和电导率可通过控制反应条件(通常为磺化剂的浓度,反应时间及温度)进行调整。

### 5.3.3 聚芳醚型聚合物

这一类DMFC的膜材料包括聚芳醚酮和聚芳醚砜。它们是DMFC膜的很好的备选材料,因为它们在氧化和还原环境中具有优良的化学稳定性并可通过控制磺化单元在聚合物中的比例来调节离子电导率。正如聚苯乙烯基膜一样,磺化可以通过直接地共聚合磺化单体(或者磺化和未磺化的混合物)或通过亲电的芳香磺化反应进行后磺化修饰来实现。这种后磺化方法,虽然简便直接,但会导致材料性能的变化,因为它不可能严格控制磺化度和精确定位磺化位点[15]。事实上,在高磺化度的情况下,这种聚合物通常会降解或变得高度易溶胀(或溶解)于水(图5.4)。

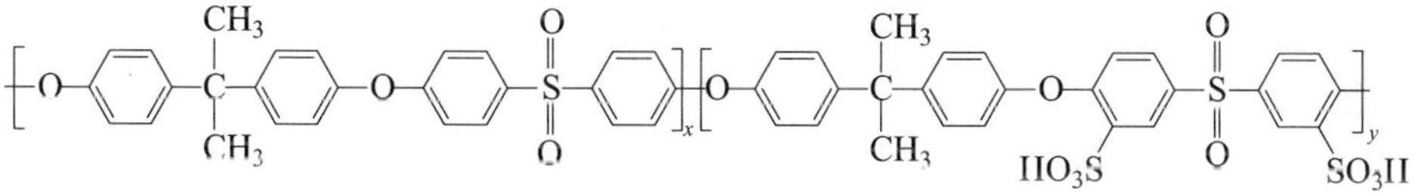

图5.4 使用直接共聚技术合成的二磺化的聚芳醚砜共聚物[16]

### 5.3.4 聚醚醚酮型聚合物

聚醚醚酮(PEEK)和聚醚醚酮酮(PEEKK)是高度稳定的聚合物,在DMFC中具有良好的应用潜力[10,17,18]。它们是半结晶的并具有优异的化学稳定性和热稳定性,可通过磺化单体的直接聚合或后磺化进行磺化,后者又会导致无法严格控制材料的磺化程度和磺化位置,但因为这是一种简单低成本的方法,仍具有一定的吸引力[19]。

磺化聚醚醚酮(SPEEK,图5.5)在DMFC中的应用潜力已被广泛地研究,特别是被制成含有其他成分(如功能化的纳米粒子)的复合膜,已有一些关于这方面的优秀综述文章报道[20]。

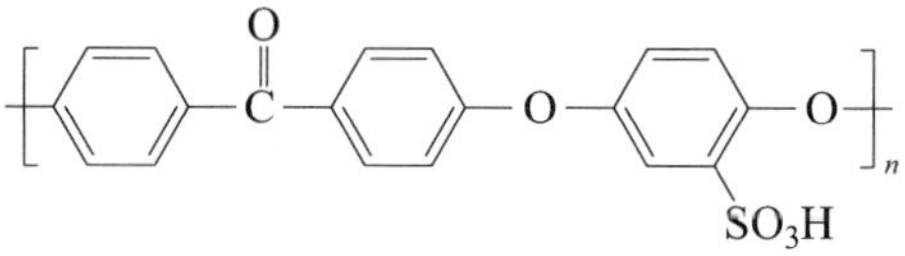

图5.5 SPEEK的重复单元

### 5.3.5 聚苯并咪唑

聚苯并咪唑类聚合物(PBI,图5.6)做为DMFC膜的研究引起了相当大的兴趣,因为它们在高于大多数其他类聚合物系统能够耐受的温度下,仍具有稳定性和质子传导性，这是由聚苯并咪唑基膜自身的特性（p*K*a≈5.5）决定的。聚苯并咪唑浸泡在酸中，如磷酸中，可形成具有良好的化学稳定性和质子传导性的单相聚合物，因此通过酸浸泡即可实现质子传导。事实上，它们不需要像全氟磺酸膜一样水化来导电，这意味着它们可以在100～200℃的温度下使用。从化学反应动力学的角度来说，这一点特别有吸引力，因为阳极和阴极动力学在这样提高的温度下大大提高。然而其缺点是，这种升高的温度不适合DMFC面向的许多便携式应用（如替换电子设备中的锂电池）。已有各种应用在高温PEM和DMFC系统中的PBI和PBI复合膜被设计出来或进行了测试[21-31]。

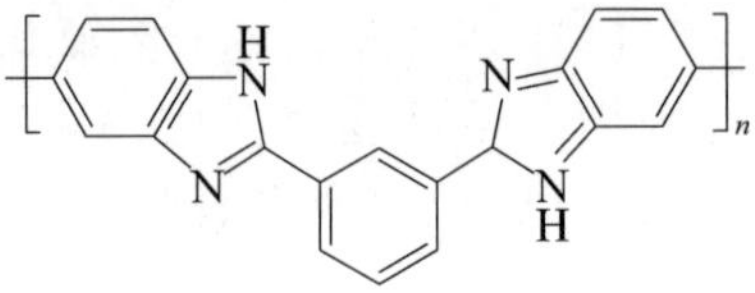

图5.6　聚苯并咪唑

## 5.3.6　聚砜和聚醚砜

聚砜（PSU）（图5.7）和聚醚砜（PES）（图5.8）是高度通用的工程聚合物，已广泛应用在包括气体分离、膜过滤、渗透蒸发和电渗析等过程中。它们具有出色的化学和机械稳定性，相对高的玻璃化温度，并且很容易从1-甲基-2-吡咯烷酮（NMP）[32]和*N*,*N*-二甲基乙酰胺（DMA）[33]等普通非质子溶剂中成膜。对于DMFC来说，PSU被认为是最容易与其他聚合物或无机组分[34-41]形成共混物的聚合物。尽管大部分PSU复合膜的质子传导性比全氟磺酸膜低，但由于减少了甲醇渗透，至少在低进料浓度时，DMFC的性能经常优于基于全氟磺酸膜的DMFC的性能[42]。

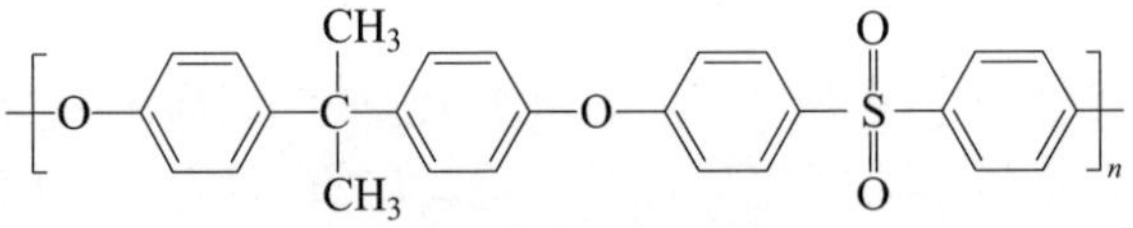

图5.7　聚砜重复单元

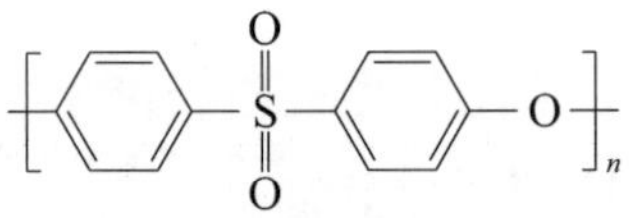

图5.8　聚醚砜重复单元

## 5.3.7　聚酰亚胺

由于聚酰亚胺高的化学和热稳定性，它们是另一类被广泛研究的应用于DMFC的聚合物[43-50]。具体来说，由于减少了环张力，六元环比五元环更稳定。

## 5.3.8　接枝聚合物电解质膜

接枝是一种制备DMFC导电膜的有趣方法，因为它不使用上一节中描述的化学修饰方法，而是将疏水不导电的聚合物暴露于辐射源中形成自由基和官能团,使得其他官能团如苯乙烯或苯乙烯磺化单体可被接枝并聚合。

Pall 公司利用这种方法生产的商用膜称为 IonClad® ,这种膜由全氟聚合物主链上接枝聚苯乙烯磺酸构成。Tricoli 等评价了 IonClad R-1010 和 R-4010 这两种膜在 DMFC 中的应用潜力,发现尽管其在 20 ~ 60℃ 的范围内的导电率与 Nafion 117 大致相同,但它们的甲醇渗透率却只有 Nafion 的 1/4,使其成为应用于 DMFC 中的很好的候选膜材料[51]。

### 5.3.9 嵌段共聚物

嵌段共聚物是一种制备 DMFC 膜的有吸引力的方法,因为它们高度可控的聚合物结构意味着离子型和非离子型结构单元可以在相同的聚合物主链上按照预定的顺序进行排列。此外,通过对嵌段长度的精细控制,微相分离也可以被控制到一定程度上。Kim 等制备了部分磺化的聚苯乙烯/聚(乙烯 - 丁烯)/聚苯乙烯(SSEBS)三嵌段共聚物(图 5.9),并评价了其质子传导性和甲醇渗透率。当嵌段共聚物的磺化度达到 34% (摩尔分数)时,可以实现类似于 Nafion 117 的质子传导率,而甲醇吸收大约只有 Nafion 117 的 1/2。较低的甲醇吸收归因于相较于 Nafion 的更精细的微结构(这是嵌段共聚物的特征),这可通过小角 X 射线散射来确定[52]。

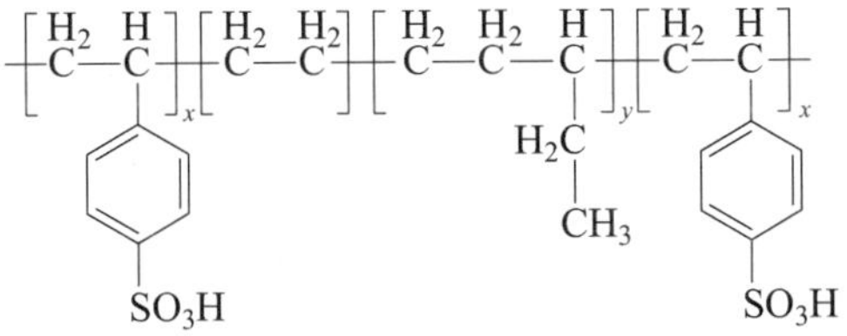

图 5.9 SSEBS 的重复单元[53]

### 5.3.10 复合聚合物膜

5.3.1 节已经指出,使用单一聚合物材料通常很难达到所需的膜性能,因此,近年来人们在合成复合聚合物膜方面做了大量的工作,以增强或改变聚合物膜的性能。

具体来说,在以 DMFC 为重点的研究中,比较集中的做法是引入填料材料,如无机纳米粒子,阻碍甲醇通过膜的通道,减小甲醇渗透。显然,这样的方法不能同时阻碍、或者至少不能以阻碍甲醇渗透的相同程度来阻碍质子的传导;然而,并非能够总是如此。许多已被报道的这类膜材料显示出显著降低甲醇渗透的特性,但同时也显示出质子传导性有所降低。明显的例外是那些无机填充材料本身是质子导体或本身具有表面酸性基团的材料,无机材料因此可以发挥阻碍主体分子运输(水和甲醇的扩散)并同时贡献额外质子传送位点的双重作用[2]。

关于全部复合膜材料的全面综述远远超出了本章的范围,有兴趣的读者可以查阅 Neburchilov 等[8]的相关综述。

## 5.4 膜性质总结

表 5.1 总结了从文献选择的 DMFC 膜,并按照一般的膜类型分类。第一项涉及 Nafion 115 和 Nafion 117,并作为基准结果与许多其他材料进行比较。必须指出的是,不同课题组之间用于测定质子传导率和甲醇渗透率的设备和方法会有变化,因此用这个表格中的数据进行绝对比较很难奏效。该表的目的仅仅是为了以简洁的方式来总结 DMFC 膜性能的当前状态。

表 5.1 膜的质子传导率和甲醇渗透率的性能数据总结

| 膜材料 | 质子传导率 /(mS/cm) | 甲醇渗透率 /($cm^2$/s) $\times 10^6$ | 文献 |
|---|---|---|---|
| Nafion | | | |
| Nafion 117 | 100 | 1.76 | [51] |
| Nafion 117 | 75 | 0.9 | [54] |
| Nafion 117 | 67 | 1.98 | [55] |
| Nafion 115 | 90 | 1.04 | [56] |
| Nafion 复合物 | | | |
| Nafion /功能化二氧化硅纳米复合物 | 6.9 | 0.358 | [2] |
| Nafion /蒙脱石纳米复合物 | 78 | 0.1 | [57] |
| 纳米二氧化硅层化 Nafion 复合物 | 77 | 0.92 | [58] |
| Nafion /聚乙烯醇共混物 | 20 | 0.65 | [59] |
| Nafion /ORMOSILS 复合物 | 19 | 1.75 | [60] |
| 其他聚合物体系 | | | |
| 磺化聚苯乙烯 | 50 | 0.52 | [12] |
| 磺化聚醚醚酮 | 70 | 0.3 | [10] |
| 磺化聚芳醚共聚物 | 100 | 0.81 | [61] |
| 复合聚酰亚胺膜 BAPS - 50 | 40 | 0.33 | [62] |
| 复合聚酰亚胺膜 BAPS - 60 | 50 | 0.45 | [62] |
| 磺化聚磷腈 | 35 | 0.13 | [54] |
| 膦化聚磷腈 | 55 | 0.14 | [54] |
| 磺化聚(苯乙烯-*b*-乙烯-*r*-丁二烯-*b*-苯乙烯)嵌段共聚物 | 45 | 2.6 | [52] |
| 磺化聚砜 | 5 | 0.06 | [56] |
| 磺化聚苯并咪唑掺杂 | 0.001 | 2.5 | [63] |

（续）

| 膜材料 | 质子传导率 /(mS/cm) | 甲醇渗透率 /($cm^2$/s) $\times 10^6$ | 文献 |
|---|---|---|---|
| 磺化聚酰亚胺 | 120 | 0.57 | [11] |
| 磺化聚苯乙烯和磺化聚(2,6-二甲基-1,4-氧化亚苯基)共混物 | 34 | 2.35 | [9] |
| 磺化聚醚醚酮酮 | 40 | 0.575 | [64] |
| 磺化酚酞型聚醚砜 | 4 | 0.210 | [65] |
| 聚(偏氟乙烯－六氟丙烯)聚合物/Nafion 共混物 | 2 | 0.2 | [66] |
| 交联的聚乙烯醇/聚(2-丙烯酰氨基-2-甲基-1-丙烷磺酸) | 90 | 0.6 | [67] |
| 磺化聚(苯乙烯)/聚(四氟乙烯)复合物 | 110 | 0.67 | [68] |
| 注:改编自参考文献[4] | | | |

## 5.5 本章小结

本章从聚合物材料化学结构的角度总结了用于直接甲醇燃料电池的众多膜材料。尽管有很多材料在某些方面表现出超越 Nafion 的性能,但 DMFC 膜领域仍需要针对关键材料进行研究。具体来说,需要开发不依赖于水化即可保持高度质子传导性的膜材料——这被认为是实现膜不渗透甲醇的唯一真正可行的手段。这类材料应该是化学稳定的,在水/甲醇溶液中不溶胀,并且在运行 DMFC 的操作条件下保持电化学稳定。

## 参考文献

[1] Yee, R. S. L., Rozendal, R. A., Zhang, K., and Ladewig, B. P. (2012) Cost effective cation exchange membranes: a review. *Chemical Engineering Research and Design*, **90** (7), 950–959.

[2] Ladewig, B. P., Knott, R. B., Hill, A. J., Riches, J. D., White, J. W., Martin, D. J., Da Costa, J. C. D., and Lu, G. Q. (2007) Physical and electrochemical characterization of nanocomposite membranes of nafion and functionalized silicon oxide. *Chemistry of Materials*, **19**(9), 2372–2381.

[3] Ladewig, B. P., Knott, R. B., Martin, D. J., Diniz da Costa, J. C., and Lu, G. Q. (2007) Nafion-MPMDMS nanocomposite membranes with low methanol permeability. *Electrochemistry Communications*, **9** (4), 781–786.

[4] Lufrano, F., Baglio, V., Staiti, P., Antonucci, V., and Arico, A. S. (2013) Performance analysis of polymer electrolyte membranes for direct methanol fuel cells. *Journal of Power Sources*, **243**, 519–534.

[5] Kreuer, K. D. (2001) On the development of proton conducting polymer membranes for hydrogen and methanol fuel cells. *Journal of Membrane Science*, **185** (1), 29–39.

[6] Mauritz, K. A. and Moore, R. B. (2004) State of understanding of Nafion. *Chemical Reviews*, **104** (10), 4535 - 4585.

[7] Savadogo, O. (1998) Emerging membranes for electrochemical systems: (I) solid polymer electrolyte membranes for fuel cell systems. *Journal of New Materials for Electrochemical Systems*, **1** (1), 47 - 66.

[8] Neburchilov, V., Martin, J., Wang, H., and Zhang, J. (2007) A review of polymer electrolyte membranes for direct methanol fuel cells. *Journal of Power Sources*, **169** (2), 221 - 238.

[9] Jung, B., Kim, B., and Yang, J. M. (2004) Transport of methanol and protons through partially sulfonated polymer blend membranes for direct methanol fuel cell. *Journal of Membrane Science*, **245** (1 - 2), 61 - 69.

[10] Gil, M., Ji, X., Li, X., Na, H., Hampsey, J. E., and Lu, Y. (2004) Direct synthesis of sulfonated aromatic poly(ether ether ketone) proton exchange membranes for fuel cell applications. *Journal of Membrane Science*, **234** (1 - 2), 75 - 81.

[11] Yin, Y., Fanga, J., Cui, Y., Tanaka, K., Kita, H., and Okamoto, K. I. (2003) Synthesis, proton conductivity and methanol permeability of a novel sulfonated polyimide from 3-(2′,4′-diaminophenoxy) propane sulfonic acid. *Polymer*, **44** (16), 4509 - 4518.

[12] Carretta, N., Tricoli, V., and Picchioni, F. (2000) Ionomeric membranes based on partially sulfonated poly(styrene): synthesis, proton conduction and methanol permeation. *Journal of Membrane Science*, **166** (2), 189 - 197.

[13] Bae, B. and Kim, D. (2003) Sulfonated polystyrene grafted polypropylene composite electrolyte membranes for direct methanol fuel cells. *Journal of Membrane Science*, **220** (1 - 2), 75 - 87.

[14] Jung, D. H., Myoung, Y. B., Cho, S. Y., Shin, D. R., and Peck, D. H. (2001) A performance evaluation of direct methanol fuel cell using impregnated tetraethyl-orthosilicate in cross-linked polymer membrane. *International Journal of Hydrogen Energy*, **26** (12), 1263 - 1269.

[15] Miyatake, K. and Hay, A. S. (2001) Synthesis and properties of poly(arylene ether)s bearing sulfonic acid groups on pendant phenyl rings. *Journal of Polymer Science, Part A: Polymer Chemistry*, **39** (19), 3211 - 3217.

[16] Harrison, W. L., Hickner, M. A., Kim, Y. S., and McGrath, J. E. (2005) Poly(arylene ether sulfone) copolymers and related systems from disulfonated monomer building blocks: synthesis, characterization, and performance - a topical review. *Fuel Cells*, **5** (2), 201 - 212.

[17] Li, L., Zhang, J., and Wang, Y. (2003) Sulfonated poly(ether ether ketone) membranes for direct methanol fuel cell. *Journal of Membrane Science*, **226** (1 - 2), 159 - 167.

[18] Kobayashi, T., Rikukawa, M., Sanui, K., and Ogata, N. (1998) Proton-conducting polymers derived from poly(etheretherketone) and poly(4-phenoxybenzoyl-1,4-phenylene). *Solid State Ionics*, **106** (3 - 4), 219 - 225.

[19] Yee, R. S. L., Zhang, K., and Ladewig, B. P. (2013) The effects of sulfonated poly (ether ether ketone) ion exchange preparation conditions on membrane properties. *Membranes*, **3** (3), 182 - 195.

[20] Liu, B., Robertson, G. P., Guiver, M. D., Sun, Y. M., Liu, Y. L., Lai, J. Y., Mikhailenko, S., and Kaliaguine, S. (2006) Sulfonated poly(aryl ether ether ketone ketone)s containing fluorinated moieties as proton exchange membrane materials. *Journal of Polymer Science, Part B: Polymer Physics*, **44** (16), 2299 - 2310.

[21] Diaz, L. A., Abuin, G. C., and Corti, H. R. (2012) Methanol sorption and permeability in Nafion and acid-doped PBI and ABPBI membranes. *Journal of Membrane Science*, **411 - 412**, 35 - 44.

[22] Haghighi, A. H., Hasani-Sadrabadi, M. M., Dashtimoghadam, E., Bahlakeh, G., Shakeri, S. E., Ma-

jedi, F. S. , Hojjati Emami, S. , and Moaddel, H. (2011) Direct methanol fuel cell performance of sulfonated poly (2,6-dimethyl-1,4- phenylene oxide)-polybenzimidazole blend proton exchange membranes. *International Journal of Hydrogen Energy*, **36** (5), 3688 – 3696.

[23] Ahmad, H. , Kamarudin, S. K. , Hasran, U. A. , and Daud, W. R. W. (2011) A novel hybrid Nafion-PBI-ZP membrane for direct methanol fuel cells. *International Journal of Hydrogen Energy*, **36** (22), 14668 – 14677.

[24] Yu, S. and Benicewicz, B. C. (2009) Synthesis and properties of functionalized polybenzimidazoles for high-temperature PEMFCs. *Macromolecules*, **42** (22), 8640 – 8648.

[25] Oono, Y. , Sounai, A. , and Hori, M. (2009) Influence of the phosphoric acid-doping level in a polybenzimidazole membrane on the cell performance of hightemperature proton exchange membrane fuel cells. *Journal of Power Sources*, **189** (2), 943 – 949.

[26] Pasupathi, S. , Ji, S. , Jan Bladergroen, B. , and Linkov, V. (2008) High DMFC performance output using modified acid-base polymer blend. *International Journal of Hydrogen Energy*, **33** (12), 3132 – 3136.

[27] Lobato, J. , Cañizares, P. , Rodrigo, M. A. , Linares, J. J. , and López-Vizcaíno, R. (2008) Performance of a vapor-fed polybenzimidazole (PBI)-based direct methanol fuel cell. *Energy and Fuels*, **22** (5), 3335 – 3345.

[28] Wycisk, R. , Chisholm, J. , Lee, J. , Lin, J. , and Pintauro, P. N. (2006) Direct methanol fuel cell membranes from Nafion-polybenzimidazole blends. *Journal of Power Sources*, **163**, 9 – 17.

[29] Wycisk, R. , Lee, J. K. , and Pintauro, P. N. (2005) Sulfonated polyphosphazene polybenzimidazole membranes for DMFCs. *Journal of the Electrochemical Society*, **152** (5), A892 – A898.

[30] Silva, V. S. , Ruffmann, B. , Vetter, S. , Mendes, A. , Madeira, L. M. , and Nunes, S. P. (2005) Characterization and application of composite membranes in DMFC. *Catalysis Today*, **104** (2 – 4), 205 – 212.

[31] Li, Q. , He, R. , Jensen, J. O. , and Bjerrum, N. J. (2003) Approaches and recent development of polymer electrolyte membranes for fuel cells operating above 100℃. *Chemistry of Materials*, **15** (26), 4896 – 4915.

[32] Kim, J. H. and Lee, K. H. (1998) Effect of PEG additive on membrane formation by phase inversion. *Journal of Membrane Science*, **138** (2), 153 – 163.

[33] Chakrabarty, B. , Ghoshal, A. K. , and Purkait, M. K. (2008) Preparation, characterization and performance studies of polysulfone membranes using PVP as an additive. *Journal of Membrane Science*, **315** (1 – 2), 36 – 47.

[34] Lufrano, F. , Baglio, V. , Di Blasi, O. , Staiti, P. , Antonucci, V. , and Aricò, A. S. (2012) Solid polymer electrolyte based on sulfonated polysulfone membranes and acidic silica for direct methanol fuel cells. *Solid State Ionics*, **216**, 90 – 94.

[35] Zhu, Y. and Manthiram, A. (2011) Synthesis and characterization of polysulfone-containing sulfonated side chains for direct methanol fuel cells. *Journal of Power Sources*, **196** (18), 7481 – 7487.

[36] Lufrano, F. , Baglio, V. , Staiti, P. , Stassi, A. , Aric, A. S. , and Antonucci, V. (2010) Investigation of sulfonated polysulfone membranes as electrolyte in a passivemode direct methanol fuel cell mini-stack. *Journal of Power Sources*, **195** (23), 7727 – 7733.

[37] Abu-Thabit, N. Y. , Ali, S. A. , and Javaid Zaidi, S. M. (2010) New highly phosphonated polysulfone membranes for PEM fuel cells. *Journal of Membrane Science*, **360** (1 – 2), 26 – 33.

[38] Lufrano, F. , Baglio, V. , Staiti, P. , Arico, A. S. , and Antonucci, V. (2008) Polymer electrolytes based on sulfonated polysulfone for direct methanol fuel cells. *Journal of Power Sources*, **179** (1),

34 – 41.

[39] Vernersson, T. , Lafitte, B. , Lindbergh, G. , and Jannasch, P. (2006) A sulfophenylated polysulfone as the DMFC electrolyte membrane: an evaluation of methanol permeability and cell performance. *Fuel Cells*, **6** (5), 340 – 346.

[40] Lufrano, F. , Baglio, V. , Staiti, P. , Arico, A. S. , and Antonucci, V. (2006) Development and characterization of sulfonated polysulfone membranes for direct methanol fuel cells. *Desalination*, **199** (1 – 3), 283 – 285.

[41] Manea, C. and Mulder, M. (2002) Characterization of polymer blends of polyethersulfone/sulfonated polysulfone and polyethersulfone/sulfonated polyetheretherketone for direct methanol fuel cell applications. *Journal of Membrane Science*, **206** (1 – 2), 443 – 453.

[42] Fu, Y. Z. and Manthiram, A. (2006) Synthesis and characterization of sulfonated polysulfone membranes for direct methanol fuel cells. *Journal of Power Sources*, **157** (1), 222 – 225.

[43] Li, Y. , Jin, R. , Cui, Z. , Wang, Z. , Xing, W. , Qiu, X. , Ji, X. , and Gao, L. (2007) Synthesis and characterization of novel sulfonated polyimides from 1,4-bis(4- aminophenoxy)-naphthyl-2,7-disulfonic acid. *Polymer*, **48** (8), 2280 – 2287.

[44] Yin, Y. , Yamada, O. , Tanaka, K. , and Okamoto, K. I. (2006) On the development of naphthalene-based sulfonated polyimide membranes for fuel cell applications. *Polymer Journal*, **38** (3), 197 – 219.

[45] Meyer, G. , Gebel, G. , Gonon, L. , Capron, P. , Marscaq, D. , Marestin, C. , and Mercier, R. (2006) Degradation of sulfonated polyimide membranes in fuel cell conditions. *Journal of Power Sources*, **157** (1), 293 – 301.

[46] Yamada, O. , Yin, Y. , Tanaka, K. , Kita, H. , and Okamoto, K. I. (2005) Polymer electrolyte fuel cells based on main-chaintype sulfonated polyimides. *Electrochimica Acta*, **50** (13), 2655 – 2659.

[47] Watari, T. , Fang, J. , Tanaka, K. , Kita, H. , Okamoto, K. I. , and Hirano, T. (2004) Synthesis, water stability and proton conductivity of novel sulfonated polyimides from 4,4′-bis(4-aminophenoxy) biphenyl-3,3′-disulfonic acid. *Journal of Membrane Science*, **230** (1 – 2), 111 – 120.

[48] Woo, Y. , Oh, S. Y. , Kang, Y. S. , and Jung, B. (2003) Synthesis and characterization of sulfonated polyimide membranes for direct methanol fuel cell. *Journal of Membrane Science*, **220** (1 – 2), 31 – 45.

[49] Guo, X. , Fang, J. , Watari, T. , Tanaka, K. , Kita, H. , and Okamoto, K. I. (2002) Novel sulfonated polyimides as polyelectrolytes for fuel cell application: 2. Synthesis and proton conductivity of polyimides from 9,9-bis(4-aminophenyl)fluorene-2,7- disulfonic acid. *Macromolecules*, **35** (17), 6707 – 6713.

[50] Fang, J. , Guo, X. , Harada, S. , Watari, T. , Tanaka, K. , Kita, H. , and Okamoto, K. I. (2002) Novel sulfonated polyimides as polyelectrolytes for fuel cell application: 1. Synthesis, proton conductivity, and water stability of polyimides from 4,4′- diaminodiphenyl ether-2,2′-disulfonic acid. *Macromolecules*, **35** (24), 9022 – 9028.

[51] Tricoli, V. , Carretta, N. , and Bartolozzi, M. (2000) Comparative investigation of proton and methanol transport in fluorinated ionomeric membranes. *Journal of the Electrochemical Society*, **147** (4), 1286 – 1290.

[52] Kim, J. , Kim, B. , and Jung, B. (2002) Proton conductivities and methanol permeabilities of membranes made from partially sulfonated polystyrene-block-poly (ethylene- ran-butylene)-block-polystyrene copolymers. *Journal of Membrane Science*, **207** (1), 129 – 137.

[53] Ganguly, A. and Bhowmick, A. K. (2008) Sulfonated styrene-(ethylene-*co*-butylene)- styrene/montmorillonite clay nanocomposites: synthesis, morphology, and properties. *Nanoscale Research Letters*, **3** (1), 36 – 44.

[54] Zhou, X. , Weston, J. , Chalkova, E. , Hofmann, M. A. , Ambler, C. M. , Allcock, H. R. , and Lvov, S. N. (2003) High temperature transport properties of polyphosphazene membranes for direct methanol fuel cells. *Electrochimica Acta*, **48**, 2173 - 2180.

[55] Elabd, Y. A. , Napadensky, E. , Sloan, J. M. , Crawford, D. M. , and Walker, C. W. (2003) Triblock copolymer ionomer membranes: Part I. Methanol and proton transport. *Journal of Membrane Science*, **217** (1 - 2), 227 - 242.

[56] Pedicini, R. , Carbone, A. , Saccà, A. , Gatto, I. , Di Marco, G. , and Passalacqua, E. (2008) Sulphonated polysulphone membranes for medium temperature in polymer electrolyte fuel cells (PEFC). *Polymer Testing*, **27** (2), 248 - 259.

[57] Song, M. K. , Park, S. B. , Kim, Y. T. , Rhee, H. W. , and Kim, J. (2003) Nanocomposite polymer membrane based on cation exchange polymer and nano-dispersed clay sheets. *Molecular Crystals and Liquid Crystals*, **407**, 15/[411] - 423/[419].

[58] Kim, D. , Scibioh, M. A. , Kwak, S. , Oh, I. H. , and Ha, H. Y. (2004) Nano-silica layered composite membranes prepared by PECVD for direct methanol fuel cells. *Electrochemistry Communications*, **6** (10), 1069 - 1074.

[59] DeLuca, N. W. and Elabd, Y. A. (2006) Nafion/poly(vinyl alcohol) blends: effect of composition and annealing temperature on transport properties. *Journal of Membrane Science*, **282** (1 - 2), 217 - 224.

[60] Kim, Y. I. , Choi, W. C. , Woo, S. I. , and Hong, W. H. (2004) Proton conductivity and methanol permeation in NafionTM/ ORMOSIL prepared with various organic silanes. *Journal of Membrane Science*, **238** (1 - 2), 213 - 222.

[61] Kim, S. Y. , Sumner, M. J. , Harrison, W. L. , Riffle, J. S. , McGrath, J. E. , and Pivovar, B. S. (2004) Direct methanol fuel cell performance of disulfonated poly(arylene ether benzonitrile) copolymers. *Journal of the Electrochemical Society*, **151** (12), A2150 - A2156.

[62] Einsla, B. R. , Yu, S. K. , Hickner, M. A. , Hong, Y. T. , Hill, M. L. , Pivovar, B. S. , and McGrath, J. E. (2005) Sulfonated naphthalene dianhydride based polyimide copolymers for proton-exchange-membrane fuel cells: II. Membrane properties and fuel cell performance. *Journal of Membrane Science*, **255** (1 - 2), 141 - 148.

[63] Pu, H. , Liu, Q. , and Liu, G. (2004) Methanol permeation and proton conductivity of acid-doped poly (*N*-ethylbenzimidazole) and poly (*N*-methylbenzimidazole). *Journal of Membrane Science*, **241** (2), 169 - 175.

[64] Li, X. , Zhao, C. , Lu, H. , Wang, Z. , and Na, H. (2005) Direct synthesis of sulfonated poly(ether ether ketone ketone)s (SPEEKKs) proton exchange membranes for fuel cell application. *Polymer*, **46** (15), 5820 - 5827.

[65] Li, L. and Wang, Y. (2005) Sulfonated polyethersulfone Cardo membranes for direct methanol fuel cell. *Journal of Membrane Science*, **246** (2), 167 - 172.

[66] Cho, K. -Y. , Eom, J. -Y. , Jung, H. Y. , Choi, N. -S. , Lee, Y. M. , Park, J. -K. , Choi, J. -H. , Park, K. -W. , and Sung, Y. -E. (2004) Characteristics of PVdF copolymer/Nafion blend membrane for direct methanol fuel cell (DMFC). *Electrochimica Acta*, **50** (2 - 3), 583 - 588.

[67] Qiao, J. , Hamaya, T. , and Okada, T. (2005) New highly proton-conducting membrane poly(vinylpyrrolidone) (PVP) modified poly (vinyl alcohol)/2-acrylamido-2-methyl -1-propanesulfonic acid (PVA-PAMPS) for low temperature direct methanol fuel cells (DMFCs). *Polymer*, **46** (24), 10809 - 10816.

[68] Shin, J. P. , Chang, B. J. , Kim, J. H. , Lee, S. B. , and Suh, D. H. (2005) Sulfonated polystyrene/PTFE composite membranes. *Journal of Membrane Science*, **251** (1 - 2), 247 - 254.

# 第6章

# 氢氧根离子交换膜及离聚物

Shuang Gu, Junhua Wang, Bingzi Zhang, Robert B. Kaspar, Yushan Yan

## 6.1 引　　言

本章聚焦在燃料电池的一个重要组成部分：氢氧根导电聚合物电解质（氢氧根交换膜（HEM）和氢氧根交换离聚物（HEI）），总结这类膜和离聚物的性能要求和工艺方法，着重论述阳离子官能团、聚合物主链和化学交联对于结构－性质关系的影响；并希望可以为今后的研究方向，特别是在燃料电池高性能聚合物电解质的设计和开发方面提供指导。虽然本章讨论的大部分内容是针对燃料电池的，但这类膜和离聚物也普遍适用于其他设备，如电解槽、溢流电池和太阳能氢气发生器。在编写本章时，引用的文献覆盖到2011年，主要是2001年—2011年的文献。

### 6.1.1 定义

HEM是指能够传导氢氧根阴离子（$OH^-$）的膜型聚合物电解质；而HEI是指黏结型聚合物电解质，它不但能够传导氢氧根阴离子，而且能够在电极催化剂层中创建三相界面。HEM和HEI已经应用于氢氧根离子交换膜燃料电池（HEMFC）中，也可以应用于许多其他电化学能量转换和存储设备中。

### 6.1.2 功能

作为离子导体，HEM和HEI实质上通过其氢氧根的传导性和化学/物理稳定性来控制HEMFC的性能和耐久性。作为支持电解质，它们促进电极动力学并且可以使用非贵重金属催化剂。

### 6.1.3 特征

与应用于质子交换膜燃料电池（PEMFC）中的常规质子交换膜（PEM）一样，

HEM 是薄膜聚合物电解质,HEMFC 因而可以实现高能量密度和高功率密度。不同之处在于导电离子的不同:HEM 传导氢氧根离子,PEM 传导质子。

HEM 与应用于碱性燃料电池(AFC)中的传导氢氧根离子的液态碱性溶液(如 KOH 水溶液)截然不同,AFC 的性能从根本上受制于二氧化碳的污染(如来自空气)[1]。在碱性溶液中,金属阳离子自由移动,而在 HEM 中,有机阳离子由共价键固定在聚合物基质中。这种固定化可以阻止碳酸盐沉淀物的形成,而生成碳酸盐沉淀物是 AFC 中溶液态电解质的主要缺点之一。虽然 HEM 中的氢氧根离子可与二氧化碳反应,生成碳酸根阴离子($CO_3^{2-}$)或碳酸氢根阴离子($HCO_3^-$),降低离子电导率,但这些抗衡离子可在燃料电池运行期间通过自净化过程完全复原成氢氧根离子[2]。此外,因为电解质是固体,在 HEM 中不存在电解液泄漏的问题。与液态碱性溶液一样,HEM 与非贵重金属催化剂兼容[3,4],这些特性表明 HEM 是有应用前景的下一代燃料电池电解质。

## 6.2 要　　求

对于高性能 HEMFC 应用来说,要求 HEM 和 HEI 具有高的氢氧根离子传导性,优异的化学稳定性,足够的物理稳定性,可控的溶解性及其他重要特性。

### 6.2.1 高氢氧根离子电导率

高氢氧根离子电导率为高效 HEM 最关键的要求,因为氢氧根离子电导率直接决定了给定厚度膜(一般为 50μm)的膜电阻。低的膜电阻在减轻电阻诱导的电池电压损失方面特别重要,尤其是在大电流密度时。HEM 通常比 PEM 具有更低的离子(氢氧根)电导率,很大程度上是因为氢氧根离子比质子固有的迁移率更低(分别为 $20.50 \times 10^{-4} cm^2 \cdot V^{-1} \cdot S^{-1}$ 和 $36.25 \times 10^{-4} cm^2 \cdot V^{-1} \cdot S^{-1}$,25℃)[5],即使在所有已知的负离子中氢氧根离子具有最高的离子迁移率。与 20℃时质子传导率约 100mS/cm 的典型商用 PEM(Nafion 212,50μm)相比,高性能 HEM 期望的氢氧根电导率应达到 50mS/cm。

除了增加氢氧根离子电导率以外,另一种降低膜电阻的方法是降低它的厚度。与 PEMFC 相比,HEMFC 实际上可以用更薄的膜来工作(如 Tokuyama 公司的产品 A901/A201,厚度为 10μm/28μm[6]),这主要是因为一般而言 HEMFC 具有较低的燃料渗透(如 $H_2$)。在 PEMFC 中,离子从阳极流向阴极,燃料渗透的方向与离子流方向相同;在 HEMFC 中,离子从阴极流向阳极,燃料渗透的方向与离子流方向相反。即使 HEM 的离子电导率只有 PEM 的 1/2,但 1/2 厚度的 HEM 会与 PEM 具有相同的膜电阻。

### 6.2.2 优良的化学稳定性

优良的化学稳定性是高性能 HEM 另一个重要的要求,因为它决定 HEM 的耐久性和操作条件(如电池温度)。这里所需要的化学稳定性包括碱、热和氧化等方面。碱性化学稳定性特别重要,这是因为在 HEM 中氢氧根阴离子本身是一种强亲核攻击离子,已被证实是膜降解的原因[7]。当电池温度升高到高性能 HEMFC 工作的适宜温度(如 60 ~ 80℃)时,需要耐久的 HEM 还具有足够的热化学稳定性[8]。与此同时,在 HEMFC 中存在的强氧化剂(通常为氧气)和高活性催化剂还要求 HEM 具有足够的氧化化学稳定性。从工程的角度来看,普遍期望 HEMFC 应用的耐久性达到 5000h。

### 6.2.3 足够的物理稳定性

与化学稳定性类似,耐久高性能 HEM 还需要具有足够的物理稳定性。物理稳定性包括尺寸和机械方面。尺寸稳定性是指在干湿循环过程中 HEM 对每一维上尺寸变化的抵抗。一般来说,尺寸稳定性由溶胀比表示:低溶胀比意味着高尺寸稳定性,反之亦然。有时,水的吸收量也可以用于描述尺寸稳定性。耐久的 HEM 需要高尺寸稳定性,因为在实际中,HEMFC 需经受频繁的开关。此外,HEM 要求足够的机械稳定性(或机械强度),例如,能够安全地组装到膜电极组件(MEA)中,可耐受膜电极中施加的用于收紧膜和电极之间界面的高压。

### 6.2.4 可控的溶解性

高耐溶剂性对于提高 HEM 的耐久性和寿命是重要的,而在低沸点的水混溶性溶剂中(如低级醇、四氢呋喃和丙酮)的特定溶解性是 HEM 应用和性能的关键[8]。低沸点(<100℃)确保了 HEI 的有机溶剂会挥发得比水快,而水混溶性保证水可以在 HEI 溶液中用作共溶剂。这两个要求旨在阻止催化剂接触纯有机溶剂,避免产生严重的安全问题和催化剂中毒。此外,良好的溶解性使 HEI 能够产生有效的三相界面以支持 HEMFC 电极中的电化学反应。

### 6.2.5 其他重要特性

低燃料渗透和低透气性对于高性能 HEM 也很重要。在燃料电池中,当燃料穿过电解质从阳极到阴极(或氧化剂从阴极到阳极渗透)时,阴极(或阳极)电位被燃料氧化(或氧化剂还原)毒害,降低了电池的整体电压。此外,HEM 必须具有低的电子电导率,以减少当电子不经过外部电路直接穿过膜从阳极到阴极时发生的内部短路。

## 6.3 制备和分类

通常来说,HEM 可通过聚合物功能化、单体聚合或膜辐射接枝等方法制备。其他方法包括化学交联或物理交联(聚合物共混、孔隙灌装和范德华相互作用调谐技术)也可以应用于制备增强的HEM。

### 6.3.1 聚合物功能化

目前,聚合物功能化是制备HEM最经常使用的方法。该方法使用商业或实验室自制的聚合物作为原料。在一般情况下,通过卤烷基化(通常为氯甲基化[9])或卤代(如溴代[10])反应在起始聚合物中引入卤代烷基。接着,通过与相应的前驱体(如叔胺[9]或叔膦[8])的季铵化反应形成阳离子官能团(如季铵或季鏻)。这种制备方法简单可靠而且灵活:几乎可以完全独立地选择聚合物骨架和阳离子官能团,保证可以得到具有独特化学结构的多种HEM。然而,对于直接功能化,很难精确控制卤烷基化(或卤化)的初始度和最终膜的功能化程度。另一缺点是,聚合物基质中的功能位点通常限于初始卤烷基化/卤化步骤中活性最强的位点。

### 6.3.2 单体聚合

单体聚合是制备HEM的一个同样重要的方法。该方法使用包含最终阳离子官能团或其前驱物的单体作为原料,然后直接聚合单体(如通过聚合不饱和键[11]或芳香酚基团[12]),合成HEM聚合物。通常,聚合通过开环或亲核缩合反应来实现。这种方法的最大优点是设计化学结构的灵活性。此外,通过调节具有官能团(或前驱体)的单体和无官能团单体的比例,在聚合步骤可以精确地控制功能化程度。这种方法的缺点是合成单体原料比较复杂。此外,官能团的选择受限,因为有些官能团在聚合反应中不兼容。

### 6.3.3 膜辐射接枝

膜辐射接枝是另一种制备HEM的方法,特别是在HEM研究的早期阶段。这种方法使用非功能化膜作为原料,大多数为聚合的氟代烯烃[13-15]。膜使用γ射线或电子束[15]照射产生聚合物自由基,然后卤代不饱和单体(主要是乙烯苄基氯)与自由基反应连接到聚合物基质中。最后,像聚合物功能化一样,聚合物通过接枝单体上的卤代烷基实现功能化。这种方法的最显著的特点是,它不需要膜合成的步骤,因为已经制备好的膜就是起始材料;膜的厚度也可以很方便地提前调整。该方法的主要缺点是材料稳定性差,由于仅使用聚合的氟代烯烃作为聚合物(这由辐射方法的性质决定),而已知C—F键在碱性介质中易于降解[16]。此外,接枝率的控制也颇具挑战性。这种方法的另一个问题是由辐射引起的膜损伤导致透气性升高。

### 6.3.4 增强方法

化学交联已被广泛用于制备高尺寸稳定性和高耐溶剂性的化学增强的

HEM。化学交联的结构和性质将在 6.6 节中讨论。

物理方法如聚合物共混[17]，孔隙灌装[18]和范德华相互作用调谐[19]也已经被用于制备物理增强的 HEM。聚合物共混是将增强材料（通常是疏水的非离子聚合物，例如聚砜）与包含阳离子官能团的聚合物混合。孔隙灌装是将功能化的聚合物灌装到增强的多孔膜基质里。

范德华相互作用调谐是将高电子密度的聚合物基质与高偶极矩的官能团（如季鏻）匹配，以提高聚合物链之间的相互作用，增加 HEM 的尺寸稳定性而不影响溶解性。

## 6.4 阳离子官能团的结构和性质

阳离子官能团一直是 HEM 化学结构的焦点，因为它通过其碱性以及密度（离子交换容量，IEC）控制氢氧根离子电导率。官能团的固有性质也决定溶解性和控制化学稳定性。当前有两种主要类型的阳离子官能团：一类基于氮原子，一类基于磷原子。

### 6.4.1 季氮基阳离子官能团

季氮基阳离子官能团有非共轭（包括四烷基铵和环烷基铵）和共轭（包括吡啶盐，胍盐和咪唑鎓）两种类型（表 6.1）。

表 6.1 HEM 阳离子官能团的化学结构

| 分类 | | 举例 |
|---|---|---|
| 季氮 | 四烷基铵[13] | $—N^+(CH_3)_3$ |
| | 环烷基铵[20-24] | $N^+$-甲基吡咯烷鎓（$H_3C$）；$N^+$-甲基哌啶鎓（$H_3C$）；$H_3C$–$N^+$（哌嗪环）$N—CH_2—N^+$（双环）N |
| | 吡啶盐[25] | 吡啶环–$N^+—CH_3$ |
| | 胍盐[26,27] | $(H_3C)N^+=C[N(CH_3)_2]_2$（$H_3C$）；三嗪环：$N(CH_3)_2$，$N(CH_3)_2$，$N^+(CH_3)_2$ |

（续）

| 分类 | 举例 | |
|---|---|---|
| 季氮 | 咪唑鎓[28,29] | 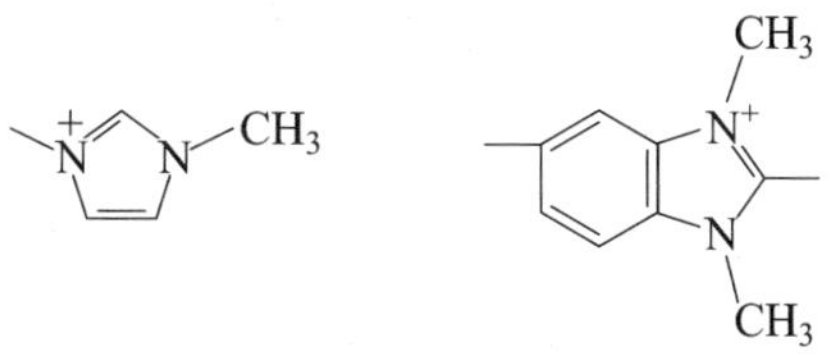 [结构式：1,3-二甲基咪唑鎓（$CH_3$）；1,2,3-三甲基苯并咪唑鎓（$N^+$，$CH_3$）] |
| 季鏻 | 三(2,4,6－三甲氧基苯基)鏻[8] | 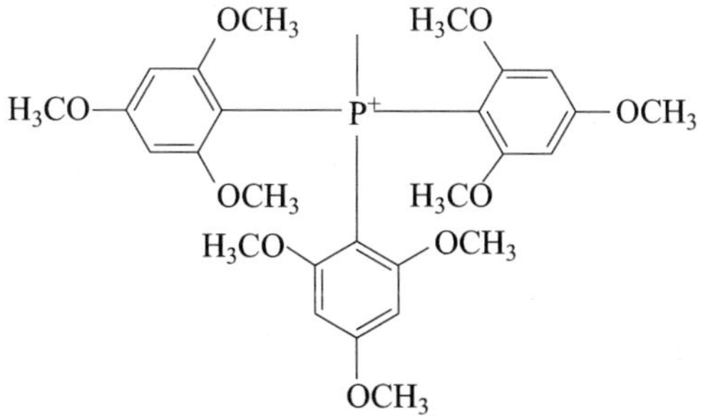 [结构式：$P^+$ 连接三个 2,4,6-三甲氧基苯基（$OCH_3$，$H_3CO$）] |

### 6.4.1.1 四烷基铵

四烷基铵是HEM和阴离子交换膜(AEM)中最常用的阳离子官能团[30]。作为最简单的四烷基铵结构,三甲基苄基季铵是HEM聚合物中最典型的阳离子官能团,其中三甲基基团作为侧基而苄基通常为聚合物基质。季氮原子带一单位正电荷,带一单位负电荷的氢氧根与其达成电荷平衡。

基于三甲基阳离子的官能团的最大优点是:它们是化学最稳定的四烷基铵阳离子[7]。在β－氢存在时,更长基团(如乙基或丙基)的引入会发生霍夫曼消除反应,这将导致严重降解。与其他阳离子的氢氧化物相比,三甲基苄基氢氧化铵的碱度适中,这会导致在给定的IEC下产生温和的氢氧根离子电导率。例如,在使用相同的聚砜基质且具有相似的均匀膜结构时,三甲基铵功能化聚合物(19mS·g/(cm·mmol)[31])的电导率或IEC归一化的氢氧根离子电导率大约是季鏻类(39mS·g/(cm·mmol)[5])的1/2(表6.2),大约是咪唑鎓类(8.4mS·g/(cm·mmol)[32])的两倍。四烷基铵阳离子官能团在低沸点的水混溶的溶剂中具有非常有限的特定溶解性,这一点妨碍了它作为高性能增溶的离聚物在电极上的应用。

### 6.4.1.2 环烷基铵

环烷基铵阳离子基团是另一种类型的铵,其中一个或多个环烷基(取代直链烷基)基团连接到氮原子上。现有的环烷基铵阳离子基团包括双环六元环系

(二氮杂二环基铵[20]和氮杂二环基铵[21]),单环六元环系(哌嗪基铵[22])和单环五元环系(吡咯烷基铵[23])。

表 6.2 基于阳离子官能团的聚合物的比溶解性和电导率

| 阳离子官能团 | | 比溶解性① | 电导率②(mS·g/mmol) |
|---|---|---|---|
| 季氮基 | 铵 | N/A | 19[31] |
| | 吡啶盐 | N/A | 0.1③[33] |
| | 胍盐 | 低级醇 | 20④[34] |
| | 咪唑鎓 | 四氢呋喃(THF) | 8.4[32] |
| 季鏻基 | 鏻 | 低级醇 | 39[5] |

① 低沸点水混溶的有机溶剂中的溶解度;
② 不交联的聚砜为骨架的 HEM 在室温下 IEC 归一化的氢氧根离子电导率;
③ 聚苯乙烯为骨架;
④ 三氟聚砜芳醚砜为骨架

在所有可用的环烷基铵中,最重要的一种是双环六元环系,由 1,4-二氮杂二环-[2,2,2]-辛烷(DABCO)构成的环烷基铵,因为它具有比传统的三甲基苄基铵更高的化学稳定性[7]。一些环烷基铵(如二氮杂二环和哌嗪基铵)具有可被季铵化的两个氮原子,所以有可能通过与来自不同聚合物链的两个卤代烷基反应来进行化学交联[35]。一般来说,环烷基氢氧化铵的溶解度和碱度与那些四烷基氢氧化铵类似。除 DABCO 基铵以外,环烷基铵应用于 HEM 存在问题,因为其中很多种表现出比三甲基苄基铵低得多的化学稳定性。

#### 6.4.1.3 吡啶盐

作为最简单的共轭阳离子体系,吡啶盐是可以同时应用于 HEM[25]和 AEM 中的第一类官能团。甲基吡啶盐是在 HEM 中使用的最简单的共轭阳离子官能团。已经确认氢氧化吡啶的化学稳定性大大小于三甲基苄基氢氧化铵,限制了其作为官能团在 HEM 中的应用。在降解过程中,吡啶盐通过亲核开环或亲核加成消除失去正电荷[36]。氢氧化吡啶的溶解度与三甲基苄基氢氧化铵类似,而氢氧化吡啶的碱度则比三甲基苄基氢氧化铵低得多,因为其在 HEM 中,即使在高 IEC 时,氢氧根离子电导率仍然非常低(对 6.05 mmol/g 的理论 IEC,氢氧根离子电导率为 0.54mS/cm;或电导率约为 0.1mS·g/(cm·mmol)[33])(表6.2)。

#### 6.4.1.4 胍盐

胍盐是在 HEM 中使用的第二类共轭阳离子官能团,最简单的形式是五甲基苄基胍盐[26]。胍盐基 HEM 在低级醇中具有良好的溶解性[26]。低级醇具有水混溶性及低沸点;所以,如在 6.2.4 节所讨论的,胍盐功能化的聚合物理论上可作为电极中的离聚物。胍盐在设计化学结构方面非常灵活,因为它有5个可自定义的基团。五甲基苄氢氧化胍的碱性已经被证明与三甲基苄基氢氧化铵非常相似,因为它们的HEM有很接近的电导率(胍盐与铵分别是20mS·g/(cm·mmol)[34]和19mS·g/(cm·mmol)[31])(表6.2)。一般情况下,氢氧化胍的化学稳定性比三甲基苄基氢氧化铵低。

#### 6.4.1.5　咪唑鎓

咪唑鎓是另一种共轭阳离子官能团[28,37],最简单的例子为甲基苄咪唑鎓。咪唑鎓功能化的HEM在另一种低沸点水混溶性溶剂四氢呋喃中具有良好的溶解性[28],同时它们不溶于醇(醇是HEMFC的重要燃料)。这样特殊并有选择性的溶解度不但使得咪唑鎓功能化的聚合物有可能作为离聚物使用,而且也可以让醇作为燃料直接在HEMFC中使用。甲基苄氢氧化咪唑的碱性大约是三甲基苄基氢氧化铵的1/2(电导率分别为8.4mS·g/(cm·mmol)[32]和19mS·g/(cm·mmol)[31])(表6.2)。通常来说,由于由两个氮原子之间环上的活泼氢引起的开环降解,咪唑鎓系被认为比铵系化学稳定性低[38]。

### 6.4.2　季鏻基阳离子官能团

HEM官能团不仅仅限于氮原子。基于三(2,4,6-三甲氧基苯基)苄基鏻的独特季鏻系已经被合成出来[5,8,19,39]。普通脂肪族或芳香族季鏻氢氧化物化学稳定性很差,而三(2,4,6-三甲氧基苯基)苄基氢氧化鏻却非常稳定,主要是因为它的9个强给电子甲氧基引起的电荷分散[5,8]。三(2,4,6-三甲氧基苯基)苄基鏻功能化的HEM的化学稳定性比三甲基苄基铵功能化的HEM高得多(例如,鏻可以承受浓度高得多的强碱性溶液,可在10M KOH溶液中浸泡2天,而铵可承受的KOH浓度仅为2mol/L)[5]。

第一个被发现可溶于低沸点水混溶溶剂(如低级醇)的HEM就是鏻基HEM[8],溶于低沸点水混溶剂是离聚物应用的必要条件。已证明鏻基离聚物可以显著提高HEMFC的电池性能(峰值功率密度增加3.5倍)[8]。

在已有的报道中,三(2,4,6-三甲氧基苯基)苄基氢氧化鏻的碱性最高。在所有的报道过的基于阳离子官能团的HEM中,其HEM具有最高的氢氧根离子电导率,通常大约是三甲基苄基铵的两倍、超过甲基咪唑鎓4倍(均在相同的聚砜聚合物基质中和均匀膜结构情况下,分别为39[5]、19[31]和8.4mS·g/(cm·mmol)[32])(表6.2)。

三(2,4,6-三甲氧基苯基)苄基氢氧化鏻官能团也具有较大的偶极矩(3.07 *D*),高于其他所有阳离子氢氧化物(1.32~2.20 *D*)[19]。聚合物可以通过官能团极化聚合物链来维持尺寸稳定性,特别是官能团与高电子密度的聚合物基质结合时(例如PPO),官能团的强偶极矩对于实现高的尺寸稳定性显得尤其重要[19]。

鏻系的可能缺点是它具有较高的相对分子质量，使得 IEC 只能达到中等水平。

## 6.5 聚合物主链的结构和性质

聚合物基质是 HEM 的骨架，其化学和物理结构深刻影响它们的性质。

### 6.5.1 化学结构

通过化学结构划分，应用于 HEM 中的聚合物基质可分为两大类，即芳香族主链聚合物和脂肪族主链聚合物（表 6.3）。一般来说，芳香族主链聚合物机械强度高、化学性质稳定，并表现出非常低的气体/燃料透过性；而脂肪族主链的聚合物更柔韧，它们可以达到超高相对分子质量，能够实现更高的尺寸稳定性。

表 6.3 HEM 聚合物基质的化学结构

| 分类 | 举例 | |
|---|---|---|
| 芳香族主链 | 聚芳醚砜[9,12,34,40-49] | 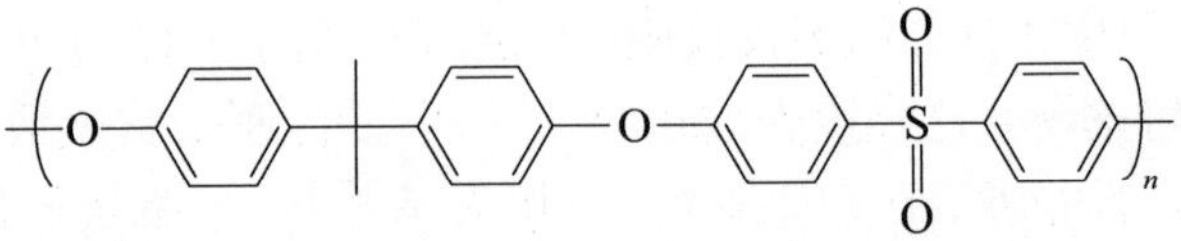 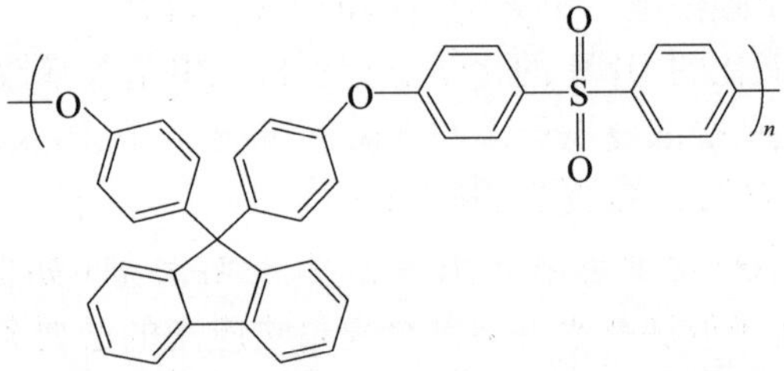 |
| | 聚芳基醚酮[50-52] | 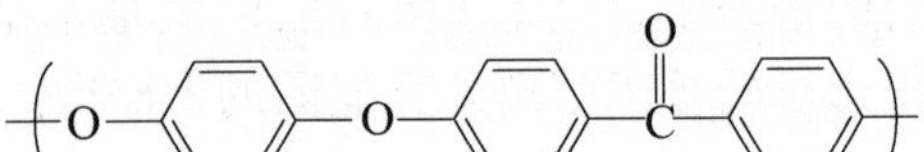 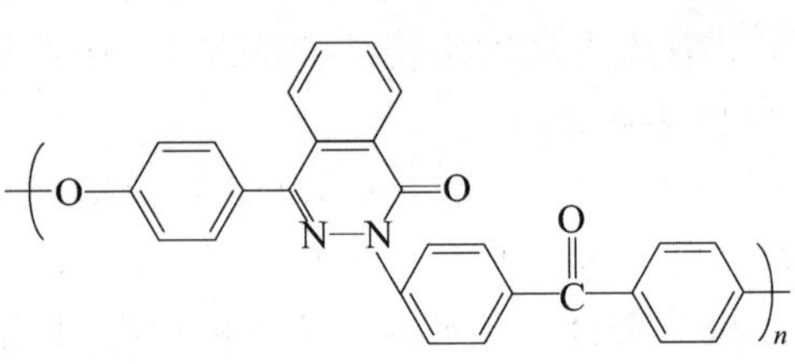 |

（续）

| 分类 | 举例 | |
|---|---|---|
| 芳香族主链 | 聚芳基醚[10,53] | 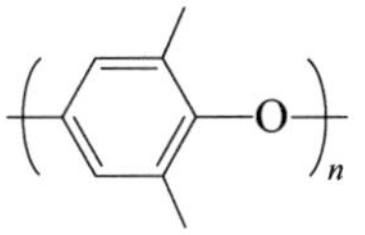 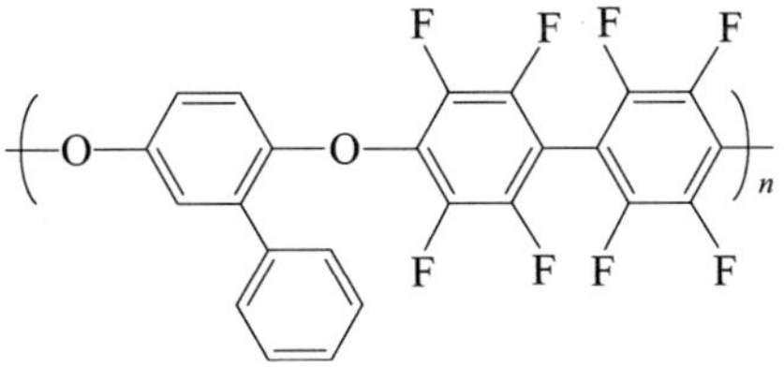 |
| | 聚亚苯基[54] | 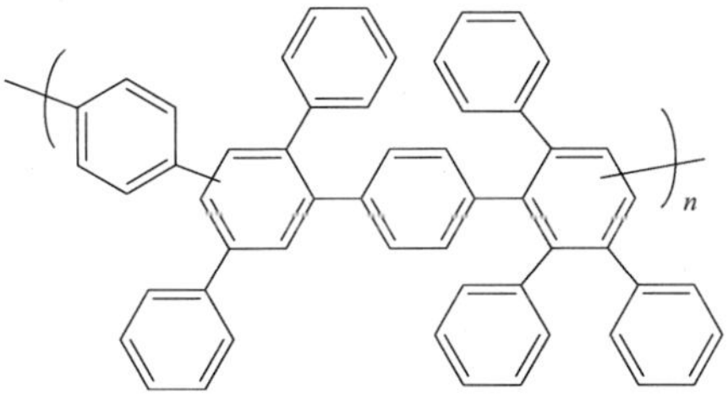 |
| | 聚芳基苯并咪唑[29,55] | 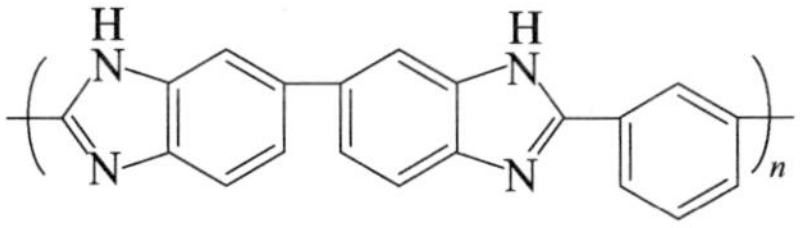 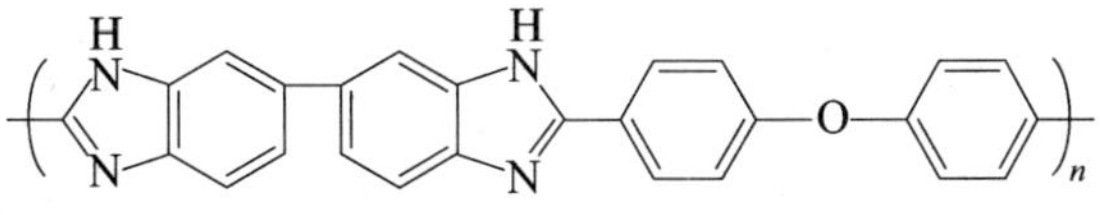 |
| | 聚芳酰亚胺[56] | 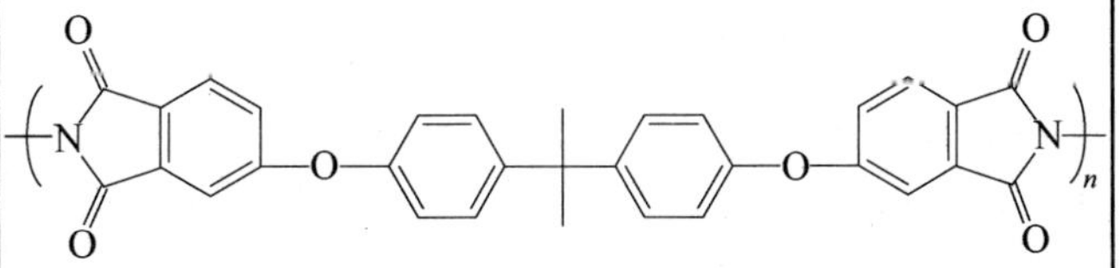 |

（续）

| 分类 | | 举例 |
|---|---|---|
| 脂肪族主链 | 含苯环[25,37,57-61] | $\text{-}[CH_2\text{—}CH(C_6H_5)]_n\text{-}$ $\text{-}[CH_2\text{—}CH(C_6H_5)]_m\text{-}[CH_2\text{—}CH(C_6H_4CH_2Cl)]_n\text{-}$<br>$\text{-}[CH_2\text{—}CH(C_6H_5)]_m\text{-}[CH_2\text{—}CH{=}CH\text{—}CH_2]_n\text{-}[CH_2\text{—}CH(C_6H_5)]_p\text{-}$ |
| | 不含苯环[11,13-15,20,21,62-71] | $\text{-}[CH_2\text{—}CH(Cl)]_n\text{-}$ $\text{-}[CH_2\text{—}CH(OH)]_n\text{-}$ $\text{-}[O\text{—}CH_2\text{—}CH(Cl)]_n\text{-}$<br>$=[CH\text{—}CH_2\text{—}CH_2\text{—}CH_2\text{—}CH_2\text{—}CH_2\text{—}CH_2\text{—}CH]_n=$<br>$\text{-}[CH_2\text{—}CH_2\text{—}CF_2\text{—}CF_2]_n\text{-}$ $\text{-}[CF_2\text{—}CF_2]_m\text{-}[CF_2\text{—}CF(CF_3)]_n\text{-}$<br>$=[CH(C_5H_8)CH]_n=$ |

### 6.5.1.1 芳香族主链聚合物

芳香族主链聚合物系列包括聚芳醚砜、聚芳基醚酮、聚芳基醚、聚亚苯基、聚芳基苯并咪唑和聚芳酰亚胺。

在高性能工程聚合物的类别中，聚芳醚砜和聚芳基醚酮是在HEM中最常用的两种聚合物基质。它们具有高的热稳定性，优异的化学稳定性，优异的机械强度，并且在许多有机溶剂中具有良好的溶解性（化学修饰的必要条件）。聚芳基醚类似于聚芳醚砜和聚芳基醚酮，但归因于更多的醚键，它通常具有较高电子密度的苯环，尤其是在使用大偶极矩的官能团时（如三(2,4,6-三甲氧基苯基)苄基季鏻），这使它作为HEM时具有更高的尺寸稳定性[19]。另一方面，它的氧化稳定性却因为活泼苯环的存在而降低，有时它的溶解度也由于聚合物链之间的强相互作用而降低。

与此相反，聚亚苯基则由于它的不活泼的苯环而具有化学氧化稳定性。但是，它却因为全部苯环结构的固有刚性而具有较差的弹性。聚芳基苯并咪唑和聚芳酰亚胺也被用作 HEM 聚合物基质。它们最大的问题是化学稳定性差,因为苯并咪唑基团和酰亚胺基团在特定条件下的碱性介质中会发生不可逆的水解。此外,聚芳基苯并咪唑在普通溶剂中溶解非常有限,很难进行化学修饰。

#### 6.5.1.2　脂肪族主链聚合物

脂肪族主链聚合物系列包括含苯环的脂肪族主链聚合物和不含苯环的脂肪族主链聚合物。含苯环的脂肪族聚合物通常含有由苯乙烯衍生的苯基乙烯片段。与不含苯环的聚合物相比,含苯环的脂肪族聚合物具有更高的热稳定性和更大的机械强度,并且苯环为化学结构的设计提供更多的选择。然而,苯基乙烯片段中活泼的 α-氢通常是 HEM 中最薄弱的环节,它对于氧化或取代反应特别敏感,这会导致膜的化学降解。

与对应的含苯环的脂肪族聚合物相比,不含苯环的脂肪族聚合物具有更高的柔韧性,并且可以达到更高的聚合物相对分子质量(在给定的 IEC 下对增加尺寸稳定性有益)。然而,全脂肪族结构导致了低机械强度和低热稳定性,并且膜中较大的自由体积会导致高的气体/燃料透过率。如果设计不够精密,引入的阳离子链接到脂肪族聚合物主链的 β-氢上的话,则会通过化学消除机制而导致严重的降解[72]。

### 6.5.2　顺序结构

聚合物基质可以通过两种方式排序,即随机或嵌段。随机聚合物是由相同(或不同)的重复单元以完全随机的方式构成。相比之下,嵌段聚合物链由不同的片段(或嵌段)组成,其中每个片段由相同的重复单元组成而相邻的片段由不同的重复单元组成。在大多数情况下,每种类型嵌段的亲水性或疏水性是变化的。随机聚合物在 HEM 中更常用,利用这种重复单元的随机链接方式可得到简单、均匀、致密的膜。

嵌段聚合物可控制 HEM 进行微相分离[73],如果优化得当可提高离子电导率和减少水的吸收[74]。然而,HEM 的嵌段结构有时会在嵌段与嵌段之间发生不兼容的难题,这可能导致气体/燃料渗透性的增加和/或机械强度的降低。这种不兼容性可以在一定程度上通过改变嵌段的长度进行控制:增加嵌段长度导致更深的相分离但也更不兼容,反之亦然[75]。因此,对于给定的嵌段聚合物系统应该有一个最佳的嵌段长度。嵌段类型的数量(两个为二嵌段,三个为三嵌段,依此类推)是嵌段聚合物的另一属性。使用更多的类型将带来更多的不兼容问题,但同时,也可能在不同类型的嵌段中产生相互调谐作用。

## 6.6　化学交联的结构和性质

化学交联一直是制备具有更好的尺寸稳定性和抗溶剂性的化学增强的

HEM 的最有效方法。键合的化学和物理结构严重影响交联的 HEM 的性质和性能。

### 6.6.1 化学结构

化学交联键合的类型有两种:基于苯环的键合和非基于苯环的键合(表6.4)。由于苯环的稳定性,芳香族的键合通常具有高机械强度和高热稳定性。它们与芳香族聚合物基质相容,也表现出高的化学稳定性,特别是当无 β-氢存在时。另一方面,包含苯环的结构难免提高聚合物的相对分子质量和尺寸,并降低了柔韧性,这会降低交联的反应活性和降低可控性。引入苯环的刚性也可能引起膜柔韧性的降低。

表 6.4 HEM 的化学交联结构

| 分类 | | 举例 |
|---|---|---|
| 基于苯环的交联 | 双链接[37,55,66] | $-C_6H_4-$　　$-C_6H_4-CH_2-$　　$-CH_2-C_6H_4-CH_2-$ |
| | 多链接[27,76] | N N N(三嗪环)　　CH$_2$ CH OH O / CH$_2$ CH OH O / OH O CH CH$_2$ / O OH CH CH$_2$ |
| 非基于苯环的交联 | 基于碳[10,35,39,48,49,77-81] | $-(CH_2)_n-$　$n$=1,2,3,4,6　　$-CH_2-C(=O)-$ |
| | 基于氮[64,82] | $-NH-(CH_2)_n-NH-$　$n$=1,2　　$-N=CH-(CH_2)_2-CH=N-$ |
| | 基于氧[65,67,83,84] | $-O-CH(CH_3)-CH_2-O-$　　$(=O)(=O)CH-(CH_2)_3-CH(O=)(O=)$ |

（续）

| 分类 | 举例 | |
|---|---|---|
| 非基于苯环的交联 | 基于硫[21,85,86] | —S—$(CH_2)_6$—S— |
| | 基于硅[59,87-90] | 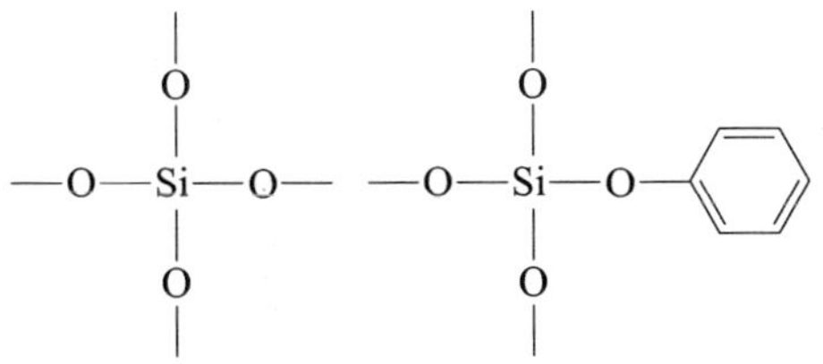 —O—Si(O—)(O—)—O—　—O—Si(O—)(O—)—O—Ph |

另一方面，非基于苯环的键合可使聚合物具有良好的柔韧性，特别是具有长的交联链时。脂肪族的键合不降低 IEC，因为它们只引入相对小的额外相对分子质量，并且同时脂肪族结构的柔韧性提供了高的交联反应活性。

### 6.6.2 物理结构

链接的物理结构通过长度、连接和链接密度影响交联的 HEM 的性质和性能。

长链接通常使膜具有良好的柔韧性但会降低材料的相容性，特别是当交联剂和聚合物链是不同的化学结构时。短链接提供更好的兼容性，并且通过增加空间位阻阻挡进攻来改进化学稳定性。例如，自交联的 HEM 通过最短的可能链接——单个亚甲基，即可改善化学稳定性和热稳定性[39,82]。

链接的连接是指每个链接里连接的数目和链接在聚合物链中的位置。链接可以是简单的双连接链接，也可以是多重连接的链接，如三重和四重连接的链接。双连接链接可以通过更方便的化学方法实现，而多重连接的链接则会更多地对 HEM 的尺寸稳定性产生影响。

在聚合物链内，链接可以连接到聚合物骨架上或阳离子官能团上。直接连接到官能团上可以保护阳离子而提高化学稳定性，但是，另一方面，它也将限制阳离子官能团的运动，阻碍期望的微相分离和减小氢氧根离子电导率。

链接密度是调谐交联的 HEM 性质的一个重要参数。在给定的 IEC 下，增加链接密度可直接提高尺寸稳定性和机械强度，但同时也会减少水的吸收，从而潜在地降低了氢氧根离子的电导率。

## 6.7 展　　望

HEM 将成为 HEMFC 的中心研究课题，因为它们的性质从根本上影响 HEMFC 的性能和耐久性。设计/探索具有较高碱性和稳定性的阳离子官能团仍是最重要的一环，因为它们是整个 HEMFC 系统的瓶颈。此外，开发/部署更好的化学和物理交联方法,将是改善尺寸稳定性的另一个重要研究方向。

## 参 考 文 献


[1] McLean, G. F., Niet, T., Prince-Richard, S., and Djilali, N. (2002) *International Journal of Hydrogen Energy*, **27**, 507 - 526.

[2] Adams, L. A., Poynton, S. D., Tamain, C., Slade, R. C. T., and Varcoe, J. R. (2008) *ChemSusChem*, **1**, 79 - 81.

[3] Varcoe, J. R., Slade, R. C. T., Wright, G. L., and Chen, Y. L. (2006) *Journal of Physical Chemistry B*, **110**, 21041 - 21049.

[4] Lu, S. F., Pan, J., Huang, A. B., Zhuang, L., and Lu, J. T. (2008) *Proceedings of the National Academy of Sciences of the United States of America*, **105**, 20611 - 20614.

[5] Gu, S., Cai, R., Luo, T., Jensen, K., Contreras, C., and Yan, Y. S. (2010) *ChemSusChem*, **3**, 555 - 558.

[6] Yanagi, H. and Fukuta, K. (2008) *ECS Transactions*, **16**, 257 - 262.

[7] Bauer, B., Strathmann, H., and Effenberger, F. (1990) *Desalination*, **79**, 125 - 144.

[8] Gu, S., Cai, R., Luo, T., Chen, Z. W., Sun, M. W., Liu, Y., He, G. H., and Yan, Y. S. (2009) *Angewandte Chemie: International Edition*, **48**, 6499 - 6502.

[9] Li, L. and Wang, Y. X. (2005) *Journal of Membrane Science*, **262**, 1 - 4.

[10] Wu, L., Xu, T. W., Wu, D., and Zheng, X. (2008) *Journal of Membrane Science*, **310**, 577 - 585.

[11] Clark, T. J., Robertson, N. J., Kostalik, H. A., Lobkovsky, E. B., Mutolo, P. F., Abruna, H. D., and Coates, G. W. (2009) *Journal of the American Chemical Society*, **131**, 12888 - 12889.

[12] Zhou, J. F., Unlu, M., Vega, J. A., and Kohl, P. A. (2009) *Journal of Power Sources*, **190**, 285 - 292.

[13] Danks, T. N., Slade, R. C. T., and Varcoe, J. R. (2002) *Journal of Materials Chemistry*, **12**, 3371 - 3373.

[14] Herman, H., Slade, R. C. T., and Varcoe, J. R. (2003) *Journal of Membrane Science*, **218**, 147 - 163.

[15] Varcoe, J. R. and Slade, R. C. T. (2006) *Electrochemistry Communications*, **8**, 839 - 843.

[16] Schulze, A. and Gulzow, E. (2004) *Journal of Power Sources*, **127**, 252 - 263.

[17] Wang, X., Li, M. Q., Golding, B. T., Sadeghi, M., Cao, Y. C., Yu, E. H., and Scott, K. (2011) *International Journal of Hydrogen Energy*, **36**, 10022 - 10026.

[18] Jung, H. M., Fujii, K., Tamaki, T., Ohashi, H., Ito, T., and Yamaguchi, T. (2011) *Journal of Membrane Science*, **373**, 107 - 111.

[19] Gu, S., Skovgard, J., and Yan, Y. S. (2012) *ChemSusChem*, **5**, 843 - 848.

[20] Agel, E., Bouet, J., and Fauvarque, J. F. (2001) *Journal of Power Sources*, **101**, 267 - 274.

[21] Stoica, D., Ogier, L., Akrour, L., Alloin, F., and Fauvarque, J. F. (2007) *Electrochimica Acta*, **53**, 1596 - 1603.

[22] Jung, M. S. J., Arges, C. G., and Ramani, V. (2011) *Journal of Materials Chemistry*, **21**, 6158 - 6160.

[23] Valade, D., Boschet, F., Roualdes, S., and Ameduri, B. (2009) *Journal of Polymer Science Part A: Polymer Chemistry*, **47**, 2043 - 2058.

[24] Faraj, M., Elia, E., Boccia, M., Filpi, A., Pucci, A., and Ciardelli, F. (2011) *Journal of Polymer*

*Science Part A*: *Polymer Chemistry*, **49**, 3437 – 3447.

[25] Huang, A. B., Xiao, C. B., and Zhuang, L. (2005) *Journal of Applied Polymer Science*, **96**, 2146 – 2153.

[26] Wang, J. H., Li, S. H., and Zhang, S. B. (2010) *Macromolecules*, **43**, 3890 – 3896.

[27] Cao, Y. C., Wang, X., Mamlouk, M., and Scott, K. (2011) *Journal of Materials Chemistry*, **21**, 12910 – 12916.

[28] Guo, M. L., Fang, J., Xu, H. K., Li, W., Lu, X. H., Lan, C. H., and Li, K. Y. (2010) *Journal of Membrane Science*, **362**, 97 – 104.

[29] Thomas, O. D., Soo, K. J. W. Y., Peckham, T. J., Kulkarni, M. P., and Holdcroft, S. (2011) *Polymer Chemistry*, **2**, 1641 – 1643.

[30] Juda, W. and Mcrae, W. A. (1950) *Journal of the American Chemical Society*, **72**, 1043 – 1044.

[31] Pan, J., Lu, S. F., Li, Y., Huang, A. B., Zhuang, L., and Lu, J. T. (2010) *Advanced Functional Materials*, **20**, 312 – 319.

[32] Zhang, F. X., Zhang, H. M., and Qu, C. (2011) *Journal of Materials Chemistry*, **21**, 12744 – 12752.

[33] Matsuoka, K., Chiba, S., Iriyama, Y., Abe, T., Matsuoka, M., Kikuchi, K., and Ogumi, Z. (2008) *Thin Solid Films*, **516**, 3309 – 3313.

[34] Kim, D. S., Labouriau, A., Guiver, M. D., and Kim, Y. S. (2011) *Chemistry of Materials*, **23**, 3795 – 3797.

[35] Schmitt, F., Granet, R., Sarrazin, C., Mackenzie, G., and Krausz, P. (2011) *Carbohydrate Polymers*, **86**, 362 – 366.

[36] Neagu, V., Bunia, I., and Plesca, I. (2000) *Polymer Degradation and Stability*, **70**, 463 – 468.

[37] Lin, B. C., Qiu, L. H., Lu, J. M., and Yan, F. (2010) *Chemistry of Materials*, **22**, 6718 – 6725.

[38] Ye, Y. S. and Elabd, Y. A. (2011) *Macromolecules*, **44**, 8494 – 8503.

[39] Gu, S., Cai, R., and Yan, Y. S. (2011) *Chemical Communications*, **47**, 2856 – 2858.

[40] Fang, J. and Shen, P. K. (2006) *Journal of Membrane Science*, **285**, 317 – 322.

[41] Hibbs, M. R., Hickner, M. A., Alam, T. M., McIntyre, S. K., Fujimoto, C. H., and Cornelius, C. J. (2008) *Chemistry of Materials*, **20**, 2566 – 2573.

[42] Wang, J. H., Zhao, Z., Gong, F. X., Li, S. H., and Zhang, S. B. (2009) *Macromolecules*, **42**, 8711 – 8717.

[43] Tanaka, M., Koike, M., Miyatake, K., and Watanabe, M. (2010) *Macromolecules*, **43**, 2657 – 2659.

[44] Yan, J. L. and Hickner, M. A. (2010) *Macromolecules*, **43**, 2349 – 2356.

[45] Zhang, Q. A., Zhang, Q. F., Wang, J. H., Zhang, S. B., and Li, S. H. (2010) *Polymer*, **51**, 5407 – 5416.

[46] Lee, K. M., Wycisk, R., Litt, M., and Pintauro, P. N. (2011) *Journal of Membrane Science*, **383**, 254 – 261.

[47] Lin, B. C., Qiu, L. H., Qiu, B., Peng, Y., and Yan, F. (2011) *Macromolecules*, **44**, 9642 – 9649.

[48] Park, J. S., Park, S. H., Yim, S. D., Yoon, Y. G., Lee, W. Y., and Kim, C. S. (2008) *Journal of Power Sources*, **178**, 620 – 626.

[49] Ni, J., Zhao, C. J., Zhang, G., Zhang, Y., Wang, J., Ma, W. J., Liu, Z. G., and Na, H. (2011) *Chemical Communications*, **47**, 8943 – 8945.

[50] Zhang, H. W. and Zhou, Z. T. (2008) *Journal of Applied Polymer Science*, **110**, 1756 – 1762.

[51] Xiong, Y., Liu, Q. L., and Zeng, Q. H. (2009) *Journal of Power Sources*, **193**, 541 – 546.

[52] Yan, X. M., He, G. H., Gu, S., Wu, X. M., Du, L. G., and Zhang, H. Y. (2011) *Journal of Mem-*

*brane Science*, **375**, 204 – 211.

[53] Zhou, J. F., Unlu, M., Anestis-Richard, I., Kim, H., and Kohl, P. A. (2011) *Journal of Power Sources*, **196**, 7924 – 7930.

[54] Hibbs, M. R., Fujimoto, C. H., and Cornelius, C. J. (2009) *Macromolecules*, **42**, 8316 – 8321.

[55] Henkensmeier, D., Kim, H. J., Lee, H. J., Lee, D. H., Oh, I. H., Hong, S. A., Nam, S. W., and Lim, T. H. (2011) *Macromolecular Materials and Engineering*, **296**, 899 – 908.

[56] Wang, G. G., Weng, Y. M., Zhao, J., Chen, R. R., and Xie, D. (2009) *Journal of Applied Polymer Science*, **112**, 721 – 727.

[57] Luo, Y. T., Guo, J. C., Wang, C. S., and Chu, D. (2010) *Journal of Power Sources*, **195**, 3765 – 3771.

[58] Luo, Y. T., Guo, J. C., Wang, C. S., and Chu, D. (2011) *Macromolecular Chemistry and Physics*, **212**, 2094 – 2102.

[59] Wu, Y. H., Wu, C. M., Xu, T. W., Yu, F., and Fu, Y. X. (2008) *Journal of Membrane Science*, **321**, 299 – 308.

[60] Zhang, C. X., Hu, J., Cong, J., Zhao, Y. P., Shen, W., Toyoda, H., Nagatsu, M., and Meng, Y. D. (2011) *Journal of Power Sources*, **196**, 5386 – 5393.

[61] Varcoe, J. R., Slade, R. C. T., and Lam How Yee, E. (2006) *Chemical Communications*, 1428 – 1429.

[62] Valade, D., Boschet, F., and Ameduri, B. (2009) *Macromolecules*, **42**, 7689 – 7700.

[63] Hong, J. H., Li, D., and Wang, H. T. (2008) *Journal of Membrane Science*, **318**, 441 – 444.

[64] Hong, J. H. and Hong, S. K. (2010) *Journal of Applied Polymer Science*, **115**, 2296 – 2301.

[65] Xiong, Y., Fang, J., Zeng, Q. H., and Liu, Q. L. (2008) *Journal of Membrane Science*, **311**, 319 – 325.

[66] Robertson, N. J., Kostalik, H. A., Clark, T. J., Mutolo, P. F., Abruna, H. D., and Coates, G. W. (2010) *Journal of the American Chemical Society*, **132**, 3400 – 3404.

[67] Wan, Y., Peppley, B., Creber, K. A. M., and Bui, V. T. (2010) *Journal of Power Sources*, **195**, 3785 – 3793.

[68] Varcoe, J. R., Slade, R. C. T., Yee, E. L. H., Poynton, S. D., Driscoll, D. J., and Apperley, D. C. (2007) *Chemistry of Materials*, **19**, 2686 – 2693.

[69] Ko, B. S., Sohn, J. Y., Nho, Y. C., and Shin, J. (2011) *Nuclear Instruments & Methods in Physics Research Section B: Beam Interactions with Materials and Atoms*, **269**, 2509 – 2513.

[70] Zhang, F. X., Zhang, H. M., Ren, J. X., and Qu, C. (2010) *Journal of Materials Chemistry*, **20**, 8139 – 8146.

[71] Kostalik, H. A., Clark, T. J., Robertson, N. J., Mutolo, P. F., Longo, J. M., Abruna, H. D., and Coates, G. W. (2010) *Macromolecules*, **43**, 7147 – 7150.

[72] Varcoe, J. R. and Slade, R. C. T. (2005) *Fuel Cells*, **5**, 187 – 200.

[73] Elabd, Y. A. and Hickner, M. A. (2011) *Macromolecules*, **44**, 1 – 11.

[74] Peckham, T. J. and Holdcroft, S. (2010) *Advanced Materials*, **22**, 4667 – 4690.

[75] Tanaka, M., Fukasawa, K., Nishino, E., Yamaguchi, S., Yamada, K., Tanaka, H., Bae, B., Miyatake, K., and Watanabe, M. (2011) *Journal of the American Chemical Society*, **133**, 10646 – 10654.

[76] Zhou, J. F., Unlu, M., Anestis-Richard, I., and Kohl, P. A. (2010) *Journal of Membrane Science*, **350**, 286 – 292.

[77] Ong, A. L., Saad, S., Lan, R., Goodfellow, R. J., and Tao, S. W. (2011) *Journal of Power Sources*,

**196**, 8272 - 8279.

[78] Wu, L. and Xu, T. W. (2008) *Journal of Membrane Science*, **322**, 286 - 292.

[79] Wang, G. G., Weng, Y. M., Chu, D., Chen, R. R., and Xie, D. (2009) *Journal of Membrane Science*, **332**, 63 - 68.

[80] Guo, T. Y., Zeng, Q. H., Zhao, C. H., Liu, Q. L., Zhu, A. M., and Broadwell, I. (2011) *Journal of Membrane Science*, **371**, 268 - 275.

[81] Wang, G. H., Weng, Y. M., Zhao, J., Chu, D., Xie, D., and Chen, R. R. (2010) *Polymers for Advanced Technologies*, **21**, 554 - 560.

[82] Pan, J., Li, Y., Zhuang, L., and Lu, J. T. (2010) *Chemical Communications*, **46**, 8597 - 8599.

[83] Wan, Y., Peppley, B., Creber, K. A. M., Bui, V. T., and Halliop, E. (2008) *Journal of Power Sources*, **185**, 183 - 187.

[84] Wang, E. D., Zhao, T. S., and Yang, W. W. (2010) *International Journal of Hydrogen Energy*, **35**, 2183 - 2189.

[85] Sollogoub, C., Guinault, A., Bonnebat, C., Bennjima, M., Akrour, L., Fauvarque, J. F., and Ogier, L. (2009) *Journal of Membrane Science*, **335**, 37 - 42.

[86] Stoica, D., Alloin, F., Marais, S., Langevin, D., Chappey, C., and Judeinstein, P. (2008) *Journal of Physical Chemistry B*, **112**, 12338 - 12346.

[87] Wu, Y. H., Wu, C. M., Xu, T. W., Lin, X. C., and Fu, Y. X. (2009) *Journal of Membrane Science*, **338**, 51 - 60.

[88] Wu, Y. H., Wu, C. M., Varcoe, J. R., Poynton, S. D., Xu, T. W., and Fu, Y. X. (2010) *Journal of Power Sources*, **195**, 3069 - 3076.

[89] Singh, S., Jasti, A., Kumar, M., and Shahi, V. K. (2010) *Polymer Chemistry*, **1**, 1302 - 1312.

[90] Tripathi, B. P., Kumar, M., and Shahi, V. K. (2010) *Journal of Membrane Science*, **360**, 90 - 101.

# 第7章

## 微生物燃料电池材料

Yanzhen Fan, Hong Liu

## 7.1 引　　言

微生物燃料电池(MFC)是一种使用微生物将存储在有机或无机物质中的化学能转换成电能的装置。近一个世纪以来,人们已经知道细菌可以产生电流[1]。然而,仅在最近的10年间,微生物燃料电池才由于在废水发电[2]、生物修复、远程传感器供电等领域的应用潜力而引起了大量的研究关注。

图7.1是MFC基本构成示意图,包括微生物、阳极、阴极和合适的隔膜。微生物氧化有机或无机物质,释放电子和质子。电子通过电路流出并在阴极上与质子和氧(或其他氧化剂)结合生成水,产生电流。质子通常在质子载体的帮助下,通过电解质溶液移动。由于微生物要求接近中性的pH条件,所以常用的离子载体有磷酸盐或碳酸盐的离子[3]。

提高功率密度和降低制造成本是微生物燃料电池实际应用的两大挑战,特别是对废水处理来说。微生物的电化学活性是一个影响微生物燃料电池功率输出的重要因素,而其他基本部件的材料在决定成本以及功率输出方面也发挥着关键的作用。电极材料的表面积、导电性、稳定性、疏水性等性质,对微生物附着、电化学反应、电子传递和收集有显著的影响。隔膜内阻的增加也可大大影响微生物燃料电池的性能[4]。聚焦于污水处理的应用潜力,本章的其余部分将简要介绍在微生物燃料电池研究中应用的各种MFC的典型结构,以及用于阳极、阴极和隔膜的常用材料,揭示目前微生物燃料电池电极材料存在的问题,并对未来发展的前景提出建议。

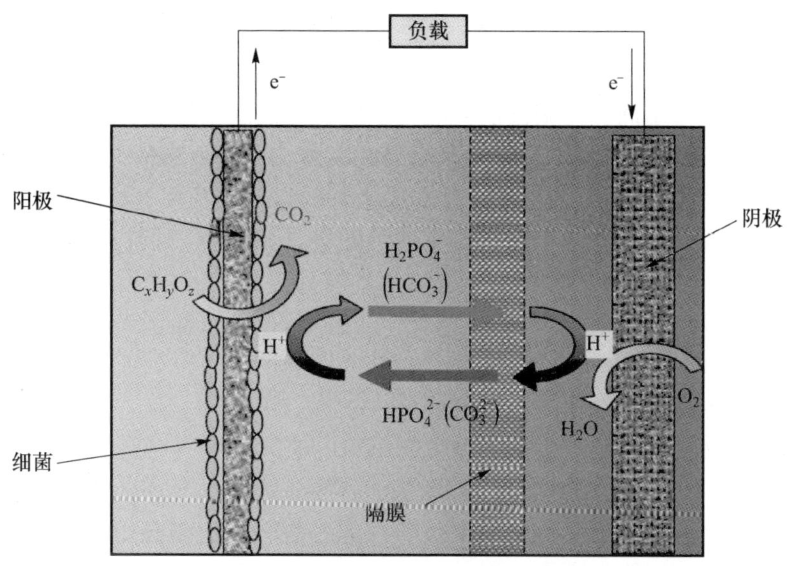


图7.1 MFC示意图：MFC包含表面覆盖生物膜的阳极、阴极、合适的隔膜。在使用磷酸盐缓冲溶液的空气阴极微生物燃料电池中，$H_2PO_4^-/HPO_4^{2-}$离子对是质子转移的主要载体；而$HCO_3^-/CO_3^{2-}$是使用碳酸氢盐缓冲溶液的微生物燃料电池中的主要质子载体

## 7.2 MFC结构

最常用的MFC结构由一个或两个腔室构成。双室或双隔间的MFC中，阳极室的细菌通过膜或盐桥与阴极室隔开。阴极的氧化剂可以是空气中的氧气，其中空气通过水鼓泡为电极提供溶解氧。这种方式不但消耗相当多的能量，而且由于氧气在水中的低溶解度和低传质速率显得相当低效。与使用空气作为氧化剂相比，其他氧化剂，如铁氰化物和高锰酸盐[5]可以大大提高MFC的性能[6]。然而，从这种类型的MFC产生的电力是不可持续的，因为氧化剂在阴极反应过程中被消耗，需要不断补充[7,8]。因此，随处可见并且近乎免费的氧气很可能是MFC实际应用中可以广泛使用的唯一氧化剂。

单室的MFC中，阳极与空气阴极通常分别位于腔室的相对侧。或者，如果使用膜电极（MEA）或布电极（CEA），阳极和阴极可位于腔室的同一侧。单室MFC可以获得比双室系统高得多的性能，这是因为与在水中相比，空气中的氧具有较高的传质速率和浓度[3]。在无膜单室MFC中，阳极和阴极在腔室的相对侧，生长在阴极上的生物膜可以作为隔膜使得扩散到阳极室的氧最小化，同时容许质子/离子的高效传递。隔膜的去除不仅降低了微生物燃料电池的成本和复杂性，同时也由于内部电阻的降低增加了功率密度[3]。

## 7.3 阳极材料

不像需要催化剂加速阳极反应的化学燃料电池的阳极，MFC 的阳极主要用作集流体，同时为生物膜生长提供可附着的表面。在这种结构中，生物膜可以作为催化剂。因此，阳极应当由适合生物膜生长的材料以及集电材料构成。

碳基材料常被用作微生物燃料电池的阳极，因为这类材料具有优异的导电性、化学稳定性、生物相容性和通用性，并且容易得到较大的表面积。各种碳材料，包括石墨棒[2]、石墨板[9]、石墨泡沫[10]、编织石墨、石墨毡[11]、石墨颗粒[12]、网状玻璃碳(RVC)[13]、粒状活性炭[14]、碳纸[15]、碳布[16]和石墨纤维刷[17]等，已经被用作微生物燃料电池的阳极电极。

非碳基材料也已被开发用作微生物燃料电池的阳极，包括各种金属：铂[18]、金[19]、钛[11]、不锈钢(SS)[20]和铜[13]等。尽管金属的电导率通常比碳材料高，但金属材料作为微生物燃料电池阳极的性能一般较差。与碳电极相比，金属电极性能差的原因可能是由于金属电极相对低的表面积和不利于生物膜生长的表面性质。铂和金材料的高成本阻碍其大规模应用。还应注意一些金属的腐蚀性和毒副作用。例如，铜和不锈钢作为微生物燃料电池阳极时会发生反应，这进一步限制了其作为 MFC 中阳极材料的应用。

### 7.3.1 块状碳材料

块状石墨板或石墨棒是最简单的 MFC 电极，因为相对便宜、易于处理、并且具有明确的表面积，因此在 MFC 研究的早期被经常使用[2]。海底微生物燃料电池中使用嵌入海洋沉积物的石墨圆盘阳极和覆盖海水的石墨圆盘阴极可以实现原位发电[14]。一些研究微生物燃料电池的先驱实验室通常使用石墨板[12,21]。Liu 等在一个单室型空气阴极微生物燃料电池中使用了 8 个石墨棒阳极，并将其应用于废水发电[2]。块状石墨电极相对平滑的表面可以更容易地计算出电极的表面积。但是，光滑表面限制了生物膜生长可获得的表面积，因此限制了 MFC 阳极的最大电流密度。用石墨棒作为微生物燃料电池阳极的最大功率密度仅为 $26mW/m^2$[2]。

### 7.3.2 粒状碳材料

与块状碳材料相比，粒状碳材料的表面积高很多。已有关于粒状石墨[22]和粒状活性炭[23,24]两种类型的粒状碳材料作为 MFC 阳极的报道。使用石墨作为管状 MFC 的阳极材料，Rabeay 等在乙酸盐和葡萄糖作为进料的基础上，净阳极室的最大功率输出分别达到 $90W/m^3$ 和 $66W/m^3$[9]。You 等在上流式空气阴极 MFC 中使用葡萄糖作底物，得到了相似的 $49W/m^3$ 的最大功率密度[25]。用醋酸钠作为底物、铁氰化钾作为氧化剂在 6-MFC 电堆中实现了更高的功率密度($258W/m^3$)[24]。尽管粒状活性炭具有更高的表面积,但粒状活性炭不如粒状石墨受欢迎,可能是由于其较低的导电性,从而降低了最大功率密度。使用粒状活性炭的微生物燃料电池的功率密度[23,26]比使用粒状石墨的低大约一到两个数量级。

## 7.3.3 纤维碳材料

纤维碳基电极,包括碳纤维、碳布、碳/石墨毡、碳纸、石墨纤维刷,由于其大的比表面积、良好的导电性、优良的物理强度和化学稳定性以及对生物膜生长有利的表面性质,因此可以说纤维碳材料作为 MFC 阳极具有最好的性能。Reimers等第一个报道了在海底微生物燃料电池中使用碳纤维,从海洋沉积物 - 海水界面获取能量[21]。Liu 等测试了碳纸和碳布在空气阴极微生物燃料电池中的性能[16,15]。与碳布相比,碳纸电极表现出较低的性能,可能是由于可供生物膜生长的表面积较低的缘故。作为低廓碳布,碳网状材料表现出比碳布阳极相似甚至更好的性能[27]。碳/石墨毡也被广泛用作 MFC 阳极[28,29]。与碳纸相比,碳毡由于具有较大的表面积、较低的电阻以及由交织纤维构成的开放式网络而显示出较高的性能[30]。Logan 等设计了一种新型石墨刷阳极,将石墨纤维卷绕于由两段钛导线扭绞构成的芯体上[17]。在与废水接种时,这种设计实现了 $73W/m^3$(基于液体体积)和 $2400mW/m^2$ 的功率密度。在类似条件下,使用石墨刷阳极的微生物燃料电池更实现了高达 $1430mW/m^2$($2.3W/m^3$)的功率密度,而使用纯碳纸电极的只有 $600mW/m^2$。

迄今为止,利用碳布阳极可以产生的最高功率密度为 $6900mW/m^2$[31] 和 $1550W/m^3$[3]。碳布的制式可能是功率密度高的主要原因。与碳毡的“砖”制式或石墨刷的“柱”制式相比,碳布的“片”制式可以更容易减少阳极和阴极之间的平均距离,从而大大降低内部电阻[31]以及增加表面积/体积比,产生显著增高的功率密度。

## 7.3.4 多孔碳材料

包括网状玻璃碳(RVC)、碳海绵和泡沫碳在内的多孔碳材料也已在微生物燃料电池中进行了测试。RVC 是低密度的多孔刚性材料,具有开放的结构和良好的导电性。Scott 等比较了碳海绵、碳布、碳纤维和 RVC 的性能[22]。使用碳海绵阳极的 MFC 达到了 $55mW/m^2$ 的最大功率密度,这是使用碳布的近两倍,高于 RVC($0.2mW/m^2$)100 倍。进一步比较石墨、海绵、纸、布、毡、纤维、泡沫和 RVC 阳极材料,也表明 RVC 的性能差,这可能与材料形态和生物膜结构所造成的可用表面积和浓度极化相关[32]。

### 7.3.5 阳极材料修饰

阳极材料修饰可以改变其表面性质,特别是改变导电性和生物相容性,从而改变它们作为 MFC 阳极的性能表现。带正电荷或负电荷的化合物附着于自然形成的如石英和砂的表面,通过改变细胞表面的静电吸引而影响细菌的粘附[33]。用氨气处理的碳布阳极增加了电极的表面电荷并且改进了 MFC 的性能[34]。Park 和 Zeikus 合成了石墨、金属(例如 $Fe^{3+}$ 和 $Mn^{4+}$)和介体化合物(如中性红)的复合材料;并且报道说,当电子介体($Mn^{4+}$ 或中性红)被引入石墨阳极时,产电增加[35]。Lowy 等[36]评估了各种修饰的石墨阳极,包括通过吸附蒽醌-1,6-二磺酸(AQDS)或1,4-萘醌(NQ)修饰的石墨、含有 $Mn^{2+}$ 和 $Ni^{2+}$ 的石墨-陶瓷复合材料、经含有 $Fe_3O_4$ 或 $Fe_3O_4$ 和 $Ni^{2+}$ 的石墨膏修饰的石墨。结果发现,这些阳极具有比纯石墨电极高 1.5 ~2.2 倍的动力学活性。

将腐败希瓦氏菌生长于金电极上,包覆单层自组装的具有羧酸官能团的不同烷基硫醇,将该电极的电流产生情况与生长相同细菌的玻碳电极进行了比较[37]。研究发现,电流的产生与单层分子链的长度和头基相关,某些头基可以增强电子与细菌的耦合,使羧酸基团与细菌细胞色素之间具有强氢键作用。这个共价连接充当了细菌和电极之间的电子通路,与玻碳相比,修饰的电极产生了显著增高的电流。石墨毡电极上使用聚乙烯亚胺黏结介体(9,10-蒽醌-2,6-二硫酸盐)和硫还原的杆菌,得到 $1.2mA/cm^2$ 的电流密度[38]。

使用导电聚合物进行阳极修饰也可提高性能。聚苯胺是经常使用的导电聚合物,可以增加 MFC 阳极的电流密度,但它也容易受到微生物的攻击和降解[39]。Schröder 等[18]报道了铂电极包覆聚苯胺的 MFC,可以实现高达 $1.5mA/cm^2$ 的电流密度。聚苯胺的修饰可改善其性能和稳定性,如氟化聚苯胺[40]、聚苯胺/碳纳米管(CNT)复合材料[41]、聚苯胺/二氧化钛复合材料[42]。

纳米材料也用来改善 MFC 阳极的性能。最近的研究已经表明,CNT 修饰的阳极提高了 MFC 的发电[43,44]。使用碳纳米管修饰的玻碳阳极,可以增加电流密度至 $(9.70 \pm 0.40)\mu A/cm^2$,比裸玻碳电极大 82 倍[45]。使用奥奈达湖希瓦氏菌 MR-1 并修饰金纳米粒子的石墨阳极产生的电流密度,比不修饰的纯石墨阳极高 20 倍;而具有类似结构的钯修饰的阳极比参照组高 0.5 ~1.5 倍[46]。使用多壁碳纳米管(MWNT)和聚电解质聚乙烯亚胺(PEI)通过层层(LBL)自组装技术来修饰碳纸电极,产生比裸碳纸阳极高 20% 的功率密度[47]。

## 7.4 阴　　极

阴极是目前 MFC 发展的主要瓶颈。高性能低成本阴极材料的开发是 MFC 技术成功应用的关键,特别是对废水处理。典型的空气阴极包括碳基层、对水侧的催化剂涂层和对大气侧的扩散层(DL)[48]。

### 7.4.1 催化剂黏结剂

在阴极的制备时经常使用黏结剂,以在阴极基底或基层表面上形成催化剂涂层。Nafion 即全氟质子交换树脂,被广泛使用。其溶液与 Pt / C 粉末混合形成糊状,然后刷涂或喷涂到碳布/纸基质上。然而,在制备 MFC 阴极使用 Nafion 溶液时,需要考虑以下几点:①Nafion 溶液相对昂贵,是 MFC 阴极总成本的主要部分;②在 MFC 的中性条件下,质子主要由阴离子或 pH 缓冲剂载体运送;Nafion 作为质子或阳离子,交换树脂会提高阴极的离子传质阻力或极化电阻。简单地将 Nafion 溶液替换成水可以改善阴极性能[49],尽管这种无黏结剂催化剂涂层的长期性能如何并不清楚。在最近的研究中也证明了阳离子交换官能团的不利影响[50]。未磺化的聚苯基砜黏结剂是非离子疏水性聚合物,在线性扫描伏安(LSV)测试中显示出最高电流和最低电荷转移电阻。在聚苯基砜黏结剂中增加磺酸基团,使离子交换容量从 0 变化至 2.54meq/g①,导致了 LSV 测试中电流响应的减小以及电荷转移电阻从 8Ω 至 23Ω 的增加。这是由于阴极黏结剂中磺酸基团存在的缘故,磺酸吸附到催化位点上阻止质子扩散到催化剂表面,从而阻碍了阴极的氧还原活性。这些结果表明,在 MFC 阴极使用非离子型黏结剂是有利的,在 MFC 中性的 pH 条件下,可以促进电荷转移和稳定性能[51]。然而,使用聚四氟乙烯(PTFE)代替 Nafion 黏结剂,却降低了最大功率密度[52]。进一步的测试表明:作为催化剂黏结剂的 Nafion – PTFE 混合物中,提高 Nafion 比例(从 0 到 100%),增加了最大功率密度(从 549mW/$m^2$ 至 1060mW/$m^2$)[53],可能原因是 PTFE 的疏水性导致它作为黏结剂时性能相对较差。与疏水的 PS – OH 黏结剂相比,增加聚苯乙烯-*b*-聚环氧乙烷黏结剂的亲水性,提高了阴极的电化学响应和 MFC 的功率密度约 15%。聚(双酚 A 环氧氯丙烷)(BAEH)是一种廉价的亲水性中性聚合物,它的性能在最初的两个循环之后为 1360mW/$m^2$ 和 630 mW/$m^2$(分别对应 0.5mg/$cm^2$ Pt 和 0.05mg/$cm^2$ Pt),比亲水的磺化 Nafion 黏结剂的性能差(1980mW/$m^2$ 和 1080mW/$m^2$ 分别对应 0.5mg/$cm^2$ Pt 和 0.05mg/$cm^2$ Pt)。然而,经过长期运行后(22 个循环,40 天),每个电池的发电却变得相似(约 1200mW/$m^2$ 和 700 ~ 800mW/$m^2$ 分别对应 0.5mg/$cm^2$ Pt 和 0.05mg/$cm^2$ Pt),可能是由于阴极生物结垢后无法通过物理清洗来完全逆转[54]。

### 7.4.2 扩散层

聚四氟乙烯被广泛用于扩散层,允许空气阴极曝气,同时防止漏水。制作微生物燃料电池阴极的一个新的简化方法是使用金属网集流体和廉价的碳/聚合物

① "meq/g"为"毫当量/g"。

扩散层——聚二甲基硅氧烷(PDMS)[55]。PDMS 可以限制通过阴极的氧的转移,改善微生物燃料电池的库仑效率。低成本的设计可以得到与碳布阴极相媲美的功率输出和高得多的库仑效率分别为(80% 和 57% )[56]。

### 7.4.3 集流体

由于微生物燃料电池中相对低的电流密度,通常使用碳布或碳纸作为基材的阴极不额外使用集流体。包含不锈钢的碳布集流体可被集成到由反应型炭黑和 Pt 催化剂的混合物以及聚二甲基硅氧烷扩散层构成的 MFC 阴极中。在这里,这些阴极的网格属性可以显著影响性能。最疏网格(30 目)制成的阴极达到(1616 ± 25)$mW/m^2$的最大功率(归一化到阴极投影表面积;(47.1 ±0.7)$W/m^3$ 基于液体体积),而最密网格(120 目)具有最低的功率密度((599 ±57)$mW/m^2$)。电化学阻抗谱表明,电荷转移和扩散阻力随网眼尺寸的增加而减小。在 MFC 测试中,阴极性能主要受反应动力学限制,而不受传质限制。随着网孔尺寸增加透氧性增加,导致扩散阻力减小。在较高的电流密度下,特别是对具有低的氧转移系数的细网孔来说,扩散成为一个限制因素。这些结果证明了用于构建 MFC 阴极的网眼尺寸的关键作用[57]。

### 7.4.4 阴极结垢

与化学燃料电池相比,由于微生物燃料电池中生物膜的生长和使用的介质溶液的复杂性,阴极结垢的问题更为严重。

Kiely 等用木质纤维素发酵的个体化终产物(乙酸、甲酸、乳酸、琥珀酸或乙醇)作为底物,运行微生物燃料电池超过一年,用以考察阴极的长期性能和细菌群落。阴极性能随时间劣化,但当除去阴极生物膜后产电增加了 26% ,而使用新的阴极后则增加了 118%[58]。对于双室微生物燃料电池中浸没在溶液中的空气阴极来说,阴极生物膜的生长和化合物规模的累积会减少通过阴极的氧扩散以及质子传质,导致在长时间运行时发电量减少。因此,在双室微生物燃料电池中恰当地控制阴极上化合物的规模和生物膜的组成,对于其长时间运行时获得稳定的电力输出是非常重要的[40]。

### 7.4.5 阴极催化剂

由于对氧还原反应的极高催化活性,铂作为阴极催化剂,广泛应用于实验室的 MFC 系统里。但是,铂的高成本降低了这种方法的吸引力。近中性的 pH 值条件也强烈影响阴极的性能,限制了微生物燃料电池的发电。大量的研究工作都集中在提高性能和降低阴极材料的成本上[44,59,60]。

#### 7.4.5.1 纳米材料修饰铂阴极

纳米材料,特别是碳纳米管,已经被用于改善铂阴极的性能。得益于碳纳米管的独特性质,铂阴极的性能可以大大提高。

将铂纳米颗粒注入单壁碳纳米管片状电极里、作为微生物燃料电池的阴极,其电流密度比电子束蒸发镀铂的阴极高出大约一个数量级。催化活性的提高可能与活性阴极层里催化剂表面积的增加有关[61]。在另一项研究中,微生物燃料电池采用碳纳米管垫状阴极,获得的最大功率密度为329mW/m$^2$,是碳布阴极的两倍以上(151mW/m$^2$)[62]。在CNT纺织阴极上通过电化学沉积Pt纳米颗粒,Pt负载量仅仅只有商用铂涂碳布阴极的19.3%,以此制作的水性阴极微生物燃料电池的最大功率密度也得到了类似的两倍提高[63]。

### 7.4.5.2 非铂金属催化剂阴极

包括铁、钴、锰和铅在内的多种金属已被研究开发用来取代Pt。在这些金属中,钴和铁经常与四甲氧基苯基卟啉(TMPP)或酞菁(Pc)一起形成大环金属络合物,在中性pH条件下其性能与Pt相当。FePc负载在KJB碳上(FePc-KJB),功率密度可达634mW/m$^2$,高于Pt阴极的593mW/m$^2$和其他金属大环络合物阴极(包括CoTMPP、FeCoTMPP、CoPc和FeCuPc)[64]。FePc和CoTMPP与铂基体系的比较显示了过渡金属基材料替代微生物燃料电池中传统阴极材料的潜力[65]。Cheng等也论证了CoTMPP阴极的性能可与Pt阴极媲美,尤其是在电流密度高于0.6mA/cm$^2$时[52]。负载量为2mg/cm$^2$的FePc基阴极微生物燃料电池的最大功率密度为550~590mW/m$^2$,这与使用含Pt量为0.5mg/cm$^2$的Pt基碳布阴极相当[66]。大环金属络合物催化剂的性能也受微生物燃料电池中化学环境的影响。在恒电流极化实验中,当缓冲溶液浓度从500mM降至50mM时,热解的FePc和CoTMPP的极限电流密度从1.5mA/cm$^2$降至0.6mA/cm$^2$(pH 3.3,$E_{cathode}=0V$)[67]。高浓度的氯离子会降低铂催化剂的性能,却能提高CoTMPP阴极的性能[64]。

Aelterman等的研究表明,有可能用铁乙二胺四乙酸(Fe-EDTA)阴极电解质代替铁氰化物[68]。尽管最大功率降低了50%,但是不需要补充阴极电解质。简单地将碳与铁螯合乙二胺四乙酸混合、在氩气气氛下热解(PFeEDTA/C),季氮铁成为氧还原反应的可能的活性位点,可以大大提高性能[48]。用PFeEDTA/C做阴极的MFC产生的最大功率密度为1122mW/m$^2$,这非常接近Pt/C阴极的1166mW/m$^2$,并且在31天的操作周期内稳定。

研究人员也研究了其他非铂金属催化剂,包括二氧化铅[69]、二氧化锰[51,70]和Co[71]。然而,它们的最大功率密度(归一化为阴极投影表面积)大约只有100mW/m$^2$或更小,这比高性能Pt空气阴极功率密度的10%还小。

### 7.4.5.3 碳阴极

碳材料,例如石墨板和石墨毡,可直接作为阴极使用,不需要额外的催化剂。

尽管高性能 Pt 催化剂阴极仍然限制微生物燃料电池的性能，但普通碳阴极的电流密度一般比含上述催化剂的阴极还要低得多。然而，最近开发的一些碳阴极表现出的性能却可以与铂催化的阴极相媲美。

最近的研究表明活性炭作为低成本阴极材料具有巨大的潜力，该研究通过冷压的方法把活性炭和 PTFE 黏结剂压到镍网集流体上。这种阴极结构不需要碳布或金属催化剂，即可制成在典型 MFC 中具有高氧还原活性的阴极[72]。AC 阴极测试得到了 $1220mW/m^2$ 的最大功率密度，而相同条件下由铂催化的碳布阴极则为 $1060mW/m^2$。其他碳材料，包括活性碳纤维毡[56]和硝酸处理过的碳粉（Vulcan XC-72R）[73]，也对微生物燃料电池中的氧还原反应表现出不错的性能。

#### 7.4.5.4 导电聚合物

导电聚合物，如聚吡咯（PPy）/聚苯胺（PANI）和介质复合材料，在微生物燃料电池的氧还原反应中表现出期望的性能。双室 MFC 中利用 PPy/蒽醌-2,6-二磺酸（AQDS）复合物修饰的阳极和阴极，将最大功率密度提高了一个数量级至 $823mW/cm^2$，同时还增加了阴极室过氧化氢的产生速率[74]。在另一项研究中，普鲁士蓝/聚苯胺（PB/PANI）修饰的阴极在酸性电解液中对氧还原表现出良好的电催化活性，显示出的性能可与铁氰化物阴极相媲美[75]。在无膜 MFC 中将聚吡咯/蒽醌-2-磺酸钠（PPy/AQS）导电聚合物修饰到不锈钢网上，得到了 $575mW/m^2$ 的最大功率密度[76]。

#### 7.4.5.5 生物阴极

漆酶已经被普遍使用在基于酶的生物燃料电池的阴极。直到最近人们才发现，细菌也可以催化微生物燃料电池中的氧还原反应。这里的生物阴极指的是以细菌为基础的阴极。

相对低的功率/电流密度是生物阴极应用的主要障碍。传质计算表明，在微生物燃料电池中氧扩散对于使用溶解氧作为电子受体的过程造成了严重的限制[77]。计算得出的石墨板生物阴极的极限电流密度为 $0.848A/m^2$，这比测得的电流密度高 3 倍，比铂空气阴极低一个数量级。然而，使用碳毡作为阴极材料制作的管状生物阴极可以得到大约 $4A/m^2$ 的最大电流密度[78]，对应的最大功率密度为 $117W/m^3$（以 1mV/s 扫描速率通过极化得到）或 $83W/m^3$（通过改变外部电阻得到）。阴极材料和快速的阴极电解液溢流速率（$6l/h$，液体保留时间小于 0.4min）可能是提高氧传质和实现这样高的电流密度的原因。石墨纤维刷作为阴极材料的 MFC 中在电流密度为 $178.6A/m^3$ 下得到略低的 $68.4W/m^3$（以 1mV/s 扫描速率通过极化得到）[79]。阴极生长藻类（小球藻）的 MFC 产出了 $5.6W/m^3$ 的最大功率密度[80]。另一个在阴极使用小球藻的 MFC 的最大电流密度为 1.0mA/$m^2$[81]。

虽然提高电流密度仍有许多工作要做，但生物阴极由于低成本和可持续的特性已吸引了大量研究者的关注。此外，生物阴极不仅可用于氧的还原也可用于许多氧化性污染物[82]。例如，铬(VI)可在生物阴极的 MFC 中被生物还原为 $Cr(OH)_3$ 沉淀物[83]。最大的 Cr(VI)还原速率为 $0.46mg_{Cr(VI)} \cdot g_{VSS}^{-1} \cdot h^{-1}$①，电流和功率密度分别达到 55.5mA/$m^2$ 和 123.4mW/$m^2$。

## 7.5 隔　　膜

隔膜是 MFC 中的一个重要组成部分，起到物理上分隔阳极和阴极的作用。研究者已经为微生物燃料电池开发了多种适合的隔膜，包括盐桥、阳离子交换膜(CEM)、阴离子交换膜(AEM)、双极性膜(BPM)、微渗透(MF)膜、超渗透(UF)膜、多孔面料和多孔材料等。

在化学燃料电池中，膜或其他导离子的隔膜需要避免燃料渗透，同时允许电极之间的带电离子进行交换。然而在微生物燃料电池中，燃料渗透不再是问题，因为阳极和阴极使用完全不同的催化剂，因而燃料/基材在阴极不会被消耗掉。在使用铁氰化钾作为电子受体的双室微生物燃料电池中仍需要阳离子交换膜，以避免有毒的铁氰化钾扩散到阳极室中，同时允许质子或其他阳离子转移到阴极室。对于其他使用铁氰化钾以外的微生物燃料电池，阴离子[84]或双极性[85]交换膜、纳米多孔聚合物滤膜[86]、超渗透膜[84]甚至 J-cloth②[87]可能是更好的选择，因为它们会维持一个相对稳定的阳极电解液 pH 值，这对保持生物活性非常关键。

微生物燃料电池隔膜的主要功能是避免阳极和阴极的直接接触(短路)，减少不希望的氧和其他物质的渗透，同时保持有效的透过隔膜的质子传质。这种昂贵、复杂的膜系统是可能被替代或去除的[87,15]。在无膜的空气阴极微生物燃料电池系统中，在阴极上生长的生物膜可以起到隔膜的作用，最小化氧扩散至阳极室，同时允许质子/离子的有效传输。隔膜的去除不但可以降低微生物燃料电池的成本和复杂性，而且由于内部电阻的降低而提高了功率密度[3]。阴极上的好氧生物膜对于氧气的积极消耗可以有效地减少通过生物膜的氧扩散。然而，生物膜也消耗衬底而导致低的库仑效率。可能的短路和氧扩散的增加制约最小电极间距至 1～2cm[88]。相对较大的电极间距不但增加了内部电阻，而且限制了电极表面积/体积比，进而影响最大体积功率密度。因此，如果需要减小电极间距，那么仍然需要在电池结构中加上隔膜。

---

① VSS：Volatile Suspended Solid，可挥发性固体悬浮物。

② J-cloth，美国生产的一种百洁布。

### 7.5.1 阳离子交换膜

可能是由于在质子交换膜燃料电池中的普及，Nafion 膜被广泛地应用于 MFC 的研究中。因为在酸性条件下，超越其他离子而对质子优先传导，Nafion 膜也称为质子交换膜。然而，使用 Nafion 或其他阳离子交换膜（CEM）的 MFC 上的电荷平衡是由占优势的阳离子（如 $Na^+$ 和 $K^+$）的传输实现的，因为这些阳离子的浓度通常比质子的浓度高 $10^5$ 倍[89]。由于通过膜的质子的低效传输，阳极 pH 值增加而阴极 pH 降低，这不仅降低了 MFC 的输出电压，还抑制了阳极微生物的活性，并导致阴极结垢。可以替代 Nafion 膜的一种低成本膜是 CMI-7000（Membranes International Inc.），由于其本身的厚度导致它比 Nafion 的透氧性低。

### 7.5.2 阴离子交换膜

在微生物燃料电池中，阴离子交换膜（AEM）通常具有比 CEM 更好的表现，因为阴离子（如磷酸盐和碳酸盐）可以促进质子传输。在含有 AEM 的 MFC 中，pH 缓冲液仍然作为质子载体。微生物燃料电池中使用的 pH 缓冲液浓度为 0.1M 量级，这比质子浓度高 $10^6$ 倍。相对较高浓度的质子载体导致通过隔膜的更高效的质子传输和更高的功率密度[3]。Kim 等报道，使用 AEM 得到了 610mW/$m^2$ 的最大功率密度，高于 Nafion（514mW/$m^2$）和 CEM（480mW/$m^2$）[84]。在使用 AEM 阴极和 200mmol/L PBS 的管状 MFC 中，功率密度可以进一步提高到 728mW/$m^2$[90]。

### 7.5.3 双极性膜

双极性膜（BPM）是阴离子交换膜和阳离子交换膜通过一个加速水分解为质子和氢氧根离子的催化中间层（下称“接合”层）层压在一起而得到的。双极性膜在微生物燃料电池中最早的应用是作为高效阴极系统应用在石墨电极上进行三价铁的还原[85]。与 AEM 和 CEM 相比，BPM 可以抑制除质子和氢氧根离子以外的离子传输，从而减少阳极室和阴极室之间的 pH 值差异。然而，水分解会增加通过膜的极化电势损失。在废水中运行的生物电化学系统中，尽管 BPM 运送质子和氢氧根离子的能力增加，并防止了阴极室的 pH 增加，但其电势损失比 CEM 和 AEM 却要大得多（分别为 0.71V，0.27V 和 0.32V）[91]。

### 7.5.4 过滤膜

过滤膜，包括 UF 和 MF 膜，可作为微生物燃料电池的隔膜使用。与离子交换膜相比，除了成本低，UF 和 MF 膜的优点还在于大多数离子可以自由地通过膜上相对较大的孔，从而降低相关的内部电阻，提高 MFC 性能[92-95]。在微生物燃料电池中膜应用的一个主要问题是氧渗透，特别是对微生物燃料电池用膜电极组件[96]。理想的隔膜应该具备控制氧渗透同时促进质子传递到阴极的能力。可以通过使用具有较小孔径的膜[97,98]或更厚的膜[99]来降低氧渗透。不幸的是，材料降低氧渗透的同时也降低了质子的传输和能量的产生。

### 7.5.5　多孔织物

微生物燃料电池中多孔织物也可以完成隔膜的主要功能，防止短路和降低氧扩散。归因于多孔的结构，它可实现电荷载体离子的高效传质，降低内阻，提高发电量。理论上，在微生物燃料电池中，任何多孔、不导电的材料均可作为隔膜使用。对材料的基本要求是不导电、耐用、低成本、高阴离子渗透性和低的氧渗透性[87]。氧渗透可以通过一层简单的织物层被有效地抑制并可生长生物膜，从而改善库仑效率，其库仑效率是没有织物层的阴极的两倍（71% 和 35%）[87]。织物层也可以隔离阴极和阳极，从而降低电极间距至 1mm 以下，使得形成布电极（CEA）结构成为可能。CEA 结构有可能大大降低内阻[31]和增加表面积/体积比，因而提高了 MFC 的体积功率密度。使用双 CEA 的 MFC 在 0.2M 碳酸氢盐缓冲溶液中实现了高达 $1550W/m^3$（$2770mW/m^2$）的功率密度[3]。由于 CEA 的低成本和多样性，可以优化透氧性和内部电阻之间的平衡。然而，需要控制好生物膜的生长，以保持它的长期稳定性。

## 7.6　展　望

虽然低功率密度和高材料成本仍然限制 MFC 技术的应用，微生物燃料电池的功率密度已在不到 10 年的时间内增加了几个数量级[100]。随着在这方面研究的逐步深入，有望进一步提高功率密度和降低材料成本。

在世界人口快速增长至约 70 亿的今天，水和能源变得越来越昂贵，而且很快就被无止境的需求所湮灭。微生物燃料电池技术使用微生物催化可生物降解的有机物直接发电，提供了一种从废水发电并同时完成废水处理的全新方法[101]。细菌可产生电流的现象人们已经知道了近一个世纪[1]，但仅在最近几年，微生物燃料电池才引起广泛的关注和研究，特别是成功地从废水中获取电力之后[2]。

在过去的 10 年里，大量的研究活动使得 MFC 的性能得到迅速提高。到现在为止，报道的最高功率密度为 $1W/m^2$ 和 $1kW/m^3$ 量级，这仍然比化学燃料电池低大约 2～3 个数量级。为了使 MFC 技术具有竞争力和可商业化，进一步改善功率密度非常必要，这就需要在微生物燃料电池的低成本和高性能材料两方面进行突破。

阳极是 MFC 目前发展阶段的一个非限制性因素。混合培养细菌的碳布阳极可以产生至少 $3mA/cm^2$ 的电流密度，而具有相同大小电极的微生物燃料电池的电流密度通常小于 1mA/cm$^2$。目前，碳布的商业价格约为几十美元；当这种材料在微生物燃料电池中大规模应用成为可能时，预计价格将会降低。阳极材料未来的开发应进一步降低成本。

由于性能和成本两方面的原因，阴极是微生物燃料电池的主要限制因素。大规模应用中使用铂太昂贵。有可能找到某些阴极材料适合于中性 pH 和相对低的电流密度。活性炭是低成本、高性能阴极材料的一个很好的例子。应特别注意寻找适用于较高电流密度下的高性能氧还原催化剂，例如 2 ~ 3mA/cm$^2$，以匹配阳极可能的电流密度。

在 MFC 的研究中往往忽略了隔膜，无论是在成本方面还是在对内部电阻的贡献方面。离子交换膜的价格大约是每平方米几百美元。成本高、性能平庸限制了其在微生物燃料电池上可能的商业化应用。而报道的产生最高体积功率密度的多孔面料，其成本可低于每平方米 1 美元。需要对厚度和控制生物膜生长进行进一步优化。

MFC 材料上的突破往往会导致反应器结构和性能上的突破。新的设计还需要新的材料。低成本和高性能电极材料的发展将极大地扩展微生物燃料电池的应用。

## 参考文献


[1] Potter, M. C. (1911) Electrical effects accompanying the decomposition of organic compounds. *Proceedings of the Royal Society of London B: Biological Sciences*, **84**, 260 – 276.

[2] Liu, H., Ramnarayanan, R., and Logan, B. E. (2004) Production of electricity during wastewater treatment using a single chamber microbial fuel cell. *Environmental Science & Technology*, **38** (7), 2281 – 2285.

[3] Fan, Y., Hu, H., and Liu, H. (2007) Sustainable power generation in microbial fuel cells using bicarbonate buffer and proton transfer mechanisms. *Environmental Science & Technology*, **41** (23), 8154 – 8158.

[4] Li, W. W., Sheng, G. P., Liu, X. W., and Yu, H. Q. (2011) Recent advances in the separators for microbial fuel cells. *Bioresource Technology*, **102** (1), 244 – 252.

[5] You, S., Zhao, Q., Zhang, J., Jiang, J., and Zhao, S. (2006) A microbial fuel cell using permanganate as the cathodic electron acceptor. *Journal of Power Sources*, **162**, 1409 – 1415.

[6] Oh, S.-E., Min, B., and Logan, B. E. (2004) Cathode performance as a factor in electricity generation in microbial fuel cells. *Environmental Science & Technology*, **38** (18), 4900 – 4904.

[7] Pham, T. H., Jang, J. K., Chang, I. S., and Kim, B. H. (2004) Improvement of cathode reaction of a mediatorless microbial fuel cell. *Journal of Microbiology and Biotechnology*, **14** (2), 324 – 329.

[8] Borole, A. P., Hamilton, C. Y., Aaron, D. S., and Tsouris, C. (2009) Investigating microbial fuel cell bioanode performance under different cathode conditions. *Biotechnology Progress*, **25** (6), 1630 – 1636.

[9] Rabaey, K., Lissens, G., Siciliano, S. D., and Verstraete, W. (2003) A microbial fuel cell capable of converting glucose to electricity at high rate and efficiency. *Biotechnology Letters*, **25** (18), 1531 – 1535.

[10] Chaudhuri, S. K. and Lovley, D. R. (2003) Electricity generation by direct oxidation of glucose in mediatorless microbial fuel cells. *Nature Biotechnology*, **21**, 1229 – 1232.

[11] Ter Heijne, A., Hamelers, H. V. M., Saakes, M., and Buisman, C. J. N. (2008) Performance of non-porous graphite and titanium-based anodes in microbial fuel cells. *Electrochimica Acta*, **53** (18), 5697 – 5703.

[12] Bond, D. R. and Lovley, D. R. (2003) Electricity production by Geobacter sulfurreducens attached to electrodes. *Applied and Environmental Microbiology*, **69**, 1548 – 1555.

[13] Kargi, F. and Eker, S. (2007) Electricity generation with simultaneous wastewater treatment by a microbial fuel cell (MFC) with Cu and Cu-Au electrodes. *Journal of Chemical Technology and Biotechnology*, **82** (7), 658 – 662.

[14] Tender, L. M., Reimers, C. E., Stecher, H. A., III, Holmes, D. E., Bond, D. R., Lowy, D. A., Pilobello, K., Fertig, S. J., and Lovley, D. R. (2002) Harnessing microbially generated power on the seafloor. *Nature Biotechnology*, **20** (8), 821 – 825.

[15] Liu, H. and Logan, B. E. (2004) Electricity generation using an air-cathode single chamber microbial fuel cell in the presence and absence of a proton exchange membrane. *Environmental Science & Technology*, **38** (14), 4040 – 4046.

[16] Liu, H., Cheng, S., and Logan, B. E. (2005) Production of electricity from acetate or butyrate using a single-chamber microbial fuel cell. *Environmental Science & Technology*, **39** (2), 658 – 662.

[17] Logan, B. E., Cheng, S., Watson, V., and Estadt, G. (2007) Graphite fiber brush anodes for increased power production in air-cathode microbial fuel cells. *Environmental Science & Technology*, **41** (9), 3341 – 3346.

[18] Schröder, U., Niessen, J., and Scholz, F. (2003) A generation of microbial fuel cells with current outputs boosted by more than one order of magnitude. *Angewandte Chemie: International Edition*, **115**, 2986 – 2989.

[19] Richter, H., McKarthy, K., Nevin, K. P., Johnson, J. P., Rotello, V. M., and Lovley, D. R. (2008) Electricity generation by *Geobacter sulfurreducens attached to gold electrodes. Langmuir*, **24**, 4376 – 4379.

[20] Dumas, C., Mollica, A., Féron, D., Basséguy, R., Etcheverry, L., and Bergel, (2007) Marine microbial fuel cell: use of stainless steel electrodes as anode and cathode materials. *Electrochimica Acta*, **53** (2), 468 – 473.

[21] Reimers, C. E., Tender, L. M., Fertig, S. J., and Wang, W. (2001) Harvesting energy from the marine sediment-water interface. *Environmental Science & Technology*, **35**, 192 – 195.

[22] Scott, K., Cotlarciuc, I., Hall, D., Lakeman, J. B., and Browning, D. (2008) Power from marine sediment fuel cells: the influence of anode material. *Journal of Applied Electrochemistry*, **38** (9), 1313 – 1319.

[23] Jiang, D. and Li, B. (2009) Novel electrode materials to enhance the bacterial adhesion and increase the power generation in microbial fuel cells (MFCs). *Water Science & Technology*, **59** (3), 557 – 563.

[24] Nam, J. -Y., Kim, H. -W., Lim, K. -H., and Shin, H. -S. (2010) Effects of organic loading rates on the continuous electricity generation from fermented wastewater using a single-chamber microbial fuel cell. *Bioresource Technology*, **101** (1 Suppl.), S33 – S37.

[25] You, S., Zhao, Q., Zhang, J., Liu, H., Jiang, J., and Zhao, S. (2008) Increased sustainable electricity generation in up-flow air-cathode microbial fuel cells. *Biosensors and Bioelectronics*, **23** (7), 1157 – 1160.

[26] Nam, J. -Y., Kim, H. -W., and Shin, H. -S. (2010) Ammonia inhibition of electricity generation in sin-

gle-chambered microbial fuel cells. *Journal of Power Sources*, **195** (19), 6428 – 6433.

[27] Wang, X., Cheng, S., Feng, Y., Merrill, M. D., Saito, T., and Logan, B. E. (2009) The use of carbon mesh anodes and the effect of different pretreatment methods on power production in microbial fuel cells. *Environmental Science & Technology*, **43** (17), 6870 – 6874.

[28] Park, D. H. and Zeikus, J. G. (1999) Utilization of electrically reduced neutral red by Actinobacillus succinogenes: physiological function of neutral red in membrane-driven fumarate reduction and energy conservation. *Journal of Bacteriology*, **181** (8), 2403 – 2410.

[29] Kim, H. J., Park, H. S., Hyun, M. S., Chang, I. S., Kim, M., and Kim, B. H. (2002) A mediator-less microbial fuel cell using a metal reducing bacterium, Shewanella putrefaciens. *Enzyme and Microbial Technology*, **30** (2), 145 – 152.

[30] In Ho, P., Gnana Kumar, G., Kim, A. R., Kim, P., and Suk Nahm, K. (2011) Microbial electricity generation of diversified carbonaceous electrodes under variable mediators. *Bioelectrochemistry*, **80** (2), 99 – 104.

[31] Fan, Y., Sharbrough, E., and Liu, H. (2008) Quantification of the internal resistance distribution of microbial fuel cells. *Environmental Science & Technology*, **42** (21), 8101 – 8107.

[32] Larrosa-Guerrero, A., Scott, K., Katuri, K. P., Godinez, C., Head, I. M., and Curtis, (2010) Open circuit versus closed circuit enrichment of anodic biofilms in MFC: effect on performance and anodic communities. *Applied Microbiology and Biotechnology*, **87** (5), 1699 – 1713.

[33] Johnson, W. P. and Logan, B. E. (1996) Enhanced transport in porous media by sediment-phase and aqueous-phase natural organic matter. *Water Research*, **30** (4), 923 – 931.

[34] Cheng, S. and Logan, B. E. (2007) Ammonia treatment of carbon cloth anodes to enhance power generation of microbial fuel cells. *Electrochemistry Communications*, **9**, 492 – 496.

[35] Park, D. H. and Zeikus, J. G. (2003) Improved fuel cell and electrode designs for producing electricity from microbial degradation. *Biotechnology and Bioengineering*, **81**, 348 – 355.

[36] Lowy, D., Tender, L. M., Zeikus, J. G., Park, D. H., and Lovley, D. R. (2006) Harvesting energy from the marine sediment-water interface II: kinetic activity of anode materials. *Biosensors and Bioelectronics*, **21**, 2058 – 2063.

[37] Crittenden, S. R., Sund, C. J., and Sumner, J. J. (2006) Mediating electron transfer from bacteria to a gold electrode via a self-assembled monolayer. *Langmuir*, **22** (23), 9473 – 9476.

[38] Adachi, M., Shimomura, T., Komatsu, M., Yakuwa, H., and Miya, A. (2008) A novel mediator-polymer-modified anode for microbial fuel cells. *Chemical Communications*, **7**, 2055 – 2057.

[39] Niessen, J., Schröder, U., Rosenbaum, M., and Scholz, F. (2004) Fluorinated polyanilines as superior materials for electrocatalytic anodes in bacterial fuel cells. *Electrochemistry Communications*, **6**, 571 – 575.

[40] Chung, K., Fujiki, I., and Okabe, S. (2011) Effect of formation of biofilms and chemical scale on the cathode electrode on the performance of a continuous two- chamber microbial fuel cell. *Bioresource Technology*, **102**, 355 – 360.

[41] Qiao, Y., Li, C. M., Bao, S. J., and Bao, Q. L. (2007) Carbon nanotube/polyaniline composite as anode material for microbial fuel cells. *Journal of Power Sources*, **170**, 79 – 84.

[42] Qiao, Y., Bao, S. J., Li, C. M., Cui, X. Q., Lu, Z. S., and Bao, J. (2008) Nanostructured polyaniline/titanium dioxide composite anode for microbial fuel cells. *ACS Nano*, **2**, 113 – 119.

[43] Sharma, T., Leel Mohana Reddy, A., Chandra, T. S., and Ramaprabhu, S. (2008) Development of carbon nanotubes and nanofluids based microbial fuel cell. *International Journal of Hydrogen Energy*, **33**, 6749 – 6754.

[44] HaoYu, E., Cheng, S., Scott, K., and Logan, B. (2007) Microbial fuel cell performance with non-Pt cathode catalysts. *Journal of Power Sources*, **171** (2), 275 – 281.

[45] Peng, L., You, S.-J., and Wang, J.-Y. (2010) Carbon nanotubes as electrode modifier promoting direct electron transfer from Shewanella oneidensis. *Biosensors and Bioelectronics*, **25**, 1248 – 1251.

[46] Fan, Y., Xu, S., Schaller, R., Jiao, J., Chaplen, F., and Liu, H. (2011) Nanoparticle decorated anodes for enhanced current generation in microbial electrochemical cells. *Biosensors and Bioelectronics*, **26** (5), 1908 – 1912.

[47] Sun, J.-J., Zhao, H.-Z., Yang, Q.-Z., Song, J., and Xue, A. (2010) A novel layer-by- layer self-assembled carbon nanotube-based anode: preparation, characterization, and application in microbial fuel cell. *Electrochimica Acta*, **55**, 3041 – 3047.

[48] Wang, L., Liang, P., Zhang, J., and Huang, X. (2011) Activity and stability of pyrolyzed iron ethylenediaminetetraacetic acid as cathode catalyst in microbial fuel cells. *Bioresource Technology*, **102** (8), 5093 – 5097.

[49] Huang, Y., He, Z., and Mansfeld, F. (2010) Performance of microbial fuel cells with and without Nafion solution as cathode binding agent. *Bioelectrochemistry*, **79** (2), 261 – 264.

[50] Saito, T., Merrill, M. D., Watson, V. J., Logan, B. E., and Hickner, M. A. (2010) Investigation of ionic polymer cathode binders for microbial fuel cells. *Electrochimica Acta*, **55** (9), 3398 – 3403.

[51] Zhang, L. X., Liu, C. S., Zhuang, L., Li, W. S., Zhou, S. G., and Zhang, J. T. (2009) Manganese dioxide as an alternative cathodic catalyst to platinum in microbial fuel cells. *Biosensors and Bioelectronics*, **24**, 2825 – 2829.

[52] Cheng, S., Liu, H., and Logan, B. E. (2006) Power densities using different cathode catalysts (Pt and CoTMPP) and polymer binders (Nafion and PTFE) in single chamber microbial fuel cells. *Environmental Science & Technology*, **40** (1), 364 – 369.

[53] Wang, X., Feng, Y., Liu, J., Shi, X., Lee, H., Li, N., and Ren, N. (2010) Power generation using adjustable Nafion/PTFE mixed binders in air-cathode microbial fuel cells. *Biosensors and Bioelectronics*, **26** (2), 946 – 948.

[54] Saito, T., Roberts, T. H., Long, T. E., Logan, B. E., and Hickner, M. A. (2011) Neutral hydrophilic cathode catalyst binders for microbial fuel cells. *Energy & Environmental Science*, **4** (3), 928 – 934.

[55] Zhang, F., Saito, T., Cheng, S., Hickner, M. A., and Logan, B. E. (2010) Microbial fuel cell cathodes with poly (dimethylsiloxane) diffusion layers constructed around stainless steel mesh current collectors. *Environmental Science & Technology*, **44** (4), 1490 – 1495.

[56] Deng, Q., Li, X., Zuo, J., Ling, A., and Logan, B. E. (2010) Power generation using an activated carbon fiber felt cathode in an upflow microbial fuel cell. *Journal of Power Sources*, **195** (4), 1130 – 1135.

[57] Zhang, F., Merrill, M. D., Tokash, J. C., Saito, T., Cheng, S., Hickner, M. A., and Logan, B. E. (2011) Mesh optimization for microbial fuel cell cathodes constructed around stainless steel mesh current collectors. *Journal of Power Sources*, **196** (3), 1097 – 1102.

[58] Kiely, P. D., Rader, G., Regan, J. M., and Logan, B. E. (2011) Long-term cathode performance and the microbial communities that develop in microbial fuel cells fed different fermentation endproducts. *Bioresource Technology*, **102** (1), 361 – 366.

[59] Lefebvre, O., Al-Mamun, A., Ooi, W. K., Tang, Z., Chua, D. H. C., and Ng, H. Y. (2008) An insight into cathode options for microbial fuel cells. *Water Science & Technology*, **57** (12), 2031 – 2037.

[60] Harnisch, F. and Schröder, U. (2010) From MFC to MXC: chemical and biological cathodes and their

potential for microbial bioelectrochemical systems. *Chemical Society Reviews*, **39** (11), 4433 – 4448.

[61] Sanchez, D. V. P. , Huynh, P. , Kozlov, M. E. , Baughman, R. H. , Vidic, R. D. , and Yun, M. (2010) Carbon nanotube/platinum (Pt) sheet as an improved cathode for microbial fuel cells. *Energy & Fuels*, **24** (11), 5897 – 5902.

[62] Wang, H. , Wu, Z. , Plaseied, A. , Jenkins, P. , Simpson, L. , Engtrakul, C. , and Ren, Z. (2011) Carbon nanotube modified air-cathodes for electricity production in microbial fuel cells. *Journal of Power Sources*, **196** (18), 7465 – 7469.

[63] Xie, X. , Pasta, M. , Hu, L. , Yang, Y. , McDonough, J. , Cha, J. , Criddle, C. S. , and Cui, Y. (2011) Nano-structured textiles as high-performance aqueous cathodes for microbial fuel cells. *Energy & Environmental Science*, **4** (4), 1293 – 1297.

[64] Wang, X. , Cheng, S. , Zhang, X. , Li, X. -Y. , and Logan, B. E. (2011) Impact of salinity on cathode catalyst performance in microbial fuel cells (MFCs). *International Journal of Hydrogen Energy*, **36** (21), 13900 – 13906.

[65] Zhao, F. , Harnisch, F. , Schröder, U. , Scholz, F. , Bogdanoff, P. , and Herrmann, (2005) Application of pyrolysed iron(II) phthalocyanine and CoTMPP based oxygen reduction catalysts as cathode materials in microbial fuel cells. *Electrochemistry Communications*, **7** (12), 1405 – 1410.

[66] Birry, L. , Mehta, P. , Jaouen, F. , Dodelet, J. -P. , Guiot, S. R. , and Tartakovsky, B. (2011) Application of iron-based cathode catalysts in a microbial fuel cell. *Electrochimica Acta*, **56** (3), 1505 – 1511.

[67] Zhao, F. , Harnisch, F. , Schröder, U. , Scholz, F. , Bogdanoff, P. , and Herrmann, (2006) Challenges and constraints of using oxygen cathodes in microbial fuel cells. *Environmental Science & Technology*, **40** (17), 5193 – 5199.

[68] Aelterman, P. , Versichele, M. , Genettello, E. , Verbeken, K. , and Verstraete, W. (2009) Microbial fuel cells operated with iron-chelated air cathodes. *Electrochimica Acta*, **54** (24), 5754 – 5760.

[69] Morris, J. M. , Jin, S. , Wang, J. Q. , Zhu, C. Z. , and Urynowicz, M. A. (2007) Lead dioxide as an alternative catalyst to platinum in microbial fuel cells. *Electrochemistry Communications*, **9**, 1730 – 1734.

[70] Zhuang, L. , Zhou, S. , Wang, Y. , Liu, C. , and Geng, S. (2009) Membrane-less cloth cathode assembly (CCA) for scalable microbial fuel cells. *Biosensors and Bioelectronics*, **24** (12), 3652 – 3656.

[71] Lefebvre, O. , Ooi, W. K. , Tang, Z. , Abdullah-Al-Mamun, Md. , Chua, D. H. C. , and Ng, H. Y. (2009) Optimization of a Pt-free cathode suitable for practical applications of microbial fuel cells. *Bioresource Technology*, **100** (20), 4907 – 4910.

[72] Zhang, F. , Cheng, S. , Pant, D. , Bogaert, G. V. , and Logan, B. E. (2009) Power generation using an activated carbon and metal mesh cathode in a microbial fuel cell. *Electrochemistry Communications*, **11** (11), 2177 – 2179.

[73] Duteanu, N. , Erable, B. , Senthil Kumar, S. M. , Ghangrekar, M. M. , and Scott, K. (2010) Effect of chemically modified Vulcan XC-72R on the performance of air-breathing cathode in a single- chamber microbial fuel cell. *Bioresource Technology*, **101** (14), 5250 – 5255.

[74] Feng, C. , Li, F. , Liu, H. , Lang, X. , and Fan, S. (2010) A dual-chamber microbial fuel cell with conductive film-modified anode and cathode and its application for the neutral electro-Fenton process. *Electrochimica Acta*, **55** (6), 2048 – 2054.

[75] Fu, L. , You, S. -J. , Zhang, G. -Q. , Yang, F. -L. , Fang, X. -H. , and Gong, Z. (2011) PB/PANI-modified electrode used as a novel oxygen reduction cathode in microbial fuel cell. *Biosensors and Bioelectronics*, **26** (5), 1975 – 1979.

[76] Feng, C. , Wan, Q. , Lv, Z. , Yue, X. , Chen, Y. , and Wei, C. (2011) One-step fabrication of mem-

braneless microbial fuel cell cathode by electropolymerization of polypyrrole onto stainless steel mesh. *Biosensors and Bioelectronics*, **26** (9), 3953 – 3957.

[77] Ter Heijne, A., Strik, D. P. B. T. B., Hamelers, H. V. M., and Buisman, C. J. N. (2010) Cathode potential and mass transfer determine performance of oxygen reducing biocathodes in microbial fuel cells. *Environmental Science & Technology*, **44** (18), 7151 – 7156.

[78] Clauwaert, P., van derHa, D., Boon, N., Verbeken, K., Verhaege, M., Rabaey, K., and Verstraete, W. (2007) Open air biocathode enables effective electricity generation with microbial fuel cells. *Environmental Science & Technology*, **41** (21), 7564 – 7569.

[79] You, S.-J., Ren, N.-Q., Zhao, Q.-L., Wang, J.-Y., and Yang, F.-L. (2009) Power generation and electrochemical analysis of biocathode microbial fuel cell using graphite fibre brush as cathode material. *Fuel Cells*, **9** (5), 588 – 596.

[80] Wang, X., Feng, Y., Liu, J., Lee, H., Li, C., Li, N., and Ren, N. (2010) Sequestration of $CO_2$ discharged from anode by algal cathode in microbial carbon capture cells (MCCs). *Biosensors and Bioelectronics*, **25** (12), 2639 – 2643.

[81] Mitra, P. and Hill, G. A. (2011) Continuous microbial fuel cell using a photoautotrophic cathode and a fermentative anode. *The Canadian Journal of Chemical Engineering*, **90** (4), 1006 – 1010.

[82] Clauwaert, P., Rabaey, K., Aelterman, P., DeSchamphelaire, L., Pham, T. H., Boeckx, P., Boon, N., and Verstraete, W. (2007) Biological denitrification in microbial fuel cells. *Environmental Science & Technology*, **41**, 3354 – 3360.

[83] Tandukar, M., Huber, S. J., Onodera, T., and Pavlostathis, S. G. (2009) Biological chromium(VI) reduction in the cathode of a microbial fuel cell. *Environmental Science & Technology*, **43** (21), 8159 – 8165.

[84] Kim, J. R., Oh, S.-E., Cheng, S., and Logan, B. E. (2007) Power generation using different cation, anion and ultrafiltration membranes in microbial fuel cells. *Environmental Science & Technology*, **41** (3), 1004 – 1009.

[85] Ter Heijne, A., Hamelers, H. V. M., De Wilde, V., Rozendal, R. A., and Buisman, C. J. N. (2006) A bipolar membrane combined with ferric iron reduction as an efficient cathode system in microbial fuel cells. *Environmental Science & Technology*, **40** (17), 5200 – 5205.

[86] Biffinger, J. C., Pietron, J., Ray, R., Little, B., and Ringeisen, B. R. (2007) A biofilm enhanced miniature microbial fuel cell using Shewanella oneidensis DSP10 and oxygen reduction cathodes. *Biosensors and Bioelectronics*, **22** (8), 1672 – 1679.

[87] Fan, Y., Hu, H., and Liu, H. (2007) Enhanced coulombic efficiency and power density of air-cathode microbial fuel cells with an improved cell configuration. *Journal of Power Sources*, **171** (2), 348 – 354.

[88] Cheng, S., Liu, H., and Logan, B. E. (2006) Increased power generation in a continuous fow MFC with advective flow through the porous anode and reduced electrode spacing. *Environmental Science & Technology*, **40** (7), 2426 – 2432.

[89] Rozendal, R. A., Hamelers, H. V. M., and Buisman, C. J. N. (2006) Effects of membrane cation transport on pH and microbial fuel cell performance. *Environmental Science & Technology*, **40**, 5206 – 5211.

[90] Zuo, Y., Cheng, S., and Logan, B. E. (2008) Ion exchange membrane cathodes for scalable microbial fuel cells. *Environmental Science & Technology*, **42** (18), 6967 – 6972.

[91] Rozendal, R. A., Sleutels, T. H. J. A., Hamelers, H. V. M., and Buisman, C. J. N. (2008) Effect of the type of ion exchange membrane on performance, ion transport, and pH in biocatalyzed electrolysis of wastewater. *Water Science & Technology*, **57**, 1757 – 1762.

[92] Zuo, Y., Cheng, S., Call, D., and Logan, B. E. (2007) Tubular membrane cathodes for scalable power generation in microbial fuel cells. *Environmental Science & Technology*, **41** (9), 3347 – 3353.

[93] Kim, J. R., Premier, G. C., Hawkes, F. R., Dinsdale, R. M., and Guwy, A. J. (2009) Development of a tubular microbial fuel cell (MFC) employing a membrane electrode assembly cathode. *Journal of Power Sources*, **187** (2), 393 – 399.

[94] Sun, J., Hu, Y., Bi, Z., and Cao, Y. (2009) Improved performance of air-cathode single-chamber microbial fuel cell for wastewater treatment using microfiltration membranes and multiple sludge inoculation. *Journal of Power Sources*, **187** (2), 471 – 479.

[95] Sun, J., Hu, Y.-Y., Bi, Z., and Cao, Y.-Q. (2009) Simultaneous decolorization of azo dye and bioelectricity generation using a microfiltration membrane air-cathode single-chamber microbial fuel cell. *Bioresource Technology*, **100** (13), 3185 – 3192.

[96] Butler, C. S. and Nerenberg, R. (2010) Performance and microbial ecology of air-cathode microbial fuel cells with layered electrode assemblies. *Applied Microbiology and Biotechnology*, **86** (5), 1399 – 1408.

[97] Zhang, X., Cheng, S., Huang, X., and Logan, B. E. (2010) The use of nylon and glass fiber filter separators with different pore sizes in air-cathode single-chamber microbial fuel cells. *Energy & Environmental Science*, **3** (5), 659 – 664.

[98] Hou, B., Sun, J., and Hu, Y.-Y. (2011) Simultaneous Congo red decolorization and electricity generation in air-cathode single-chamber microbial fuel cell with different microfiltration, ultrafiltration and proton exchange membranes. *Bioresource Technology*, **102** (6), 4433 – 4438.

[99] Watson, V. J., Saito, T., Hickner, M. A., and Logan, B. E. (2011) Polymer coatings as separator layers for microbial fuel cell cathodes. *Journal of Power Sources*, **196** (6), 3009 – 3014.

[100] Logan, B. E. and Regan, J. M. (2006) Microbial fuel cells: challenges and applications. *Environmental Science & Technology*, **40**, 5172 – 5180.

[101] Logan, B. E., Aelterman, P., Hamelers, B., Rozendal, R., Schröeder, U., Keller, J., Freguiac, S., Verstraete, W., and Rabaey, K. (2006) Microbial fuel cells: methodology and technology. *Environmental Science & Technology*, **40** (17), 5181 – 5192.

# 第 8 章

## 生物电化学系统

Falk Harnisch, Korneel Rabaey

## 8.1 生物电化学系统和生物电催化

生物电化学系统(BES)利用生物催化剂催化阴极或阳极反应,或同时催化两个反应。因此,生物电催化可以看作增加给定电极反应的反应速率且适当减小反应过电势的过程。如图 8.1 所示,3 种不同的生物电催化从单组酶(如脱氢酶[1])或多组酶到细胞器(如线粒体[2,3])再到整个微生物细胞。后者被称为微生物的生物电催化,因为微生物能自生,对多电子反应常常具有专项催化性能,且具有能够转化复杂的物质或混合物等优点,在过去数年里引起了越来越多的关注。本章重点介绍微生物的生物电催化。

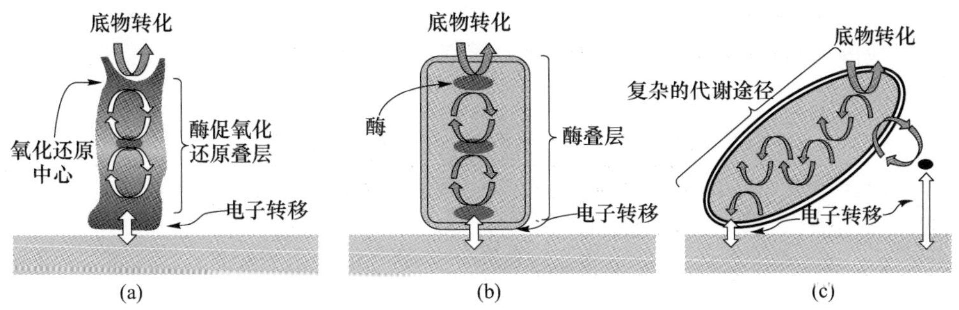


图 8.1 不同类型的生物电催化剂:(a) 酶;(b) 细胞器;(c) 微生物

首先,我们将讨论传统的化学反应和微生物的生物电催化反应之间的异同;其次,介绍细胞外的电子转移(EET)的主要机制和该技术的利用;最后介绍生物学上和电化学上用来分析微生物电子转移和识别鉴定介质、蛋白质和微生物的方法。

## 8.2 微生物的生物电催化本质

微生物电催化剂的微生物是活的细胞。因此,它们需要给自己的代谢和增殖提供大量的能量。细胞消耗能量生成两种主要的能量载体 ATP 和 NAD(P)H。为了生成能量,微生物内部要建立电势梯度,这将导致外部电势的损失。在进行生物电催化时,获得的能量来自能量不相同的给电子体和接受电子体的相互反应。以醋酸在生物膜上的氧化为例,作为典型的阳极生物膜可提供 0.4V 的电势(本章提供的电势如未特别说明均为此值),并通过控制杆菌的种类控制反应,生物膜上的醋酸氧化反应如下:

$$C_2H_4O_2 + 2H_2O \longrightarrow 2CO_2 + 8H^+ + 8e^- \tag{8.1}$$

(式(8.1)及后面所有的反应式与推导公式,为了简便都写成电中性结构)

醋酸氧化为 $CO_2$ 的生物标准电势 $E^{0\prime} = -0.290V$;需要注意的是电势会受到溶液中局部浓度和氧化还原产物的影响[4]。可推导的最大电势差 $\Delta E$ 为 0.690V,相当于电子转移时最大能量差为 66.6kJ/mol($\Delta G_{max} = 66.6kJ/mol$,依据 $\Delta G = -nF\Delta E$ 计算)。实际上,微生物转移电子时阳极的电势没有 −0.290V,而是约为 −0.150V(相对 SHE,以 Ag/AgCl 为参比电极时为 −0.350V)。所以,阳极的最终电子转移电势差在 0.550V 之内。微生物生物电催化氧化醋酸真实得到的最大能量为 $\Delta G_{real} = 53.1kJ/mol$。从微生物的角度来看,还剩余 $\Delta G_{microbio} = 13.5kJ/mol$ 的能量可以去开发。我们知道微生物的催化在电子给体和受体中得到的能量是不同的,因为反应的驱动因素不同。所以,严格来讲"生物电催化"这样讲可能是不准确的,但是在文献中这样表述很方便所以就这样表达,在本章中也这样表达。

与传统的贵金属电催化相比,低温燃料电池中微生物电催化的优缺点如表 8.1 所列。

表 8.1 微生物电催化剂和化学电催化剂催化低温燃料电池的工作条件

| 参数 | 常见电化学催化剂 | 微生物生物电化学催化剂 |
|---|---|---|
| 工作温度 | 常温到高温 | 控温 15~75℃ |
| 燃料底物 | 简单分子,如:$H_2$、甲醇 | 简单到复杂分子,或混合底物 |
| 纯度 | 有要求 | 无要求 |
| 寿命 | 有限,会腐蚀 | 自我更新 |
| 成本 | 相当大 | 可忽略 |
| 催化剂体积 | 小(纳米级) | 相当大(单细胞在微米水平,生物膜在 10~15μm) |
| 媒介要求 | 纯水或溶剂 | 含有营养物质(N,P,微量元素)和维生素的水性混合物 |
| 选择性 | 通常较低 | 可能高达 100% |

## 8.3 微生物电子转移机制

如上所述,微生物从最大可利用的能量中提取一定比例的能量用于自身代谢。如果发生阳极生物电催化,产生的能量差会低于基质含有的能量。与传统的电催化体系类似,微生物电催化体系的主要能量损失发生在生物电催化剂和电极界面上(图8.2)。

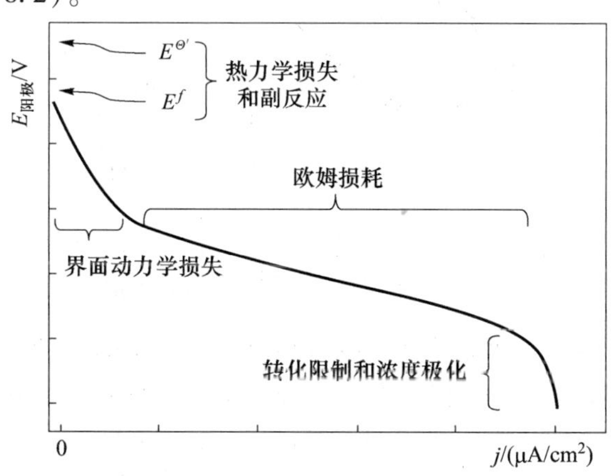


图8.2 微生物的生物阳极极化曲线

如图8.2所示,主要的能量损失包括热力学以及动力学损失,微生物的新陈代谢损失和在电极界面上电子转移动力学损失,实例见文献[6、7]详细论述。下面将讨论微生物和电极之间的电子传递途径。值得注意的是,大部分研究主要集中在生物电催化作用的阳极,而作为电子受体的阴极,很少有人去了解它的微生物阴极反应机制[8]。

从1911年,Potter[9]首次提出到20世纪90年代中期,人们普遍认为氧化还原反应的物质在微生物和电极间的电子穿梭要外加物质促进转化,这些外加物质介质通常是有毒的有机染料如中性红[10]或亚甲基蓝。然而,这些内源性物质的添加是不必要的[11],我们现在开始认识到直接电子转移(DET)(参见8.3.1节)和间接电子转移(MET)机制是基于初级和次级代谢产物,并不是添加的氧化还原介质(MET)(参见8.3.2节)。

### 8.3.1 直接电子转移

常见的直接电子转移机制要求微生物和最终电子受体有直接的接触。如图8.3所示,直接电子转移可以分为两种形式:通过膜上的氧化还原蛋白结合和通过细菌菌毛——“纳米线”连接。近期研究表明,两种连接方式可以相互协调来实现电子高效转移。

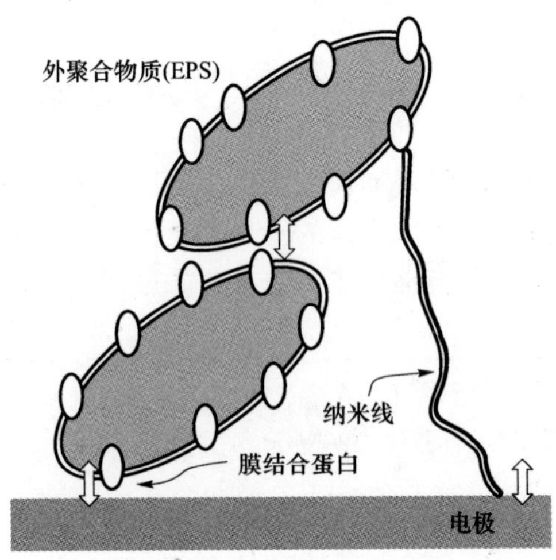


图 8.3　微生物与电极通过膜蛋白质结合及通过纳米线连接进行直接电子转移模拟图

通过膜蛋白连接实现电子直接传递看似是最直接的途径,所有的细胞和阳极接触并通过膜上的结合点传递电子。许多微生物具有能使膜结合的酶,并能够让微生物进行直接电子传递,典型的菌种有地杆菌和希瓦氏菌属。因为地杆菌有多种多样的用于限制个体电势带宽的 c 型细胞色素,使得在不同的条件下都能进行电子转移[13];希瓦氏菌似乎是因为聚亚铁血红素这种复合体的单克隆抗体使它有较宽的电位区间(关于这种复合体的综述可以参照文献[14])。因为地杆菌相对于其他的微生物可以实现更高的电流密度,这使它更适合进行微生物的生物电催化。一些单层膜结构的微生物像红育菌,使电流密度受到严重限制(如只有 $3\mu A/cm^2$)[15],其他不是单层膜的物种能产生显著的更高的电流密度,这涉及阳极上的生物膜是多层结构的事实。这也强调直接电子转移只是形成电流的一种方式。

在 2005 年,微生物菌毛或类菌毛结构在 EET 上的传递作用首次被报道。过去对于纳米菌毛线的解释多在横向导电上[16,17],最近的解释多为纵向导电[18,19]。这些纳米线使微生物的膜和外面相联系,电子可在膜连接位点内穿梭(如细胞色素)并通过菌毛最终传导给电子受体(图 8.3)[16,17]。报道指出,这种方式可以使电流密度比上述的 DET 利用膜蛋白连接传导理论高出几个数量级,目前实现的最大电流密度为 $3mA/cm^2$(表 8.2)[20]。关于电子转移方式的许多机理问题和复杂的生物膜所起的作用依然有待解决[21]。需要注意的是,电流密度不仅仅和纳米线的存在有关,也和复杂的细胞色素有关,两者都和细菌膜有关[22,23]。可能是细胞色素允许电子从纳米线转移到电极,或者说纳米线可将电子从电池内传送到电极表面。

关于间接的电子转移,通过初级代谢产物的氧化,可以实现微生物和电极材料在电池悬浮液中不进行直接的相互作用，直接电子转移需要微生物细胞和电极表面有接触。接触的连接并不是永久性的，一些菌种拥有向电子受体运动的能力，如同在悬浊液中的间接电子转移[24]。

此外，有越来越多的证据表明外聚合物质（EPS）的矩阵结构也可能在微生物的电子转移中发挥重要作用[25]。

表 8.2　不同的电极材料与其最大电流的研究

| 阳极材料 | 细菌来源 | 培养基质 | $j_{max}$/(μA/cm$^2$) | 文献 |
|---|---|---|---|---|
| 玻璃碳 | 地杆菌 | 醋酸盐 | 100 | [56] |
| 石墨纤维刷 | 预处理的 MFC | 醋酸盐 | 800 | [57] |
| 多晶碳 | 污水 | 醋酸盐 | 920 | [55] |
| 碳纸 | 污水 | 醋酸盐 | 1350 | [55] |
| 聚苯胺粘在石墨上 | 稻田土壤 | 淀粉、蛋白胨、鱼汁 | 2400 | [58] |
| 电纺碳纤维垫 | 污水 | 醋酸盐 | 3000 | [20] |
| 注：如温度等测试条件，以上测试在实际操作时是不同的 | | | | |

## 8.3.2　间接电子转移（MET）

间接电子转移是通过微生物的初级代谢产物和次级代谢产物来实现的[26]，这两个概念在图 8.4 进行了介绍。次级代谢产物大多是不能进行电子可逆传送的氧化产物，而初级代谢产物大多能进行可逆电子传送。细胞内外的电子穿梭取决于代谢产物。

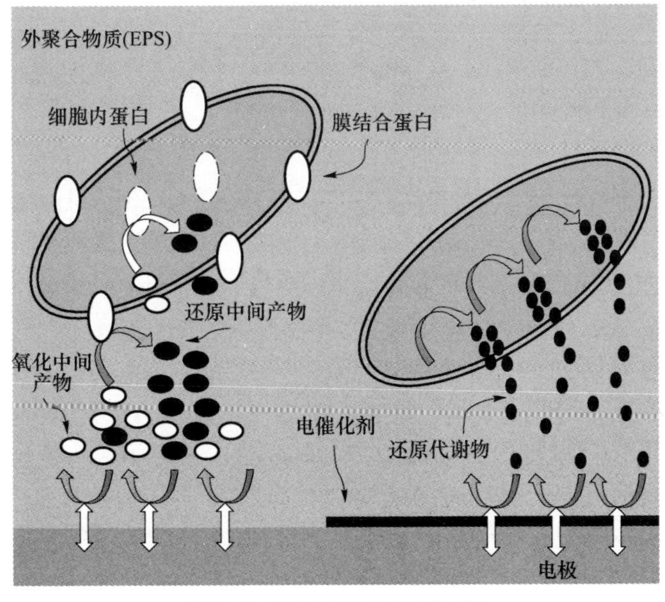


图 8.4　间接电子转移机理图

#### 8.3.2.1 以次级代谢产物为基础的间接电子转移

鉴于添加外源氧化还原物质在EET研究上的进行[27]，目前对微生物次级代谢产物转移电子的研究越来越多。次生代谢产物的生成不是细菌主要的代谢方式，因此它的合成对于细菌来说并不容易。一些次生代谢产物如异氰氨酸[28]或2-氨基-3-羧基-1,4-萘醌[29]允许电子在厚的生物膜中转移（甚至可以在浮游细胞和阳极间转移电子），从而导致微生物燃料电池（MFC）阳极性能的提高。有几种微生物（如假单细胞）[30]已经被报道具有这种转移电子能力。有趣的是，有越来越多的证据表明运输电子的化合物不仅在生成该物质的同类微生物中起作用，在其他种类的微生物中也起作用（甚至更高效）[30]。在细菌相互作用的复杂环境下，需要进一步观察运输电子的化合物是否通过信号物质如群体感应来实现微生物之间的相互联系[31]。最后，我们可以认为一些化合物如腐殖质，是微生物在自然条件下或在实验室EET试验下降解有机体的结果[32]。

#### 8.3.2.2 以初级代谢产物为基础的间接电子转移

初级代谢产物和微生物的新陈代谢有直接的联系，它们主要来源于微生物的无氧呼吸和发酵，产氢和生成硫化物就是很好的例子。代谢生成的化合物从微生物的角度看基本是废弃物，但却可以起到调节细胞内的氧化还原平衡的作用。为了研究阳极微生物的生物电化学体系，这些化合物需要在电催化电极表面被氧化。该阳极概念已应用于在铂聚合物的“三明治”电极上的葡萄糖发酵过程中产生的氢气的氧化上。在之后的研究中，非贵金属电极催化剂取代了贵金属①材料，并且被认同的氧化反应不只是氢气的氧化，还有如甲酸、乳酸等小相对分子质量酸的氧化[33-35]。此外，不同种类的硫单质的利用[36-38]可以把电子转移的概念进行分类，需要注意硫物种在生物电化学体系里能进行硫化物和硫单质的可逆转换。

## 8.4 由生理机理到工艺转变的微生物电化学体系

自20世纪Potter的初步调查结果[9,40]以来，关于利用微生物分解发电的研究陆续见报。从美国国家航空航天局实行“阿波罗”计划并专注于人的粪便和尿液利用开发开始，微生物的电化学催化研究的关注度逐渐增多[41,42]。直到太阳能光电板在宇宙飞船上表现出优良的电能供给性能，对微生物电化学催化的关注才有所下降。20世纪90年代，Wilkinson首次报道了用生物电催化体系为机器人供电的研究，此类机器人被称为食肉机器人[43]。基于自然界如污水中自我形成的具有电催化作用的生物膜[11,44,45]被系统有意识的驱动来产生能量的

① 根据原著，此处应译为“非贵金属”，但结合本书上下文，此处应为“贵金属”。译者注。

发现,进一步促进了该研究工作的进展。在很长一段时间里这一体系只在微生物燃料电池中被应用(图8.5(a))[46]。过去一段时间的主要任务是在微生物发酵中引入新的应用[47,48],在2005年实现了微生物的协助产氢[49]。从那时起,许多概念和应用被提出,包括生物降解电池[50]、微生物脱盐电池[51]、生物光电化学电池[52]等许多研究进展[7]。这些现在都归纳于微生物电化学系统中[53]。微生物电合成是一个新的方法[54],这种方法中微生物不仅在阳极发生反应,在阴极也进行化合物的生成和累积,微生物电合成电池的主要结构如图8.5(b)所示。

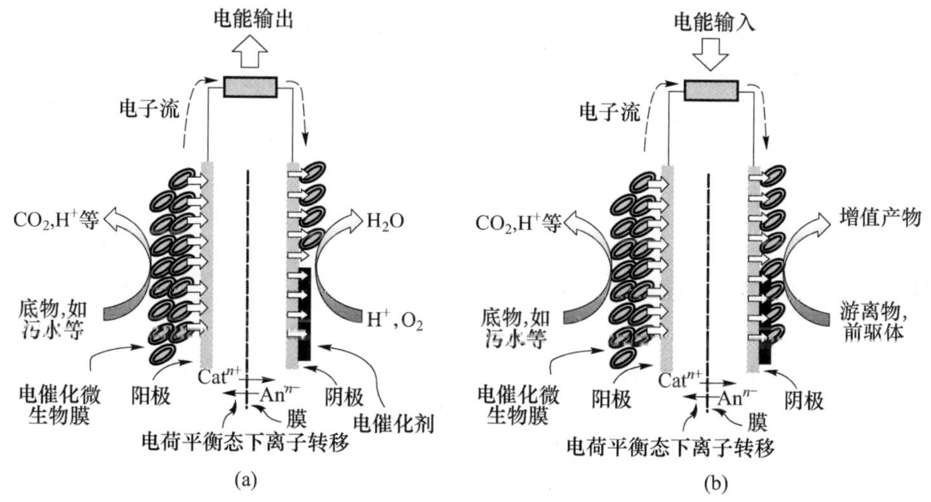


图8.5 (a)微生物燃料电池模拟图和(b)微生物电合成电池阳极处理污水阴极产生生化药品模拟图

在大多数情况下,生物电化学体系是通过生物膜附着在电极上来实现生物电催化反应。因此,电极扮演了双重角色,即作为生物膜基底,又作为电池或电子最终转移体的反应界面。由此可见,微生物的表面积既不是其几何学面积,也不是实际的通过气体吸附来测定的表面面积[55]。到目前为止,还不能准确测定和生物学相关的表面积。表8.2给出了不同电极材料用电活性微生物膜分别测定的最大几何学电流密度。

值得注意的是,从废水工程的角度来看,库仑效率和相关有机物的移动通常被认为是与最佳几何学电流密度有重要关系的参数。因此,三维电极材料,如石墨颗粒经常被采用[59]。关于体积电流密度的确定,通过确定反应器的体积比确定反应电极的面积更方便。无论怎样,任何研究报告都应该包括膜电极基于表面积转换后的参数[60]。

## 8.5 BES技术的应用潜力

近年来,生物电化学体系由于其自身潜在的应用前景有了巨大的进展。直到2005年研究重点聚焦到微生物燃料电池的发电方面[26,61]。能源作为低价值的商品,将燃料电池投入实际应用,除了要有有利可图的市场应用,如远程电能输送装置,还要使微生物燃料电池实现整体造价低,并具有可接受的体积和性能[62]。此外,它要求高效处理废水、生成少的沉淀产物的同时,还能去除其中的营养物质。

考虑到电能的价值,产生增值产品是较好的途径。产生氢气能带来适度的价值,要产生氢氧化钠[65]和双氧水[66]对反应体系有更复杂的要求,这会提高竞争力并且不需要进行完整的污水处理。对于未来在生物炼油市场中,生产生化试剂开发中间价值商品是很好的机会[54]。然而,这种体系对二次效益有要求[51]。

生物电化学体系中的生物降解和生物传感是完全不同的情况。在生物降解时,生物电化学体系能控制降解或固化过程,例如,硝基苯[67]和铀[68]。生物电化学体系可以为遥感器同时提供感应能力和电能。

## 8.6 微生物电化学体系和微生物电催化剂的表征

微生物生物电催化剂表征的方法和技术的发展是最活跃和重要的领域之一。可以分为两种不同类型的方法,即电化学方法(见8.5.1节)和生物方法(见8.5.2节)。前者通常专注于电子转移的识别和表征,后者致力于生物电催化剂的识别、分析、空间分布以及所处的环境。

### 8.6.1 电化学方法

#### 8.6.1.1 极化曲线

极化曲线是最为普遍的电化学分析技术,可以对单个电极或整个生物电化学体系进行极化曲线分析。它是用来测定微生物燃料电池最大功率的流行手段。图8.6所示为微生物燃料电池电势的典型极化曲线和推导功率曲线。从能量的角度分析,极化曲线和功率曲线能提供一个特定的生物电化学体系相关的重要限制和瓶颈因素等信息,如电池整体的内阻抗值及最大功率所在的具体位置。如图8.2单个电极的极化曲线或许能提供更进一步的关于电化学催化和电势瓶颈的信息。当记录极化曲线时采用合适的扫描速率避免功率或电流过大是非常重要的。

#### 8.6.1.2 伏安法

为了分析生物电催化时微生物的具体电子转移机理,伏安测试方法被应用。伏安测试法有好几种形式(如方波伏安法),但主要的是循环伏安法[69-71]。图8.7所示为地杆菌[72]以醋酸作为电子源的阳极不旋转和旋转循环伏安法。

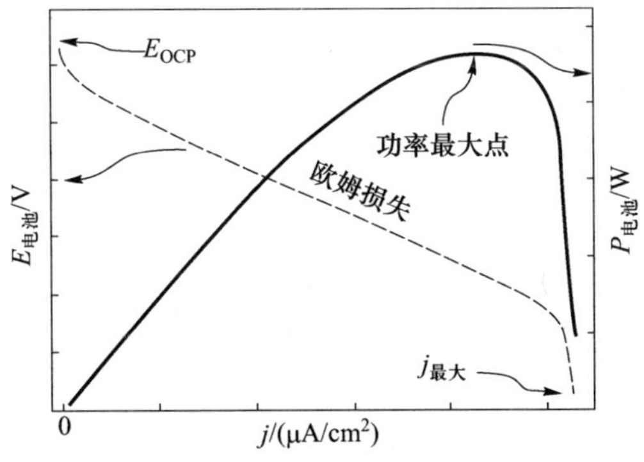


图 8.6　燃料电池极化和功率曲线

图 8.7(a)为基质耗尽且不旋转时阳极生物膜的循环伏安图。从图可以清楚看见4个氧化还原电对,代表可能的电子转移位点,它们的表观电位在图上已分别标出并命名为 $E_1^f \sim E_4^f$。相对比的是基质以醋酸作为给电子供体在旋转条件下的CV图,图显示为典型的电催化S型图(图8.7(b))。对图谱进行一阶微分可以清楚看到 $E_2^f$ 和 $E_3^f$ 是和微生物的电子转移有关的,而 $E_1^f$ 和 $E_4^f$ 与生物电催化活性无关。结果表明微生物从醋酸得到电子并转移给阳极的电势为 -0.350V(相对于Ag/AgCl电极,相对于SHE电极为 -0.150V)。需要对电化学数据进行深入的动力学和热力学分析,包括速率常数等,为了解释具有高复杂度的膜表征,分子或酶以传统模式的电化学应用是有困难的,需要新的模型和程序(见文献[73]关于动力学)。

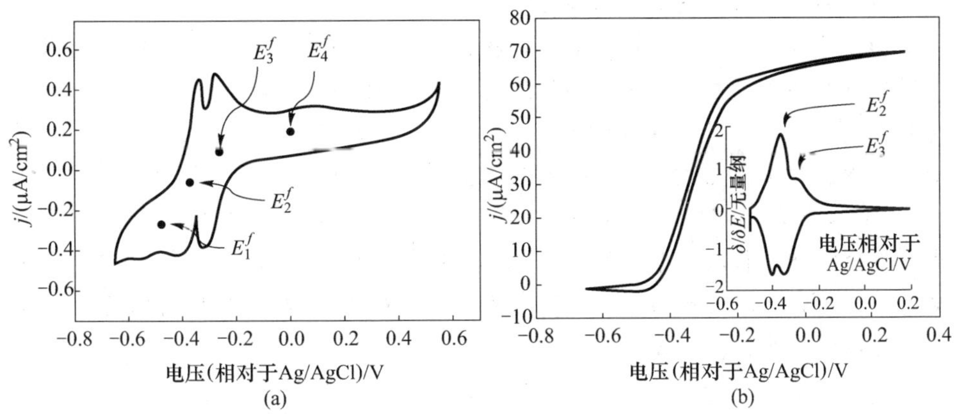


图 8.7　(a)地杆菌生物膜在基质耗尽时,在石墨杆电极上相对于Ag/AgCl电极0.2V时的循环伏安曲线(不旋转),$E_1^f$ 到 $E_4^f$ 指示生物膜被检测到的4个氧化还原电对;(b)相同的生物膜在提供醋酸基质下的循环伏安曲线(旋转),插图为循环伏安曲线的一阶导数图,可以看到 $E_2^f$ 和 $E_3^f$ 与微生物的电子转移有关(所有的扫描速率为1mV/s)(数据依据文献[69])

#### 8.6.1.3 光谱电化学分析和其他分析技术

如前面所述，动态电化学分析如循环伏安法可以用来分析反应的热力学，截至目前，以它来对生物电催化电极的结构动力学进行分析仍很有限。通过纯粹的电化学方法来鉴定氧化还原过程产生的物质是不可能的。尽管光谱分析技术能鉴定从溶液中纯化的介质，但目前低浓度的介质溶液，甚至对直接电子转移起作用的蛋白质的鉴定仍不可行。结合光谱分析和电化学分析解释光谱电化学能提供我们想要的分析信息。首先，光谱电化学实验是电活性微生物地杆菌悬浊液电池的衰减全反射表面增强红外光谱(ATR-SEIRAS)测试[74]和紫外可见光谱测试[75]，两者能对氧化还原活性蛋白质细胞色素进行常规鉴定。使用表面增强共振拉曼散射(SERRS)对活性阳极膜进行基本光谱电化学研究，能准确鉴定催化中心的膜转移蛋白，发现地杆菌生物膜是由 6 个低旋转的亚铁红素组成并以两个组氨酸作为轴向配体[76]。

进一步的被用于整体生物电化学体系和单个电极研究的技术包括电化学阻抗谱[77]、间歇电流法[78]和塔菲尔法[79]等。

### 8.6.2 生物学方法

多年来，大量的微生物分子技术被广泛应用到阳极生物膜的研究中。但由于各种方法的发展速度的差别性，目前还没有全面系统的综述介绍，因此下面将对最近的研究结果作简单介绍和说明。

对于电极表面的微生物群落的分析，许多方法的存在和方法的使用均取决于问题的所在。高通量测序可为基础的方法提供最佳的信息水平，这是合理的。然而，这会引起测量成本增加的问题。一旦群落的数据被收集，可以比简单的成分计算多样性关系参数推导出更多的信息，如文献[80]所述。需要注意的是基于 DNA 的分析不能提供活性相关信息，典型的 16SrDNA 对群体进行分析不能得到新陈代谢的有效信息，特定单种微生物或混合培养物的生物电化学催化电势也不能得到。结合转录组学和蛋白质组学，纳米二次离子质谱分析法利用缺失突变体或原位成像可能在未来被应用。

需要强调的关键问题是生物膜在电极上是几微米厚的三维结构，研究中需保持这种结构状态。典型的分子分析要包括生物膜组成和活性，生物膜在不同条件下会分层[81]，在混合种群系统提供不同种类的有机体膜的层化会很明显，或者在纯培养体系中通过改变转录组和蛋白质组。关于微生物种群分析，共焦激光扫描显微镜(CLSM)结合荧光原位杂交(FISH)是群落分析的重要组成部分[82]，而纯培养物中特定基因的原位成像或生物膜二次抽样分析对理解膜层化有帮助。

在这些方法的范畴,免疫胶体金标记在处理电子转移上有重要价值。例如,它确定了细胞色素能和"纳米线"建立联系[17],在地杆菌生物膜上细胞色素释放并累积在电极表面附近。在更广泛的背景下,如今常见的重要的成像技术如SEM和TEM在理解具有电活性的生物膜体系的结构上不能忽视。所需的样品制备过程同样需要特别小心,因为原生的生物膜状态可能被破坏。

## 8.7 本章小结

生物电化学体系的领域在迅速扩大,导致在生物能源、生物制品、生物降解和生物传感等方面有广泛的应用。由于它和多学科相关的性质,新的发展需要从各学科整合技术,同时为每个特定领域的数据报告提供必要的技术。下面是进一步发展的关键瓶颈。

(1) 更好地理解和设计电池电极的界面;

(2) 在保持生物膜电活性的条件下研究新型高效廉价的材料(电极、膜等);

(3) 建立适当的工艺参数和特征(如反应器的几何图形)实现生产;

(4) 研究和设计微生物电催化剂使它具有多功能性、稳定性、高效性。

这章不仅介绍了应用,也让读者对微生物在自然状态和实验状态下的功能有更好的理解。

## 致谢

衷心感谢德国博士后奖学金和澳大利亚研究奖学金的支持;感谢澳大利亚研究理事会,昆士兰大学微生物合成中心以及基金(BOF10/MRP/005)的支持。

## 参考文献

[1] Cracknell, J. A., Vincent, K. A., and Armstrong, F. A. (2008) Enzymes as working or inspirational electrocatalysts for fuel cells and electrolysis. *Chemical Reviews*, **108**, 2439 - 2461.

[2] Moehlenbrock, M. J., Toby, T. K., Waheed, A., and Minteer, S. D. (2010) Metabolon catalyzed pyruvate/air biofuel cells. *Journal of the American Chemical Society*, **132**, 6288 - 6289.

[3] Arechderra, R. and Minteer, S. D. (2008) Organelle-based biofuel cells: immobilized mitochondria on carbon paper electrodes. *Electrochimica Acta*, **53**, 6698 - 6703.

[4] Madigan, M. T., Martink, J. M., and Parker, J. (1999) *Brock Biology of Microorganisms*, Prentice Hall International, Inc.

[5] Katz, E., Shipway, A. N., and Willner, I. (2003) Biochemical fuel cells, in *Handbook of Fuel Cells: Fundamentals, Technology and Applications* (eds W. Vielstich, H. A. Gasteiger, and A. Lamm), John Wiley & Sons, Inc., Hoboken.

[6] Schröder, U. and Harnisch, F. (2010) Electrochemical losses defining BES performance, in *Bioelectrochemical Systems: From Extracellular Electron Transfer to Biotechnological Application* (eds K. Rabaey, L. Angenent, U. Schröder, and J. Keller), IWA Publishing, London.

[7] Harnisch, F. and Schrøder, U. (2010) From MFC to MXC: chemical and biological cathodes and their potential for microbial bioelectrochemical systems. *Chemical Society Reviews*, **39**, 4433 – 4448.

[8] Rosenbaum, M., Aulenta, F., Villano, M., and Angenent, L. T. (2011) Cathodes as electron donors for microbial metabolism: Which extracellular electron transfer mechanisms are involved? *Bioresource Technology*, **102** (1), 324 – 333.

[9] Potter, M. C. (1912) Electrical effects accompanying the decomposition of organic compounds. *Proceedings of the Royal Society B: Biological Sciences*, **84**, 260 – 276.

[10] Park, D. H. and Zeikus, J. G. (2000) Electricity generation in microbial fuel cell using neutral red as an electronophore. *Applied and Environmental Microbiology*, **66** (4), 1292 – 1297.

[11] Kim, B. H., Park, D. H., Shin, P. K., Chang, I. S., and Kim, H. J. (1999) Mediator-less biofuel cell.

[12] Lovley, D. R. (2008) The microbe electric: conversion of organic matter to electricity. *Current Opinion in Biotechnology*, **19**, 1 – 8.

[13] Lovley, D. R. (2006) Bug juice: harvesting electricity with microorganisms. *Nature Reviews Microbiology*, **4**, 497 – 508.

[14] Hartshorne, R. S., Reardon, C. L., Ross, D., Nuester, J., Clarke, T. A., Gates, A. J., Mills, P. C., Fredrickson, J. K., Zachara, J. M., Shi, L., Beliaev, A. S., Marshall, M. J., Tien, M., Brantley, S., Butt, J. N., and Richardson, D. J. (2009) Characterization of an electron conduit between bacteria and the extracellular environment. *Proceedings of the National Academy of Sciences of the United States of America*, **106** (52), 22169 – 22174.

[15] Chaudhuri, S. K. and Lovley, D. R. (2003) Electricity generation by direct oxidation of glucose in mediatorless microbial fuel cells. *Nature Biotechnology*, **21** (10), 1229 – 1232.

[16] Reguera, G., McCarthy, K. D., Mehta, T., Nicoll, J. S., Tuominen, M. T., and Lovley, D. R. (2005) Extracellular electron transfer via microbial nanowires. *Nature*, **435**, 1098 – 1101.

[17] Gorby, Y. A., Yanina, S., McLean, J. S., Rosso, K. M., Moyles, D., Dohnalkova, A., Beveridge, T. J., Chang, I. S., Kim, B. H., Kim, K. S., Culley, D. E., Reed, S. B., Romine, M. F., Saffarini, D. A., Hill, E. A., Shi, L., Elias, D. A., Kennedy, D. W., Pinchuk, G., Watanabe, K., Ishii, S. I., Logan, B., and Nealson, K. H., and Fredrickson, J. K. (2006) Electrically conductive bacterial nanowires produced by *Shewanella oneidensis strain MR*-1 *and other microorganisms. Proceedings of the National Academy of Sciences of the United States of America*, **103**, 11358 – 11363.

[18] Malvankar, N. S., Vargas, M., Nevin, K. P., Franks, A. E., Leang, C., Kim, B. -C., Inoue, K., Mester, T., Covalla, S. F., Johnson, J. P., Rotello, V. M., and Tuominen, M. T., and Lovley, D. R. (2011) Tunable metallic-like conductivity in microbial nanowire networks. *Nature Nanotechnology*, **6** (9), 573 – 579.

[19] El-Naggar, M. Y., Wanger, G., Leung, K. M., Yuzvinskya, T. D., Southame, G., Yang, J., Lau, W. M., Nealson, K. H., and Gorby, Y. A. (2010) Electrical transport along bacterial nanowires from *Shewanella oneidensis* MR-1. *Proceedings of the National Academy of Sciences of the United States of America*, **107** (42), 18127 – 18131.

[20] Chen, S., Hou, H., Harnisch, F., Patil, S. A., Carmona-Martinez, A. A., Agarwal, S., Zhang, Y., Sinha-Ray, S., Yarin, A. L., Greiner, A., and Schröder, U. (2011) Electrospun and solution blown

threedimensional carbon fiber nonwovens for application as electrodes in microbial fuel cells. *Energy & Environmental Science*, **4**, 1417 – 1421.

[21] Cao, B. E. A. (2011) Extracellular polymeric substances from Shewanella sp. HRCH-1 biofilms: characterization by infrared spectroscopy and proteomics. *Environmental Microbiology*, **13** (4), 1018 – 1031.

[22] Nevin, K. P., Kim, B. -C., Glaven, R. H., Johnson, J. P., Woodard, T. L., Methé, B. A., DiDonato, R. J. J., Covalla, S. F., Franks, A. E., Liu, A., and Lovley, D. R. (2009) Anode biofilm transcriptomics reveals outer surface components essential for high density current production in *Geobacter sulfurreducens fuel cells*. *PLoS One*, **4** (5), e5628.

[23] Mehta, T., Coppi, M. V., Childers, S. E., and Lovley, D. R. (2005) Outer membrane *c*-type cytochromes required for Fe(III) and Mn(IV) oxide reduction in *Geobacter sulfurreducens*. *Applied and Environmental Microbiology*, **71** (12), 8634 – 8641.

[24] Harris, H. W., El-Naggar, M. Y., Bretschger, O., Ward, M. J., Romine, M. F., Obraztsova, A. Y., and Nealson, K. H. (2010) Electrokinesis is a microbial behavior that requires extracellular electron transport. *Proceedings of the National Academy of Sciences of the United States of America*, **107** (1), 326 – 331.

[25] Magnuson, T. S. (2011) How the xap locus put electrical "Zap" in Geobacter sulfurreducens biofilms. *Journal of Bacteriology*, **193** (5), 1021 – 1022.

[26] Rabaey, K. and Verstraete, W. (2005) Microbial fuel cells: novel biotechnology for electricity generation. *Trends in Biotechnology*, **23** (6), 291 – 298.

[27] Schröder, U. (2007) Anodic electron transfer mechanisms in microbial fuel cells and their energy efficiency. *Physical Chemistry Chemical Physics*, **9**, 2619 – 2629.

[28] Rabaey, K., Boon, N., Verstraete, W., and Höfte, M. (2005) Microbial phenazine production enhances electron transfer in biofuel cells. *Environmental Science & Technology*, **39** (9), 3401 – 3408.

[29] Hernandez, M. E., Kappler, A., and Newman, D. K. (2004) Phenazines and other redoxactive antibiotics promote microbial mineral reduction. *Applied and Environmental Microbiology*, **70** (2), 921 – 928.

[30] Pham, T. H., Boon, N., Aelterman, P., Clauwaert, P., De Schamphelaire, L., Vanhaecke, L., De Maeyer, K., Höfte, M., and Verstraete, W., and Rabaey, K. (2008) Metabolites produced by Pseudomonas sp. enable Gram positive bacterium to achieve extracellular electron transfer. *Applied Microbiology and Biotechnology*, **77**, 1119 – 1129.

[31] Venkataraman, A., Rosenbaum, M., Arends, J. B. A., Halitschke, R., and Angenent, L. T. (2010) Quorum sensing regulates electric current generation of Pseudomonas aeruginosa PA14 in bioelectrochemical systems. *Electrochemistry Communications*, **12** (3), 459 – 462.

[32] Benz, M., Schink, B., and Brune, A. (1998) Humic acid reduction by Propionibacterium freudenreichii and other fermenting bacteria. *Applied and Environmental Microbiology*, **64** (11), 4507 – 4512.

[33] Harnisch, F., Schröder, U., Quaas, M., and Scholz, F. (2008) Electrocatalytic and corrosion behaviour of tungsten carbide in pH neutral electrolytes. *Applied Catalysis B: Environmental*, **87**, 63 – 69.

[34] Rosenbaum, M., Zhao, F., Quaas, M., Wulff, H., Schröder, U., and Scholz, F. (2007) Evaluation of catalytic properties of tungsten carbide for the anode of microbial fuel cells. *Applied Catalysis B: Environmental*, **74**, 262 – 270.

[35] Rosenbaum, M., Zhao, F., Schröder, U., and Scholz, F. (2006) Interfacing electrocatalysis and biocatalysis using tungsten carbide: a high performance noble-metal-free microbial fuel cell. *Angewandte Chemie: International Edition*, **45**, 6658 – 6661.

[36] Rabaey, K., van den Sompel, K., Maignien, L., Boon, N., Aelterman, P., Clauwaert, P., de Schamphelaire, L., Pham, H. T., Vermuelen, J., Verhaege, M., and Lens, P., and Verstraete, W. (2006)

Microbial fuel cells for sulfide removal. *Environmental Science & Technology*, **40**, 5218 – 5224.

[37] Zhao, F. , Rahunen, N. , Varcoe, J. R. , Roberts, A. J. , Avignone-Rossa, C. , Thumser, A. E. , and Slade, R. C. T. (2009) Factors affecting the performance of microbial fuel cells for sulfur pollutants removal. *Biosensors & Bioelectromagnetics*, **24** (7), 1931 – 1936.

[38] Zhao, F. , Rahunen, N. , Varcoe, J. R. , Chandra, A. , Avignone-Rossa, C. , Thumser, A. E. , and Slade, R. C. T. (2008) Activated carbon cloth as anode for sulfate removal in a microbial fuel cell. *Environmental Science & Technology*, **42** (13), 4971 – 4976.

[39] Dutta, P. K. , Keller, J. , Yuan, Z. G. , Rozendal, R. A. , and Rabaey, K. (2009) Role of sulfur during acetate oxidation in biological anodes. *Environmental Science & Technology*, **43**, 3839 – 3845.

[40] Potter, M. C. (1931) Measurement of the electricity liberated during the downgrade reactions of organic compounds. *Nature*, **127** (3206), 554 – 555.

[41] Bean, R. C. , Inami, Y. H. , Basford, P. R. , Boyer, M. H. , Shepherd, W. C. , Walwick, E. R. , and Kay, R. E. (1964) Study of the fundamental principles of bioelectrochemistry. NASA Final Technical Report (NASA N64 – 25906), p. 107.

[42] Canfield, J. H. , Goldner, B. H. , and Lutwack, R. (1963) Utilization of human wastes as electrochemical fuels. NASA Technical Report (N66 23679), p. 63.

[43] Wilkinson, S. (2000) "Gastrobots"-benefits and challenges of microbial fuel cells in food powered robot. *Autonomous Robots*, **9**, 99 – 111.

[44] Kim, B. , Chang, I. , Hyun, M. , Kim, H. , and Park, H. (2001) An electrochemical method for enrichment of microorganism, a biosensor for analysing organic substance and BOD.

[45] Rabaey, K. , Boon, N. , Siciliano, S. D. , Verhaege, M. , and Verstraete, W. (2004) Biofuel cells select for microbial consortia that self-mediate electron transfer. *Applied and Environmental Microbiology*, **70** (9), 5373 – 5382.

[46] Rabaey, K. , Angenent, L. , Schrøder, U. , and Keller, J. (eds) (2010) *Bioelectrochemical Systems: From Extracellular Electron Transfer to Biotechnological Application*, IWA Publishing, London.

[47] Emde, R. and Schink, B. (1990) Enhanced propionate formation by Propionibacterium freudenreichii subsp. freudenreichii in a 3-electrode amperometric culture system. *Appliedand Environmental Microbiology*, **56**, 2771 – 2776.

[48] Hongo, M. and Iwahara, M. (1979) Electrochemical studies on fermentation. 1. Application of electro-energizing method to L-glutamic acid fermentation. *Agricultural and Biological Chemistry*, **43**, 2075 – 2081.

[49] Rozendal, R. and Buisman, C. J. N. (2005) Process for producing hydrogen. Patent WO2005005981.

[50] Aulenta, F. , Canosa, A. , De Roma, L. , Reale, P. , Panero, S. , and Rossetti, S. , and Majone, M. (2009) Influence of mediator immobilization on the electrochemically assisted microbial dechlorination of trichloroethene (TCE) and *cis*-dichloroethene (*cis* – DCE). *Journal of Chemical Technology and Biotechnology (Oxford, Oxfordshire*: 1986), **84**, 864 – 870.

[51] Cao, X. , Huang, X. , Liang, P. , Xiao, K. , Zhou, Y. , Zhang, X. , and Logan, B. E. (2009) A new method for water desalination using microbial desalination cells. *Environmental Science & Technology*, **43** (18), 7148 – 7152.

[52] Rosenbaum, M. , He, Z. , and Angenent, L. (2010) Light energy to bioelectricity: photosynthetic microbial fuel cells. *Current Opinion in Biotechnology*, **21**, 259 – 264.

[53] Rabaey, K. , Rodríguez, J. , Blackall, L. , Keller, J. , Batstone, D. , Verstraete, W. , and Nealson, K. H. (2007) Microbial ecology meets electrochemistry: electricity driven and driving communities. *ISME Journal*, **1** (1), 9 – 18.

[54] Rabaey, K. and Rozendal, R. A. (2010) Microbial electrosynthesis: revisiting the electrical route for microbial production. *Nature Reviews Microbiology*, **8**, 706 – 716.

[55] Liu, Y., Harnisch, F., Schröder, U., Fricke, K., Climent, V., and Feliu, J. M. (2010) The study of electrochemically active mixed culture microbial biofilms on different carbon-based anode materials. *Biosensors & Bioelectronics*, **25**, 2167 – 2171.

[56] Katuri, K. P., Kavanagh, P., Rengaraj, S., and Leech, D. (2010) Geobacter sulfurreducens biofilms developed under different growth conditions on glassy carbon electrodes: insights using cyclic voltammetry. *Chemical Communications*, **46**, 4758 – 4760.

[57] Logan, B. E., Cheng, S., Watson, V., and Estadt, G. (2007) Graphite fiber brush anodes for increased power production in air-cathode microbial fuel cells. *Environmental Science & Technology*, **41**, 3341 – 3346.

[58] Zhao, Y., Watanabe, K., Nakamura, R., Mori, S., Liu, H., Ishii, K., and Hashimoto, K. (2010) Three-dimensional conductive nanowire networks formaximizing anode performance in microbial fuel cells. *Chemistry: A European Journal*, **16**, 4982 – 4985.

[59] Rabaey, K., Clauwaert, P., Aelterman, P., and Verstraete, W. (2005) Tubular microbial fuel cells for efficient electricity generation. *Environmental Science & Technology*, **39** (20), 8077 – 8082.

[60] Foley, J. M., Rozendal, R. A., Hertle, C. K., Lant, P. A., and Rabaey, K. (2010) Life cycle assessment of high-rate anaerobic treatment, microbial fuel cells, and microbial electrolysis cells. *Environmental Science & Technology*, **44** (9), 3629 – 3637.

[61] Logan, B. E., Hamelers, B., Rozendal, R., Schröder, U., Keller, J., Freguia, S., Aelterman, P., Verstraete, W., and Rabaey, K. (2006) Microbial fuel cells: methodology and technology. *Environmental Science & Technology*, **40** (17), 5181 – 5191.

[62] Rozendal, R. A., Hamelers, H. V. M., Rabaey, K., Keller, J., and Buisman, C. J. N. (2008) Towards practical implementation of bioelectrochemical wastewater treatment. *Trends in Biotechnology*, **26** (8), 450 – 459.

[63] Tender, L. M., Reimers, C. E., Stecher, H. A., Holmes, D. E., Bond, D. R., and Lovley, D. R. (2002) Harnessing microbially generated power on the seafloor. *Nature Biotechnology*, **20**, 821 – 825.

[64] Logan, B. E., Call, D., Cheng, S., Hamelers, H. V. M., Sleutels, T. H. J. A., Jeremiasse, A. W., and Rozendal, R. A. (2008) Microbial electrolysis cells for high yield hydrogen gas production from organic matter. *Environmental Science & Technology*, **42**, 8630 – 8640.

[65] Rabaey, K., Bützer, S., Brown, S., Keller, J., and Rozendal, R. A. (2010) High current generation coupled to caustic production using a lamellar bioelectrochemical system. *Environmental Science & Technology*, **44** (11), 4315 – 4321.

[66] Rozendal, R., Leone, E., Keller, J., and Rabaey, K. (2009) Efficient hydrogen peroxide generation from organic matter in a bioelectrochemical system. *Electrochemistry Communications*, **11** (9), 1752 – 1755.

[67] Mu, Y., Rozendal, R. A., Rabaey, K., and Keller, J. (2009) Nitrobenzene removal in bioelectrochemical systems. *Environmental Science & Technology*, **43**, 8690 – 8695.

[68] Gregory, K. B. and Lovley, D. R. (2005) Remediation and recovery of uranium from contaminated subsurfaceenvironments with electrodes. *Environmental Science & Technology*, **39**, 8943 – 8947.

[69] Fricke, K., Harnisch, F., and Schröder, U. (2008) On the use of cyclic voltammetry for the study of anodic electron transfer in microbial fuel cells. *Environmental Science & Technology*, **1** (1), 144 – 147.

[70] Marsili, E., Rollefson, J. B., Baron, D. B., and Hozalski, R. M., and Bond, D. R. (2008) Microbial

biofilm voltammetry: direct electrochemical characterization of catalytic electrode-attached biofilm. *Applied and Environmental Microbiology*, **74** (23), 7329 – 7337.

[71] Srikanth, S., Marsili, E., Flickinger, M., and Bond, D. R. (2007) Electrochemical characterization of Geobacter sulfurreducens cells immobilized on graphite paper electrodes. *Biotechnology and Bioengineering*, **99** (5), 1065 – 1073.

[72] Harnisch, F., Koch, C., Patil, S. A., Hübschmann, T., Müller, S., and Schröder, U. (2011) Revealing the electrochemically driven selection in natural community derived microbial biofilms using flowcytometry. *Energy & Environmental Science*, **4** (4), 1265 – 1267.

[73] Torres, C. I., Marcus, A. K., Lee, H.-S., Parameswaran, P., Krajmalnik-Brown, R., and Rittmann, B. E. (2010) A kinetic perspective on extracellular electron transfer by anode-respiring bacteria. *FEMS Microbiology Reviews*, **34**, 3 – 17.

[74] Busalmen, J. P., Esteve-Nunez, A., Bernú, A., and Feliu, J. M. (2008) C-type cytochromes wire electricity-producing bacteria to electrodes. *Angewandte Chemie*, **47**, 4874 – 4877.

[75] Nakamura, R., Ishii, K., and Hashimoto, K. (2009) Electronic absorption spectra and redox properties of C type cytochromes in living microbes. *Angewandte Chemie*, **121**, 1634 – 1636.

[76] Millo, D., Harnisch, F., Patil, S. A., Ly, K. H., Schröder, U., and Hildebrandt, P. (2011) In situ spectroelectrochemcial investigation of electrocatalytic microbial biofilms by surface-enhanced resonance Raman spectroscopy. *Angewandte Chemie: International Edition*, **50**, 2625 – 2627.

[77] He, Z. and Mansfeld, F. (2009) Exploring the use of electrochemical impedance spectroscopy in microbial fuel cell studies. *Energy & Environmental Science*, **2**, 215 – 219.

[78] Schröder, U. and Harnisch, F. (2009) Electrochemical losses defining BES performance, in *Bioelectrochemical Systems: From Extracellular Electron Transfer to Biotechnological Application* (eds K. Rabaey, L. Angenent, U. Schröder, and J. Keller), International Water Association.

[79] Lowy, D. A. (2010) Importance of Tafel plots in the investigation of bioelectrochemical systems, *in Bioelectrochemical Systems: From Extracellular Electron Transfer to Biotechnological Application* (eds K. abaey, L. Angenent, U. Schröder, and J. Keller), IWA.

[80] Marzorati, M., Wittebolle, L., Boon, N., Daffonchio, D., and Verstraete, W. (2008) How to get more out of molecular fingerprints: practical tools for microbial ecology. *Environmental Microbiology Reports*, **10**, 1571 – 1581.

[81] Franks, A. E., Nevin, K. P., Glaven, R., and Lovely, D. R. (2010) A novel approach for spatial analysis of global gene expression within a *Geobacter sulfurreducens currentproducing biofilm. The ISME Journal*, **4** (4), 509 – 519.

[82] Read, S. T., Dutta, P., Bond, P. L., Keller, J., and Rabaey, K. (2010) Initial development and structure of biofilms on microbial fuel cell anodes. *BMC Microbiology*, **10**, 98.

[83] Franks, A. E., Nevin, K. P., Glaven, R. H., and Lovley, D. L. (2010) Microtoming coupled to microarray analysis to evaluate the spatial metabolic status of Geobacter sulfurreducens biofilms. *ISME Journal*, **4**, 509 – 519.

[84] Inoue, K., Leang, C., Franks, A. E., Woodard, T. L., Nevin, K. P., and Lovley, D. R. (2010) Specific localization of the ctype cytochrome OmcZ at the anode surface in current-producing biofilms of *Geobacter sulfurreducens. Environmental Microbiology Reports*, **3** (2), 211 – 217.

# 第9章

# 微流体燃料电池材料

Seyed Ali Mousavi Shaegh, Nam – Trung Nguyen

## 9.1 引　言

便携式电子设备性能的增加和运行时间不断延长的需求，促进了高能量密度能源的研究与开发。许多研究仅仅专注于电池功率密度，但是电池工艺技术的最新进展并不能满足能源与能源消费者之间的能量差距，并且这种能量差距在接下来将变得更大[1]。

此外，无线网络传感器的出现，不仅可以应用于生物、环保、军事和安检，也为长时间运行的可靠性电源打开了新的市场。并且，实现/开发新设备，如小型无人机或者是智能昆虫机器人和智能窃听器[2]都依赖小型电源的存在。

金属氢化物如硼氢化钠，与甲醇和大多数碳氢燃料的能量密度均比商业的电池技术能量密度更高[3]。微型燃料电池可以认为是一种合适的能源应用设备。微型燃料电池可以应用于可再充电电池的混合系统以提高整个系统的灵活性和可靠性。一个微型燃料电池通常由一个发动机、辅助系统、燃料箱和助燃剂容器组成。若燃料体积和发动机容量的比值达到最大化，燃料在辅助设备上的功耗或者助燃剂运送和发动机功率的校正也都明显减少的话，这些元件的性能会比电池更优越[3]。

与气态氢相比，液体燃料电池具有高能量密度和高安全性等特点。目前，研究热门的燃料电池种类包括直接甲醇燃料电池（DMFC）[4]、直接甲酸燃料电池[5]、通过从金属氢化物中置换氢来运行的质子交换膜燃料电池（PEMFC）[6]和无膜微流体燃料电池[7]。

常规质子交换膜燃料电池小型化到微型质子交换膜燃料电池时，必须考虑到它的设计和操作方面的注意事项。质子交换膜由全氟磺酸溶液制成，作为质子交换膜燃料电池的核心部分，它需要在特定的条件下，为质子导电性提供足量的水。电池运行的动态条件进一步增加了燃料电池发动机和辅助系统的复杂性。另外,由于全氟磺酸膜纳米孔结构对水的吸收和保留导致了膜的膨胀和收缩,这可能给燃料电池的组装带来问题。此外,目前微型燃料电池的制造技术是用于制备凹槽或电极(如在有机玻璃上激光加工[8],硅刻蚀[9])的精细加工技术和制备膜电极组件(MEA)的传统方法和材料的组合。因此,在所需材料和制造工艺以及微型燃料电池的特征尺寸和现有技术之间并不匹配。为了解决这种不匹配,进行了许多研究,通过引入新材料,制造一个整体式的燃料电池。例如在介孔硅中形成质子传导膜,且与精细加工技术相兼容[1,10,11]。

与基于微机电系统(MEMS)方法的小型化传统电源类似,微/纳流控技术可为能量转换系统提供新方法,并且能够充分利用缩减流体系统所出现的特定现象。

由于对流质量传递在层流的情况下不稳定,在低雷诺数条件下,许多不同物质不同浓度的物质流可并排通过微孔道。根据不同的佩克莱特数及对流扩散的相对重要性显示,液体流可以单独通过孔道。分散混合的两种物质流通过液-液界面会导致浓度梯度。利用控制微流体接口这种特性,许多应用如萃取、分子分离[13-15]、精细加工以及在微通孔道内的精细加工[16]和微光流体透镜等都已实现[17]。

费里诺等[18]提出了基于两种物质流在微孔道里分层的无膜燃料电池的概念。由图9.1可知,氧化剂和燃料的两种物质流被引入具有集成的微孔道电极作为电化学反应和集电活动的有效区域。阳极和阴极电解液都支持液态电解质,以便于整个通道的离子传导。来自氧化燃料或减少的氧化剂的离子通过浓度梯度迁移穿过孔道,同时电子通过外电路到达阴极电极一侧。

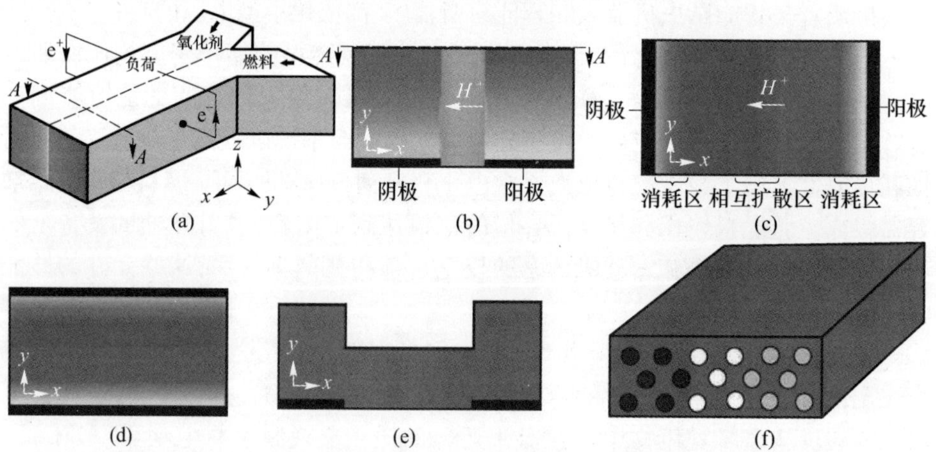


图9.1 溢出设计图。(a)具有两种物质流的LFFC通过Y形孔道的原理图;(b)两电极都在底部的两物流平行穿过通道的截面原理图;(c)垂直于侧壁电极的液-液界面区,阳极、阴极损耗边界及相互扩散区域的截面图;(d)上下电极组通道的截面图;(e)两电极都在通道凹槽底部的截面图;(f)石墨棒为电极的通道截面图[19](版权2011,Elsevier)

因为燃料和氧化剂可以平行通过孔道,作为两股物质流分离器的膜已经没有存在的必要性[18]。两物流的相互扩散区域被限制在通道中心的界面宽度。为了避免燃料和氧化剂的交叉影响,电极对之间的距离应优化到同时保持通过孔道的欧姆损耗和泵激能量的最小值。

在本章中,对无膜层流为基础的燃料电池(LLFC)的操作的基本原理首次进行详解。并且,就设计和开发制造无膜 LFFC 技术,电极流体结构的效应及其在电池性能上的安排进行讨论。随后,详细介绍有关无膜 LLFC 的燃料、氧化剂和电解质。最后,将对材料的限制条件和选择性进行一些讨论。

研究证明,无膜 LLFC 能够为生物燃料电池提供一个合适的平台[20]。生物燃料电池超出了本章的讨论范围,但是它们也可以从这些技术发展中受益。为了获得更多有关无膜 LLFC 的设计要素和性能限制因素的知识,读者可以参阅一些综述类文章[7,20]。

## 9.2 基本原理

无膜 LFFC 遵循有膜燃料电池的基本电化学原理。它们最主要的区别是膜可以作为电荷传输介质和对电极的隔板,在微孔道创建一个密闭的液-液界面。通常情况下,微孔道定义为孔道特征尺寸小于 1mm 且大于 1μm[21],并且在微孔道的流体称为微流体。微流体的基本原理和应用参见文献[22]。

与比例因子 $R$ 相比,微流体的表面积与体积之比为[$(R^2/R^3)=R^{-1}$],比值随着电池的微型化而减少。微流体系统可以利用界面性能的依赖性开发一系列更广泛的应用[23]。随着流体系统尺寸的减少,在低雷诺数条件下就可以建立层流态。在这种情况下,基于表面的影响(包括表面张力或黏度)能够控制基于体积的影响,提供新的微观现象,例如,可以获得在微孔道里具有双层物质流的密闭液-液界面。

由于液体在微孔道的连续性和层状性质[21],质量守恒定律的流体流动服从连续性方程:

$$\frac{\partial\rho}{\partial t}+\nabla\cdot(\rho\boldsymbol{u})=0 \tag{9.1}$$

流体密度($\rho$)是常数,它导致了不可压缩的流体条件,$\nabla\boldsymbol{u}=0$。纳维-斯托克斯(Navier-Stokes)方程的解可以用来确定三维速度场($u$):

$$\rho\left(\frac{\partial\boldsymbol{u}}{\partial t}+\boldsymbol{u}\cdot\nabla\boldsymbol{u}\right)=-\nabla p+\mu\nabla^2\boldsymbol{u}+\boldsymbol{f} \tag{9.2}$$

式中:$\rho$ 为压力;$f$ 为单位体积的体积力。

如图 9.1(a)所示,把燃料和氧化剂两物流分别引入孔道,它们在具有液-液界面的孔道里相遇,形成并行的双层流体。该界面成为燃料和氧化剂两物流的分隔膜。集流体和表面涂有适当催化层的电极在孔道壁面即电化学反应发生的区域组装。为了使两电极间进行电荷转移,燃料和氧化剂溶液都应该具有离子导电性,可以通过在两物质流中加入支持电解质获得,支持电解质包含氢氧化物或者水合氢离子如氢氧化钾或硫酸的稀溶液。

在给定的温度和压力下,在式(9.3)给出的燃料和氧化剂方程组中,一个确定的氧化还原反应电池的理论开路电势由式(9.4)即能斯特方程决定:

$$v_A A + v_B B \leftrightarrow v_C C + v_D D \tag{9.3}$$

$$E^0(T,P) = E^0 + \frac{R_{uT}}{nF}\ln\left(\frac{a_A^{v_A} a_B^{v_B}}{a_C^{v_C} a_D^{v_D}}\right) \tag{9.4}$$

式中:$E^0$ 为标准状态下的平衡电势;$a$ 为每种物质的活度。

在水溶液中,活度用浓度来表示。由于在燃料和氧化剂两流之间没有膜,所以两个流的 pH 值可以单独修饰,以增加半电池电势。理论电池电势会因为阳极和阴极的活化损失($\eta_{a,a}$ $\eta_{a,c}$)、欧姆损耗($\eta_r$)、阳极和阴极质量运输损失($\eta_{m,a}$,$\eta_{m,c}$)以及其他损失 $\eta_x$ 而降低。

$$E_{\text{cell}} = E^0(T,P) - \eta_{a,a} - |\eta_{a,c}| - \eta_r - \eta_{m,a} - |\eta_{m,c}| - \eta_x \tag{9.5}$$

式中:$\eta_x$ 为燃料/氧化剂通过电解液到达对面电极时的交叉效应或者是电池内部短路引起的更多理论平衡开路电位偏离能斯特平衡电压。

在水溶液中的电氧化反应不活泼,如甲酸和甲醇的反应,它的活化损失一般都比质子交换膜燃料电池高并且电池电位下降得更快。

欧姆损耗主要是由电解质的离子电阻率($R_{\text{electrolyte}}$)和电极外电阻($R_{\text{external}}$)以及电池连接时的电流($i$)造成,即

$$\eta_r = i \cdot (R_{\text{electrolyte}} + R_{\text{external}}) \tag{9.6}$$

在两电极间的支持电解质的欧姆电阻主要取决于阳极到阴极间电荷运输的长度及其横截面积和离子电导率:

$$R_{\text{electrolyte}} = \frac{d}{\sigma A} \tag{9.7}$$

在相对湿度为 100% 以及标准膜厚度为 50 ~ 200μm 的室温条件下,作为质子交换膜的标准全氟磺酸溶液的透过电导率为 0.1S/cm$^2$。与此相比,作为一种常见的支持电解质 0.5mol/L 的硫酸溶液的电导率为 0.2S/cm$^2$。在无膜 LLFC 中,阳极到阴极的间距通常为 0.5 ~ 1.5mm,这导致了它比 PEM 燃料电池总的欧姆损失更高。

低浓度的氧化剂或燃料是传质损失的主要来源。由于电极过渡层补给缓慢,电池在低流速下运行产生高电流密度,而传质损失对高电流密度下的电池电势的降低至关重要。

当电化学反应发生时,燃料和氧化剂分别在对应的电极被消耗从而产生电流。电流密度分布由巴特勒 - 沃尔默方程模拟实验,根据参考浓度下给定电极

的交换电流得出($C_{i,\mathrm{ref}}$):

$$J_0 = i_0\left(\frac{C_i}{C_{i,\mathrm{ref}}}\right)^{\beta_i}\left[\exp\left(\frac{\alpha_a F_\eta}{RT}\right) - \exp\left(-\frac{\alpha_c F_\eta}{RT}\right)\right] \tag{9.8}$$

式中:$C_i$ 为物质的浓度;"$i$"指燃料或氧化剂;$\beta_i$ 为物质基本电荷转移步骤的反应级数;$\alpha_a$,$\alpha_c$ 为阳极和阴极的电荷转移系数;$R$ 为摩尔气体常数;$T$ 为工作温度;$F$ 为法拉第常数;$\eta$ 为表面超电势,有

$$\eta = \phi_s - \phi_e = E^0(T,P) \tag{9.9}$$

其中:$\phi_s$,$\phi_e$ 分别为电极和电解质的电势。

电极的物质浓度分布取决于物质的扩散对流运输,并且可以通过质量守恒方程计算出来:

$$\nabla\cdot(D_i\ \nabla C_i + C_i\boldsymbol{u}) = S_i \tag{9.10}$$

式中:$D_i$ 为物质"$i$"的扩散系数;$S_i$ 为物质"$i$"在电化学反应中阳极与阴极的净化学反应速率,也可表示每立方米消耗物质的速率:

$$S_i = \frac{J_0}{nF} \tag{9.11}$$

增加燃料/氧化剂的流量以加速电极耗损区的补给,在低燃料利用率为代价下使电池电流和能量密度最大化。燃料的利用率为

$$\varepsilon_{\mathrm{fuel}} = \frac{J}{n\mathrm{F}v_{\mathrm{fuel}}} \tag{9.12}$$

式中:$v_{\mathrm{fuel}}$为燃料供应速率(mol/s)。

低燃料利用率可能会使燃料/氧化剂系统需要资源回收,这将使得燃料电池系统更复杂。

## 9.3　无膜 LFFC 的设计和使用的材料

具有支持电解质的两个流之间的界面代表一个虚拟膜,可提供离子导电介质和有效地控制反应物在细胞间的混合。总之,微流体系统的无膜 LFFC 具有以下优点:

(1) 消除膜,从而减少了电池的尺寸和提高电池设计和制造的灵活性,包括小型化[18,25]。

(2) 由于两种物质流是通过单一通道,一些设计考虑消除了燃料和氧化剂输送系统的应用,简化了密封和包装的要求。此外,燃料和氧化剂两物流的组成可以单独进行量身定制,以最大限度地提高阳极和阴极的反应动力学[26]。

(3) 不存在与膜相关的问题如水管理、膜污染和破坏等[18,27]。与质子交换膜燃料电池不同,没有聚合物膜中 Pt 的溶出。由于两物质流都包含液体电解质,电子导体、电荷传输媒介以及反应物供给通道(这三者称为三相界面)的精确建

立比质子交换膜燃料电池更简单。

(4) 无膜 LFFC 作为电源,与其他微流体系统兼容(如芯片实验室设备)。

流体架构对无膜 LFFC 的性能发挥起着重要的作用。根据使用电极的体系结构和组成,无膜 LFFC 可分为:①流式平面电极;②三维多孔导流型电极;③无膜 LFFC 与吸气式阴极。所有的设计都是利用常见的燃料和氧化剂,也可以用常规的方法达到相同的精度水平。所有的设计,碱性或酸性电解质都可以适用。

表 9.1 列出了用于制备燃料电池的材料和催化剂。为了得到可靠的燃料电池,需要考虑不同的材料和加工因素。需要注意的设计因素如下[28]:

(1) 形成通道所需的载体基质和设计的几何形状。

(2) 催化剂的结构和用沉积法负载催化剂的载体。

(3) 电极和通道结构的组装方法,形成一个液体密封的电池,且在接口处有仪表以及燃料和氧化剂的输送系统。

如图 9.1 所示,主要由一个主体微孔道组成的无膜 LFFC 可以覆盖在不透液的支撑结构中。共层流在孔道里面相遇有两种形式,即并排流和垂直分层流。并排流是垂直的燃料 - 氧化剂界面,而垂直分层流是水平的液液界面,电极放在底部(图 9.1(b)),侧壁(图 9.1(c)),以及顶部和底部(图 9.1(d))。另外,电极也可以放置在有凹槽的孔道(图 9.1(e)),以防大量的气泡或者一排电极被放在单一的孔道(图 9.1(f))。

如图 9.1 所示,设计的溢流装置使得燃料和氧化剂可以流经平面电极的表面。只有小部分的燃料和氧化剂流靠近催化层参与电化学反应。由于缺乏有效的对流传质,边界层低浓度反应物的损耗会蔓延到两电极。为了提高超流设计中燃料的利用率,改良设计过的电极应用于孔道,以增大电化学反应的有效面积。如图 9.1(f)所示,在孔道中心的石墨棒是电绝缘的[27]。

### 9.3.1 流动结构与流程设计

通常,催化剂和集流体的电极被置放于通道的两边。在一些并排共层流的无膜 LFFC 设计中,它的通道结构是由聚酯二甲基硅烷(PDMS)复制成形制得,然后再和电极一起密封到一个固体基质中[47,48]。PDMS 模型利用了基于软光刻技术的程序[51]。软光刻技术是一种图案转移技术。“软”一词指的是一种弹性印章,其表面有图案浮雕。软光刻技术对于传递微图案有两种基本技术,即微接触印刷技术和复制成形技术。复制成形技术通常用来制造微通道。这种技术始于薄层光刻胶的堆积作用,如硅片上的 SU8 胶或者利用旋转涂布的玻片。给定黏度下的光刻胶厚度由旋转涂布的旋转速度和程度决定。厚度层决定了 PDMS 的孔道深度。为了使光刻胶稳定存在于基片上,在电热板上或炉子里低温(70 ~ 90℃)加热。通道的图案应该印制在塑料还是玻璃的模具上取决于通道的最佳特征尺寸。这种光掩模与基片对齐,然后再用紫外光照射一定时间。像

表 9.1 微流体燃料电池设备的对料概览

| 设计 | 燃料/氧化剂 | 电解液 | 阳极催化剂 | 阴极催化剂 | 通道 | 集流体 |
|---|---|---|---|---|---|---|
| Ferrignod 等[18] | 钒氧化物(Ⅱ)(1M)<br>钒氧化物(Ⅴ)(1M) | 硫酸(25%) | 无 | 无 | PDMS/SU8 | 用电子束蒸镀机使金沉积在碳上 |
| Kjeang 等[27] | 钒氧化物(Ⅱ)(1~2M)<br>钒氧化物(Ⅴ)(1~2M) | 硫酸(1~2M) | 无 | 无 | PDMS/缩醛树脂 | 石墨棒作为电极 |
| Kjeang 等[29] | 钒氧化物(Ⅱ)(2M)<br>钒氧化物(Ⅴ)(2M) | 硫酸(2M) | 无 | 无 | PDMS | Toray 碳纸作为电极 |
| Kjeang 等[30] | 钒氧化物(Ⅱ)(2M)<br>钒氧化物(Ⅴ)(2M) | 硫酸(4M) | 无 | 无 | PDMS | Toray 碳纸作为溢流多孔电极集流体 |
| Tominaka 等[25] | 甲醇/空气 | 硫酸(0.5M)或硫酸钠(0.5M) | 电沉积铂-钌合金 | 电沉积钯-钴合金作为氧气电化学还原反应的选择性催化剂 | 硅胶 | 用电子蒸镀机把200nm金沉积在30nm钛上 |
| Morales-Acosta 等[31] | 甲酸(0.1M,0.5M)/溶解氧 | 硫酸 | 0.7mg/cm² 钯负载在MWCNT上或1.9mg/cm² 钯负载在VXC-72上 | 1.9mg/cm² 钯负载在VXC-72上 | 1mm PMMA | VXC-72作为集流体 |
| Jayashree 等[32] | 甲酸(1M)/空气 | 硫酸(0.5M) | 10mg/cm² 钯黑纳米颗粒涂在石墨上 | 0.35mg/cm² 铂和另外2mg/cm² 铂涂在Toray碳纸上作为吸气式阴极 | 2mm PMMA | 石墨阳极侧板 |
| Jayashree 等[33] | 甲醇(1M)/空气 | 硫酸(0.5M)或氢氧化钾(1M) | 10mg/cm² 铂-钌原子质量比为50:50的合金涂在石墨上 | 0.35mg/cm² 铂和另外2mg/cm² 铂涂在Toray碳纸上作为吸气式阴极 | 2mm PMMA | 石墨阳极侧板 |

（续）

| 设计 | 燃料/氧化剂 | 电解液 | 阳极催化剂 | 阴极催化剂 | 通道 | 集流体 |
|---|---|---|---|---|---|---|
| Brushett 等[34] | 甲酸(1M)、甲醇(1M)、乙醇(1M)、硼氢化钠(1M)、肼(3M)<br>空气 | 硫酸(0.5M) | 钯黑<br>铂/钌黑<br>铂/钌黑<br>铂黑<br>铂/碳，全部催化剂以 $10mg/cm^2$ 涂在石墨上 | $2mg/cm^2$ 铂黑涂在 Toray 碳纸上作为吸气式阴极 | 2mm PMMA | 石墨阳极侧板 |
| Whipple 等[35] | 甲醇(0.1～15M)<br>空气 | 硫酸(1M) | 在碳纸上负载 $10mg/cm^2$ 的 Pt/Ru | $2mg/cm^2$ 铂黑或 $Ru_xSe_y$ 作为甲醇催化剂涂在 Toray 碳纸上作为吸气式阴极 | 2mm PMMA | 石墨阳极侧板 |
| Hollinger 等[36] | 甲醇(0.063M，0.125M，0.25M，1M)<br>空气，氧气 | 硫酸(1M) | $10mg/cm^2$ 铂－钌原子比为 50∶50，涂在碳纸上 | $2mg/cm^2$ 铂－碳，质量比为 50∶50 的合金粉刷在 Sigracet 碳纸上作为吸气式阴极，阴极上在催化剂层上有一薄层 Nafion 溶液 | 150μm 聚酰亚胺薄膜聚碳酸酯分离器 | 石墨阳极侧板 |
| Mitrovski，Nuzzo 等[37] | 氢气<br>空气 | 硫酸(5M)或氢氧化钠(2.5M) | 电沉积铂或钯在 100nm 的铂上 | 电沉积铂或钯在 100nm 的铂上 | PDMS | 100nm 铂通过电子束蒸发沉积在 15nm 的钛上 |
| Mitrovski 等[38] | 氢气<br>氧气 | 硫酸(0.1M)或氢氧化钠(0.1M 和 1M) | 电沉积铂或钯在 100nm 的铂上 | 电沉积铂或钯在 100nm 的铂上 | PDMS | 100nm 铂通过电子束蒸发沉积在 15nm 的钛上 |

（续）

| 设计 | 燃料/氧化剂 | 电解液 | 阳极催化剂 | 阴极催化剂 | 通道 | 集流体 |
|---|---|---|---|---|---|---|
| Li 等[39] | 甲酸（0.5M）溶解氧或过氧化氢（0.01M） | 硫酸（0.1M） | 铂钌原子比率为50∶50，湿法喷涂在金上 | 铂黑湿法喷涂在金上 | PMMA | 喷溅金 |
| Mousavi Shaegh 等[40] | 甲酸（1M）<br>空气 | 硫酸（0.5M） | 铂黑 | 铂黑纳米粒子 | PMMA | Toray 碳纸 |
| Choban 等[41] | 甲酸（2.1M）<br>溶解氧 | 硫酸（0.5M） | 使用电沉积得到的铂黑 | 使用电沉积得到的铂黑 | PDMS/聚氨酯 | 在 2.8～5nm 的铬上喷溅 25～150nm 的金 |
| Choban 等[26] | 甲醇<br>溶解氧 | 硫酸或<br>氢氧化钾（1M） | 无载体铂钌质量比为 50∶50 的合金纳米粒子 | 无载体的铂黑纳米粒子 | 石墨/聚碳酸酯板 | 石墨板 |
| Choban 等[42] | 甲醇（1M）<br>溶解氧 | 硫酸（0.5M） | $2\sim4mg/cm^2$ 无载体铂钌质量比为 50∶50 的合金或无载体 Pt 纳米粒子和无载体 Ru 纳米粒子的 50∶50 混合物或无负载 Pt 纳米粒子滴涂在石墨板上 | $2\sim4mg/cm^2$<br>无载体铂纳米粒子滴涂在石墨板上 | 石墨/聚碳酸酯板 | 石墨板作为集流体 |
| Cohen 等[28] | 甲醛（0.5M）<br>溶解氧 | 硫酸（0.1M） | 把 20nm 的钯和 50nm 的铂或 50nm 的钯和 100nm 的铂负载在 200nm 的铬上再蒸发在聚酰亚胺薄膜上（卡普顿） | 把 20nm 的钯和 50nm 的铂或 50nm 的钯和 100nm 的铂负载在 200nm 的铬上再蒸发在卡普顿薄膜上 | 硅 | 和催化剂层相结合 |

（续）

| 设计 | 燃料/氧化剂 | 电解液 | 阳极催化剂 | 阴极催化剂 | 通道 | 集流体 |
|---|---|---|---|---|---|---|
| Cohen 等[43] | 溶解氢<br>溶解氧 | 氢氧化钾(0.1M)<br>或硫酸(0.1M) | 50nm 的钯和 50nm 的铂蒸发在卡普顿膜上 | 50nm 的钯和 50nm 的铂蒸发在卡普顿膜上 | 硅 | 和催化剂层相结合 |
| Lópezontesinos 等[44] | 甲酸(1M)<br>高锰酸钾<br>(60mM,144mM) | 硫酸(0.5M) | 在铬/金合金上电沉积钯 | 无 | 硅 | 铬/金 |
| Hasegawa 等[45] | 过氧化氢(0.75M)<br>过氧化氢(0.75M) | 氢氧化钠(0.75M)或<br>硫酸(0.375M) | 喷溅铂 | 喷溅铂 | 玻璃 | 和催化剂层相结合 |
| Sun 等[46] | 甲酸(2.1M)<br>高锰酸钾(0.144M) | 硫酸(0.5M) | 通过电子束蒸发<br>40nm 铂和 5nm 钛 | 通过电子束蒸发<br>40nm 铂和 5nm 钛 | PDMS | 和催化剂层相结合 |
| Kjeang 等[47] | 甲酸(1M)<br>过氧化氢(1～3M) | 磷酸盐<br>(1～3M;pH 6～8)<br>磷酸盐<br>(1～2M;pH 0～1) | 在金上电沉积钯 | 在金上电沉积钯 | PDMS | 100nm 金 |
| Kjeang 等[48] | 甲酸(1.2M)<br>次氯酸钠(0.67M) | 氢氧化钠(2.8M) | 电沉积钯在碳纸上 | 电沉积钯在碳纸上 | PDMS | Toray 碳纸 |
| Salloum 等[49] | 甲酸(0.04M)<br>高锰酸钾(0.01M) | 硫酸(0.5～1M) | 铂黑纳米粒子 | 铂黑纳米粒子 | PMMA | Toray 碳纸 |
| Sung，Choi 等[50] | 甲醇(2M)<br>过氧化氢(0.05M) | 高锰酸钾(0.2M) | 电沉积氢氧化镍<br>在 20nm 的铬和<br>300nm 的金上 | 电沉积氢氧化银<br>在 20nm 的铬和<br>300nm 的金上 | 玻璃 | 铬/金 |

SU8胶这种负电荷光刻胶,在紫外光照射下,暴露在紫外光的部分会产生交联,在之后的加热中会更加稳定。然后,基片会完全沉浸在显像剂中,没有被紫外照射的部分会被除掉,胶片中只剩下主要的部分,这部分可以用于好几个复制模具。软聚合物如PDMS,一种基体树脂的混合物和固化剂,倒入模具,然后抽真空,在置于炉子或电热板上加热几个小时。加热后,孔道结构就会从模板中脱除。孔道开放区域可以通过黏结玻璃或PDMS进行可逆或不可逆密封。以玻璃或PDMS作为密封基质,采用氧气等离子体处理孔道会导致不可逆结合。另外,氧气等离子体处理可以同时提高孔道表面的亲水性和浸润性。

PDMS的一些特性使其对制备微流体设备非常有用,包括容易制备(快速成型、密封、与用户接口)、在紫外可见光区透明、化学惰性、低极性、低电导率、有弹性以及制备成本低[52]。

由于氢气和氧气在PDMS中具有较高的溶解度和渗透性(在35℃下$D_{H_2} = 1.4 \times 10^{-4} cm^2/s, D_{H_2} = 34 \times 10^{-5} cm^2/s$)[53],气体反应物可以通过PDMS薄层提供给一对被含有硫酸或氢氧化钠作为电解液的孔道分隔的电极[37,38]。这样,运行于溶解的氢气/氧气[38]和氢气/空气[37]的电池的能量密度大约为$0.7 mW/cm^2$,原因是溶解在PDMS的氢气的渗透速率受到了限制。

石墨可以用来制作孔道的基底材料。Choban等[26]用石墨作为基底材料实现了3种功能:集流体、催化剂支撑材料和结构材料。如图9.2(a)所示,石墨平板并排放置,合理的间距决定了通道的宽度。在组装之前,催化剂浆料被涂布在石墨平板上。为了使通道密封,用1mm厚的聚碳酸酯平板和PDMS薄膜做垫片。

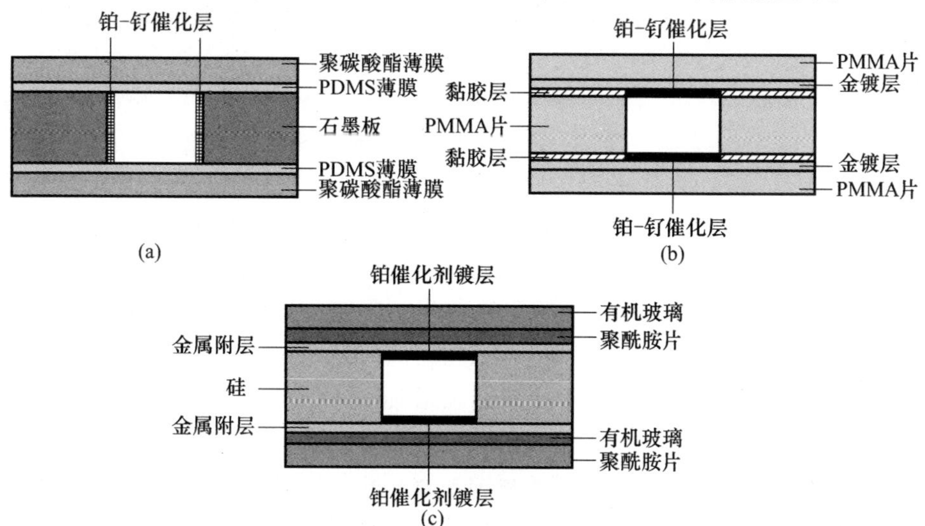


图9.2 无膜LFFC的不同层组载体基质的结构。(a)垂直电极和垂直液-液界面;(b)具有横向电极且基于聚甲基丙烯酸甲酯的无膜LFFC;(c)具有横向溢流电极且基于硅基衍生物微孔道的无膜LFFC[19](版权2011,Elsevier)

聚甲基丙烯酸甲酯(PMMA)是另一种经常用于微流控芯片和微型燃料电池的聚合物材料[4]。PMMA 是一种热塑性聚合物,通常是线性连接,在高于玻璃转化温度时软化[8]。PMMA 拥有非晶结构,在可见光谱中具有高达 92% 的透光率。这种材料同样也具有其他一些优良性质如低摩擦因数、高化学电阻和良好的电绝缘性。PMMA 所有的这些特性和性能使其成为微流体设备,尤其是与化学应用有关的设备[8]的一种良好的基底材料。

PMMA 基底可用多种方法微加工,如 X 射线照射,后加工,热压成型和激光加工[8]。对于激光加工,它的微孔道的横截面积取决激光束的形状、移动速度、激光功率和基底材料的热扩散率。激光能量呈正态分布,从而使得以 PMMA 为基底的通道横截面积的形状也成正态分布状[8]。

Li 等[39]利用红外二氧化碳激光器在 PMMA 上刻画孔道以制作无膜燃料电池。3 张 PMMA 片材用来创建通道(图 9.2(b))。使用激光在 1mm 的中间片层产生一个狭缝,并且用胶黏剂使其夹在顶部和底部的 PMMA 两模板之间。底层作为电极支撑和液封盘。在 PMMA 片层的表面涂有一层 100nm 的 Au 单质,以减少接触电阻。这一 Au 单质层相当于集流体。PMMA 片层的表面用合适的砂纸(1200 细砂)进行机械打磨以改善其表面粗糙度和 Au 单质层在基底的附着力。出口和入口的孔在 PMMA 的顶部。Pt - Ru 和 Pt 催化剂浆料分别喷涂在 Au 单质层以制得阳极和阴极电极,负载量为 4.5mg/$cm^2$。

刚性微孔道同样也是利用光刻加工法在硅胶上合成[28,43]。因为硅胶很稳定,不会受热变形,所以硅胶孔道和基底可以用来研究随温度变化的燃料电池的氧化反应性能[28]。如图 9.2(c)所示,利用电子束蒸镀技术将 Pt 镀在金属粘附层上。作为集流体的沉积金属薄膜的电导率取决于沉积的厚度、工艺和沉积速率。为了加深对薄层材料和精细加工相应流程的了解,读者可以参考文献[54]。利用标准光刻法对硅胶的处理,刚性微孔道的设计可以受益于其简单的合成方法,参数设置的多样化以及微孔道尺寸的最佳化[28]。然而,整个制造工艺可能比其他聚合物兼容工艺花费更多一些。

为了在硅胶上制造通道,通道的图案可被转移到表面覆盖有光阻材料以作为牺牲层的硅片上。在制造过程中,牺牲层中不需要的部分被除去,开放的区域允许硅刻蚀以生成通道。值得一提的是硅刻蚀的制造工艺多种多样,包括外延、溅射、化学法和物理气相沉积,可以在硅胶上得到想要的图层或得到薄膜集流体和电极。干湿法化学刻蚀可以用来制作复杂的几何形状的通道,电化学腐蚀法则可以用来制作多孔硅通道。读者可以参考文献[54,55]以了解更多有关以硅胶为基底的制造工艺。

由于在阳极和阴极之间有不间断的电氧化和电还原反应的进行,通过通道的染料和氧化剂的浓度将逐渐减少。由于对流传质的缺乏,使得催化活性面积的反应物得不到补充，在催化剂覆盖的电极上就会形成一个消耗边界层[40]，如图 9.3 所示。

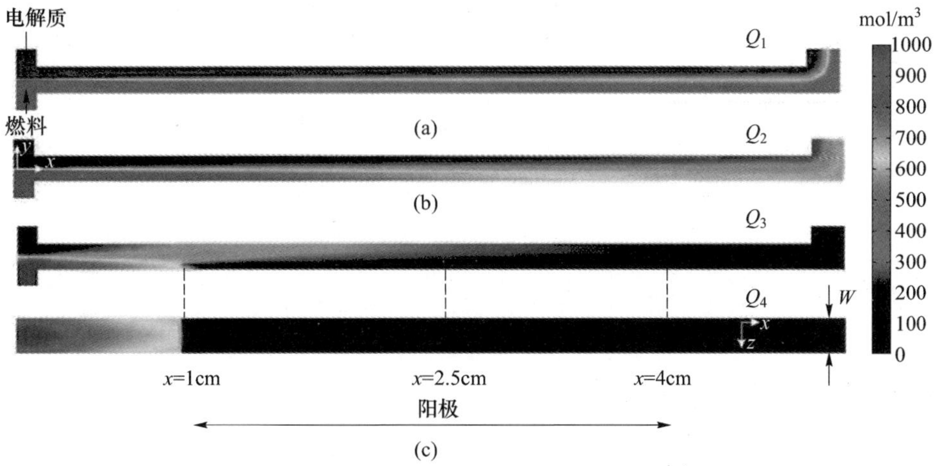


图 9.3　无膜 LFFC 中甲酸浓度的数值模拟，阳极接入口浓度为 1M。(a) $Q_1 = 500\mu L/min$，最大电流密度为 68mA/cm；(b) $Q_2 = 100\mu L/min$，最大电流密度为 58mA/cm；(c) $Q_3 = 10\mu L/min$，最大电流密度为 40mA/cm[40]（版权 2010，英国物理学会）

如果有效补充的过渡层不存在，那么在电极活性面积上的反应物浓度将急剧减少。在这种情况下，只有离电极最近几毫米范围内的物质浓度对电流贡献最大，并且电极的长度也受到一定限制。另外，为了控制扩散混合和附加损失，必须要限制通道的长度[57]。

我们采用损耗层的无源化控制增大最大电流的方法而不是使用附加能量激发的方法研究并联连续电极对于电流密度与燃料利用的影响[58,59]。Lim 等[58]提议每一个电极可以由一排微型电极组成。把一个电极的长度分成两个或更多区域以得到更短的电极，使其间距相当于它们长度的 3 倍，这样就能够避免损耗层不断增加的厚度。因此，与具有相同活性面积的单一电极装置相比，这种方法使最大功率密度提高了 25%。

## 9.3.2　溢流设计的流体构造学与装配

在溢流设计装置中，反应物流穿过包括催化活性面积在内的三维多孔电极，如图 9.4 所示。Salloum 等[49]提出了一种具有圆盘电极的对流传送无膜 LFFC（图 9.4(a)）。载体基质由 PMMA 制成。硫酸中的甲酸作为燃料流从圆盘的中心入口引入，通过覆盖在碳纸上的催化剂纳米颗粒发生氧化反应，碳纸充当多孔阳极。硫酸中高锰酸钾作为氧化剂流，从中心入口的外环被引入。氧化剂减少，然后与被氧化过的燃料混合，且在阳极和阴极之间保持 2mm 的间距，以防止因氧化剂回流导致的短路。

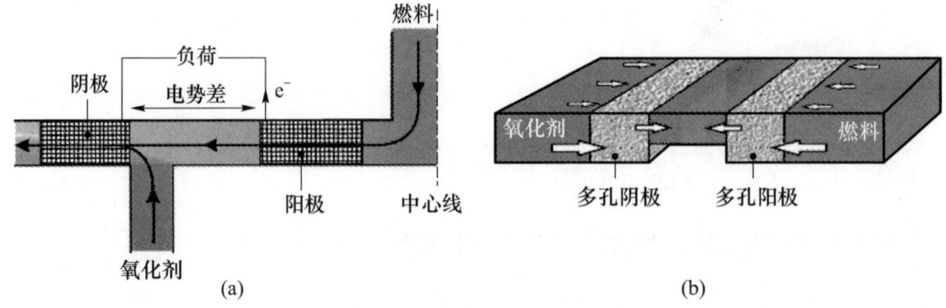


图 9.4　溢流设计。(a)无膜 LFFC 的放射状流体构造;(b)具有“多重入口”概念的三维多孔电极[19](版权 2011,Elsevier)

由对流运输的影响而导致的离子迁移可使燃料的利用率增加到 58%,并且能够独立控制燃料和氧化剂的流速。实验结果表明通过增加燃料流速,燃料的利用率从 58%、流速为 100μL/min 降到 4%、流速为 5mL/min。增大支持电解质硫酸的浓度可以使最大功率密度从 1.5mW/$cm^2$ 提高到 3mW/$cm^2$,这是因为减少了电荷传输损失。

“多重入口”的概念是由 Sinton 等提出来的[30]。流通式设计没有利用补充新鲜原料的流体网络入口,整个电池由 PDMS 制成。如图 9.4(b)所示,在 PEMFC 中作为气体扩散介质的亲水性多孔纤维碳纸,被剪成带状并放置在两个隔室。孔隙率为 78% 的高孔隙率碳纸(HGP-H 90,购置于 Toray)为电化学反应提供了大的比表面积,提高了燃料利用率,因为它提高了扩散/对流传质[30]。运行全钒氧化还原物种的电池,在流速为 300μL/min 下可得到的最大功率密度为 131mW/$cm^2$。

## 9.3.3　吸气式阴极 LFFC 的流体构造学与装配

水性介质中,作为氧化剂的溶解氧的传质和扩散($2\times10^{-5}cm^2/s$)限制了基于溶解氧运行的无膜 LFFC 的性能。另外,在水介质(2～4mM)中低浓度的溶解氧不能给阴极的损耗边界层提供充分的反应物补充。在通道侧壁利用气体扩散电极(GDE)可以促进燃料电池接受空气中高浓度的氧(10mM),比在水介质中的扩散率(0.2$cm^2$/s)高出 4 个数量级。

具有吸气式阴极的无膜 LFFC 的概念是由 Kenis 等在 2005 年提出的[32]。如图 9.5(a)和(b)所示,GDE 由 Toray 公司的碳纸涂覆含有铂黑纳米颗粒的催化剂制成,并应用于电池顶层作为吸气式阴极。电池的通道是由 PMMA 组成,通道和石墨板紧贴在一起,石墨板上覆盖一层钯黑纳米颗粒作为溢流阳极。

由于氧化剂流在这个设计中被舍弃了,需要电解液流来分隔燃料流,避免其直接暴露在阴极。因此,为了避免燃料的交叉损失和催化剂中毒,当 1M 的甲酸作为燃料时,就需要 0.5M 的硫酸同时添加至燃料和电解液中以促进通道中的

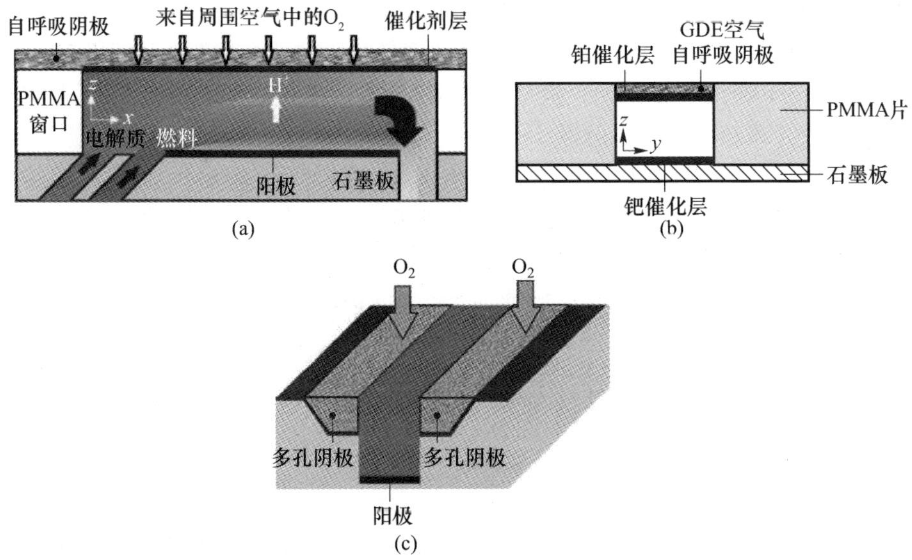


图 9.5　吸气式阴极无膜 LFFC。(a) 通道中燃料和电解液流的安排示意图[32]；(b)对图(a)中通道的截面和载体基质的描述；(c)具有吸气式阴极的整体设计的电池[19](版权 2011,Elsevier)

质子传导。在这种情况之下,当燃料和电解液的流速之比为 1∶1 且流量为 300mL/min时,可以得到最大电流密度为 130mA/cm$^2$,功率密度为 26mW/cm$^2$,最大输出功率为 16mW。阳极和阴极的电势可以通过参比电极获得,这也表明氧化剂浓度不是限制吸气式无膜 LFFC 性能的来源[33]。

Tominaka 等[25]提出了以甲醇为燃料的整体设计。整个电池,就是一个容器,顶部开口,由硅胶制成一个整体,省去了用来分隔燃料流和阴极的电解液。钯 - 钴(Pd - Co)合金通过电沉积在 Au 单质薄层上,作为电极或者集流器。在这个设计中,全部是无源燃料和氧化剂传输系统,氧气通过多孔阴极供给。2M 的甲醇溶液包含硫酸在内作为电解液,滴加到微孔道内。这种燃料电池的开路电压和最大纯功率分别能够达到 0.5V 和 1.4μW。

## 9.3.4　性能比较

为了进行定量比较,基于 3 种电极设计的不同的无膜 LFFC 的性能如表 9.2 和表 9.3 所列。

由于燃料型电池对电池动力学有明显的影响,只对以甲酸或甲醇为燃料的无膜 LFFC 进行了对比讨论。表 9.2 列出了一些以甲酸或甲醇为燃料的 3 种不同流体结构的最大功率密度的设计特征。

表 9.3 显示以甲酸或甲醇为燃料且具有溢流设计结构的无膜 LFFC 的功率密度的数量级范围为 0.1 ~ 1mW/cm$^2$,而流过设计的范围则为 1 ~ 10mW/cm$^2$。

溢流设计的低性能主要归因于阴极氧化剂的浓度过低以及电极损耗层增长。流过设计的性能的发展与通过多孔和三维电极使得活性部位的反应物得到有效补给从而加快了催化反应有关。吸气式无膜 LFFC 具有高性能的主要原因是阴极较高的氧气(来自空气)传送速率。由于来自空气的氧气通过 GDE 传输,使得在活性部位氧气的浓度几乎恒定,因而使这一阶段的反应得到优化。

表 9.2 在甲酸或甲醇中运行的无膜层流燃料电池的性能

| 设计 | 燃料/氧化剂 | 电解液 | 电极结构 | 最大电流密度/($mA/cm^2$) | 最大功率密度/($mW/cm^2$) | 设计特点 |
|---|---|---|---|---|---|---|
| Chohan 等[26] | 甲醇(1M)<br>溶解氧 | 硫酸或氢氧化钾(1M) | 流动电极 | 40 | 12 | 混合介质运行 |
| Chohan 等[42] | 甲醇(1M)<br>溶解氧 | 硫酸(0.5M) | 流动电极 | 8 | 2.8 | — |
| Cohen 等[28] | 甲酸(0.5M)<br>溶解氧 | 硫酸(0.1M) | 流动电极 | 1.5 | 0.18 | 堆叠的微孔孔道 |
| Chohan 等[41] | 甲酸(2.1M)<br>溶解氧 | 硫酸(0.5M) | 流动电极 | 0.8 | 0.17 | |
| Li 等[39] | 甲酸(0.5M)<br>溶解氧 | 硫酸(0.1M) | 流动电极 | 1.5 | 0.58 | 激光加工 |
| Morales - Acosta 等[31] | 甲酸(0.1M, 0.5M)<br>溶解氧<br>甲酸(0.04M) | 硫酸 | 流动电极 | 11.5 | 3.3 | Pd/MWCNT 作催化剂 |
| Salloum 等[49] | 高锰酸钾(0.01M)<br>甲酸(1.2M) | 硫酸(0.5~1M) | 溢流电极 | 5 | 2.8 | 连续的径向流 |
| Kjeang 等[48] | 次氯酸钠(0.67M) | 氢氧化钠(2.8M) | 溢流电极 | 230 | 52 | 三维立体结构的电极 |
| Jayashree 等[32] | 甲酸(1M)<br>空气 | 硫酸(0.5M) | 吸气式电极 | 130 | 26 | — |
| Jayashree 等[33] | 甲醇(1M)<br>氢氧化钾(1M)空气 | 硫酸(0.5M) | 吸气式电极 | 120 | 17 | — |
| Brushett 等[34] | 甲酸(1M)<br>空气 | 硫酸(0.5M)或氢氧化钾(1M) | 吸气式电极 | 130 | 26 | — |
| Hollinger 等[36] | 甲醇(1M)<br>氧气 | 硫酸(1M) | 吸气式电极 | 655 | 70 | 在阴极加入 80℃的甲醇并通入氧气 |

（续）

| 设计 | 燃料/氧化剂 | 电解液 | 电极结构 | 最大电流密度 /($mA/cm^2$) | 最大功率密度 /($mW/cm^2$) | 设计特点 |
|---|---|---|---|---|---|---|
| Whipple 等[35] | 甲醇(0.1～15M) 空气 | 硫酸 (0.5M) | 吸气式电极 | 62 | 4 | 使用 $Ru_xSe_y$ 作阴极抗甲醇催化剂 |
| 注:转自文献[19],版权 2011,Elsevier | | | | | | |

表 9.3　以甲酸或甲醇为燃料的最大限度运行的电池的 3 种不同流体结构的设计特点

| 设计 | 最大功率密度 /($mW/cm^2$) | 设计特点及注解 |
|---|---|---|
| 流动电极 | | |
| Choban 等[26] | 12,每个入口的流速 3mL/min | 并排流,Y 形,通道的长度为 29mm,高度为 1mm,宽度为 0.75mm,铂/钌和铂黑纳米颗粒的 Nafion 溶液分别应用于石墨板上作为阳极和阴极,燃料利用率小于 10% |
| 溢流电极 | | |
| Kjeang 等[48] | 52,电压为 0.4V,每个入口的流速 60μL/min | 并排流,T 形,通道的长度为 12mm,高度为 0.3mm,宽度为 3mm,电沉积钯和金在条形碳纸上作为溢流设计电池的阳极和阴极,在峰值功率密度,燃料利用率为 85%,在低电压下为 100% |
| 吸气式电极 | | |
| Hollinger 等[36] | 70,每个入口的流速 0.3mL/min | 垂直分层流,T 形,通道的长度为 48mm,宽度为 3.3mm,铂/钌和铂黑纳米颗粒的 Nafion 溶液刷在碳纸上作为电极,石墨板充当集流体,作为多孔分隔器应用于燃料 - 电解液的界面。运行于 80℃ 下,氧气供应速度为 50mL/min,阴极上的 Nafion 热压薄膜可以减轻燃料的交叉影响 |
| 注:转自文献[19],版权 2011,Elsevier | | |

## 9.4　燃料、氧化剂和电解液

### 9.4.1　燃料类型

无膜 LFFC 燃料的选择范围很广。关于燃料和氧化剂的选择,这里有许多可供选择的燃料,包括氢气($H_2$)[37,43,60]、甲醇($CH_3OH$)[26,33]、乙醇($C_2H_5OH$)[34]、甲酸[28,46]、过氧化氢($H_2O_2$)[45]、钒的氧化还原物种[18,29]、硼氢化钠($NaBH_4$)[34]和肼($N_2H_4$)[34]。

在这些水性燃料中,能量密度分别为 2.08kW·h/L 和 4.69kW·h/L 的甲酸和甲醇在无膜 LFFC 中的应用引起了更多关注,因为它们容易得到且电催化作用易于研究。甲酸/氧气燃料电池具有高的理论电动势(1.45V),而甲醇的相应的电动势为 1.2V。

Cohen 等[28]测试了以甲酸为燃料的无膜 LFFC,用 Pt 作催化剂,使甲酸和溶解氧结合被氧化:

$$HCOOH \longrightarrow 2H^+ + 2e^- + CO_2, \quad E^0 = 0.22V \tag{9.13}$$

$$4H^+ + O_2 + 4e^- \longrightarrow 2H_2O, \quad E^0 = 1.23V \tag{9.14}$$

然而,使用甲酸也有一些缺点,例如,CO 会使 Pt 催化剂中毒。但是整个系统很容易控制,电池也有很大的开路电势和高电化学效益[28]。

通常,钒氧化还原电对溶解于支持电解液中,如硫酸,$V^{2+}/V^{3+}$ 作阳极电解液,$VO_2^+/VO^{2+}$ 作阴极电解液。在无膜 LFFC 中,钒氧化还原电对被用来形成全钒的燃料/氧化剂系统[18,27,30,61]。基于以下阳极和阴极上的氧化还原反应和标准电极电势,无膜 LFFC 可以正常工作:

$$V^{3+} + e^- \rightleftharpoons V^{2+}, \quad E^0 = -0.496V \text{ 相对于 SCE} \tag{9.15}$$

$$VO_2^+ + 2H + e^- \rightleftharpoons VO^{2+} + H_2O, \quad E^0 = 0.750V \text{ 相对于 SCE} \tag{9.16}$$

对微流体燃料电池[20,30]来说,这种氧化还原反应组合的优点如下:①在反应速率和传输特性方面给电化学半电池提供了很好的协调性;②具有较高的溶解度和接近 5.4M 的高氧化还原浓度;③具有高的开路电压(在相同 pH 下,接近 1.7V),因为其与正式的氧化还原电势存在很大的不同;④使用纯碳电极,没有用贵金属催化剂来加快反应速率。

钒氧化还原燃料电池的能量密度受钒氧化还原物种溶解度的限制。为了解决这个问题,Kjeang 等[48]提出了一种新的碱性微流体燃料电池,以甲酸和次氯酸盐氧化剂为基础。反应物溶液来自甲酸和次氯酸钠溶液,这两种溶液都是易得到且稳定的高浓度的液体,从而导致了燃料电池系统具有较高的整体能量密度。甲酸盐的氧化反应和次氯酸盐的还原反应分别在多孔钯和金电极的碱性介质中进行。结果表明,在低的过电位下产生快速动力学,同时通过碳酸盐的吸附防止 $CO_2$ 的生成。

Brushett 等[34]在吸气式流体结构的电池中,以 3M 的肼为燃料,0.5M 的硫酸作电解液。在这个以肼为燃料的无膜 LFFC 中,室温下最大功率密度为 $80mW/cm^2$ 已经实现,与此同时,阳极 Pt 电极上催化剂的负载量为 $1mg/cm^2$,燃料和电解液的流速比为 0.3mL/min。此外,在先前提到的相同条件下,用流速为 50 标准 mL/min 的氧气而不是静止的空气输送方法来检测肼,峰值功率密度和最大电流密度没有大的改变,这表明这种无膜 LFFC 组不受氧气传输的限制。

当其安全性不是主要问题时,这种直接肼酸性无膜 LFFC 有望做成实用型的微型电源。肼/$O_2$ 燃料电池的理论开路电压为 1.56V。另外，肼/$O_2$ 完全电解氧化的最后产物为氮气和水，使得肼燃料电池可以成为环境友好零排放的能量转换器[34]。同样，Brushett 等[34]在阳极 Pt 电极上利用溶于 1M 的氢氧化钾的 1M 的硼氢化钠溶液作为燃料。然而，能量密度为 9.29kW·h/kg 的硼氢化钠在酸性介质中是不稳定的，但是在碱性介质中则有较高的稳定性。在吸气式阴极上，燃料和电解液（1M KOH）的流速比为 0.3mL/min 时，测得最大峰值功率密度为 101mW/$cm^2$。这种高性能主要归因于提高了 Pt 对于硼氢化钠阴离子氧化反应的电催化活性。

## 9.4.2　氧化剂类型

氧化剂可以是水中的溶解氧[26,39,42,62]、空气[25,32-35]、过氧化氢[45,47]、钒氧化物物种[18,27,29]、高锰酸钾[46,49]和次氯酸钠[48]。

从根本上说，阴极半电池电动力学比阳极上的要慢。氧还原反应不活泼。氧还原反应需要许多单个步骤和有效的分子重组才能完成，因此，大多数激活过电压损失都发生在阴极。在标准温度和大气压下（298K，1atm），在铂电极上氧还原反应的交换电流密度比氢的氧化反应低 6 个数量级。在溶解氧作为氧化剂时，阴极低的电动力学反应和低氧浓度导致了较低的功率密度。为了改善电解液中溶解氧传质的限制，可以使用高溶解度的氧化剂替代溶解氧。

少数使用液体氧化剂的燃料电池一般会有更高的功率输出。Choban 等[62]用 0.144M 高锰酸钾替代了 0.5M 氧饱和的硫酸溶液做氧化剂，用 2.1M 甲酸做燃料。结果表明，用高锰酸钾做氧化剂的电池运行的电流密度增大了一个数量级，这是因为在水介质中高锰酸钾的溶解度较高[41]。

Li 等[39]利用 0.1M 氧饱和的硫酸溶液作氧化剂，溶于 0.1M 硫酸的 0.5M 甲酸溶液作燃料。电池的最大功率密度为 0.58mW/$cm^2$，这是由于从氧化剂流到阴极的氧气供应不足引起的。

同样地，气态空气在吸气式电池结构中作为氧化剂[34,40]。由于空气中氧气的浓度（0.2 $cm^2$/s）比在水介质中（$2\times10^{-5}$ $cm^2$/s）[32]高 4 个数量级，因而吸气式阴极的设计可以提供更高的功率密度和功率输出。

## 9.4.3　电解液类型

在大多数无膜 LFFC 设计中，燃料和氧化剂都是溶解在电解液中，并且通常都是用注射泵引入通道中。添加电解液到物质流的主要原因是为了提高离子电导以减少在阳极和阴极间的欧姆损失。例如，在无膜 LFFC 的两物流中加入硫酸，可以使质子源更接近阴极，并且可以维持在阴极消耗的质子的浓度梯度[62]。Choban 等[62]的实验结果表明用 0.5M 的硫酸作支持电解液，可以得到的最大电流密度为 0.9mW/$cm^2$，而在最大体积流速为 0.8mL/min，以水作电解液的 LFFC

上,最大电流密度仅为 0.2mW/cm²。

无膜使得电池在酸碱两种介质中都可以运行,同样在"混合介质"如阴极在酸性介质而阳极在碱性介质中也可以运行,反之亦然[26]。这种介质的灵活性使得设计者可以单独设计阴极和阳极物质流的组成,以使单个电极的动力学和整个电池电势达到最优化[26]。此外,操作者可以自由地将无膜 LFFC 在全酸、全碱或者混合介质模型中运行。

电解液的 pH 值对单个电极上的反应动力学以及氧化或还原反应发生电极上的电极电势起作用[26]。强酸或强碱都是代表性的电解液,如硫酸或氢氧化钾,其中分别包括快速移动的水合氢离子或氢氧根离子[20]。在碱性介质中运行的燃料电池会加快使阳极催化剂中毒的 CO 的电化学氧化反应,并且能够提高阴极的氧还原动力学[26]。然而,在有膜燃料电池中,由于碳酸盐形成物的存在导致膜堵塞,电池的长期稳定性受到限制,并且限制了这些碱兼容的膜在液体燃料电池运行中的使用[26]。

Bruhsett 等[34]分别在酸性($H_2SO_4$)和碱性条件(KOH)下,测试了以乙醇和甲醇为燃料运行的吸气式无膜 LFFC 的性能。以甲醇和乙醇为燃料的电池测试显示了它们分别在碱性介质(1.2V 和 0.7V, 17.2mW/cm² 和 12.1mW/cm²)和酸性条件(0.93V 和 0.41V, 11.8mW/cm² 和 1.9mW/cm²)下的开路电压和最大功率密度的比较。相比于酸性介质,在碱性介质中提高的性能是由于增强了乙醇的氧化动力学和氧还原反应动力学。

Choban 等[26]研究了在酸性、碱性以及酸/碱介质中,在 1M 甲醇和溶解氧条件下运行的无膜 LFFC 的电池电势和功率输出。结果表明,研究过程中电池电势和功率输出在酸性和碱性介质中都受到了阴极的限制,这是因为溶液中氧浓度过低。在酸性介质中,氧气的溶解度大约为 1mM,而在碱性介质中大约降低 25%,使得其性能在电流-电压特性曲线中提前下降[26]。同样,与碳酸盐形成物也没有关系,因为系统中形成的任何碳酸盐都会被流体快速清除掉。

在酸性阳极流与碱性阴极流结合的电池中,最大开路电压(OCP)为 0.38V,但是也有观察到一个小于 0.1V 的开路电压,这是因为阴极和阳极上的过电势。换言之,在甲醇氧化和氧还原反应中释放的能量大部分已经被应用于水的电离反应。在这种组合中,与原反应相结合的电解反应不能产生大量的能量[26]。

在碱性阳极和酸性阴极的条件下,两种电流的组合产生了令人满意的高理论开路电压 2.04V。但是,实际开路电压为 1.4V,归因于氧还原和甲醇氧化反应的慢动力学。同样,混合介质的功率密度比那些为全酸性介质或者全碱性介质的无膜 LFFC 要高。

大多数观察到的混合介质型电池的额外功率密度是由电化学酸碱中和反应提供,在阴极的质子减少,阳极甲醇的氧化反应中氢氧根离子被消耗。因此,当与不同无膜 LFFC 构型相比较时,$H_2SO_4$ 和 KOH 的消耗必须要包括在内。

当电池在碱性阳极和酸性阴极条件下运行时,$OH^-$ 和 $H^+$ 在阳极和阴极消耗的速率为一分子甲醇转移 6 个电子。最大理论能量密度(基于 1M 的甲醇和空气中氧气的反应,同时消耗 6 当量的 $H_2SO_4$ 和 KOH)为 495W · h/kg,比以全碱性和全酸性为介质且只消耗甲醇的 LFFC 的能量密度理论值(6000W · h/kg)低得多[26]。

Hasegawa 等[45]把混合介质应用于以过氧化氢为燃料和氧化剂的微流体燃料电池中,酸碱介质分别为 $H_2SO_4$ 和 NaOH。这种设计产生的功率密度高达 $23mW/cm^2$。而设计的缺点为阴极的过氧化氢会自发分解产生气泡,氧气的逸出可能会妨碍共层流的交界面。

这个设计的总反应为

$$H_2O_2(aq) + HO_2^-(aq) + 2H^+(aq) + OH^-(aq) \longrightarrow O_2 + 3H_2O \quad (9.17)$$

其中涉及了 $H_2O_2$ 的歧化反应以及 $H^+$ 和 $OH^-$ 的结合反应。在酸/碱性两极电解液界面的 $SO_4^{2-}$ 和 $Na^+$ 的电荷相互抵消以至于反应式(9.17)可以持续进行。因此,这种燃料电池的产物为水、氧气和盐。

## 9.5 本章小结

本章讨论了无膜 LFFC 的设计,制造工艺和性能。设计创新包括新的流体结构和使用新材料调整电极在通道中的位置,这些都是为了加强单一电池的性能以及提高在单一流通通道中燃料的利用率而提出的。为了在实际应用中得到足够多的功率,电池堆叠是不可避免的。面向商业化前景,包括成本和耐久性,除了无膜 LFFC 的创新流体设计和优化叠加外,替代材料的开发有巨大的研究潜力。当然,燃料和氧化剂的最优选择也需要解决。在酸性或碱性介质中,电池的新电极和具有高电导率和长久耐用的集流体仍需不断的研究与开发。同样,电池比表面积的电阻率(ASR)组合包括电解液、阳极和阴极的电阻率都应该减少以提供更高的功率密度。石墨的抗腐蚀能力不同于其他金属材料,但是它的精细加工会增加总的成本。在电化学催化剂方面,无膜 LFFC 一般可以受益于电化学催化剂材料的不断改进,如直接液体燃料电池的耐中毒纳米催化剂。整个燃料电池系统应该在极低的自耗下运行。而精细加工和微流体技术可以提供被动的或极低能耗的燃料/氧化剂传输系统。

## 参考文献

[1] Moghaddam, S., Pengwang, E., Jiang, Y. B., Garcia, A. R., Burnett, D. J., Brinker, C. J., Masel, R. I., and Shannon, M. A. (2010) An inorganic-organic proton exchange membrane for fuel cells with a

controlled nanoscale pore structure. *Nat Nano*, **5** (3), 230 – 236.

[2] Moghaddam, S. et al. (2010) An enhanced microfluidic control system for improving power density of a hydride-based micro fuel cell. *Journal of Power Sources*, **195** (7), 1866 – 1871.

[3] Moghaddam, S. et al. (2008) A selfregulating hydrogen generator for micro fuel cells. *Journal of Power Sources*, **185** (1), 445 – 450.

[4] Nguyen, N. T. and Chan, S. H. (2006) Micromachined polymer electrolyte membrane and direct methanol fuel cells: a review. *Journal of Micromechanics and Microengineering*, **16** (4), R1 – R12.

[5] Ha, S., Dunbar, Z., and Masel, R. I. (2006) Characterization of a high performing passive direct formic acid fuel cell. *Journal of Power Sources*, **158** (1), 129 – 136..

[6] Moghaddam, S. et al. (2010) An enhanced microfluidic control system for improving power density of a hydride-based micro fuel cell. *Journal of Power Sources*, **195** (7), 1866 – 1871.

[7] Mousavi Shaegh, S. A., Nguyen, N. T., and Chan, S. H. (2011) A review on membraneless laminar flow-based fuel cells. *International Journal of Hydrogen Energy*, **36** (9), 5675 – 5694.

[8] Chan, S. H., Nguyen, N. T., Xia, Z., and Wu, Z. (2005) Development of a polymeric micro fuel cell containing asermicromachined flow channels. *Journal of Micromechanics and Microengineering*, **15** (1), 231 – 236.

[9] Kelley, S. C., Deluga, G. A., and Smyrl, W. H. (2000) A miniature methanol/air polymer electrolyte fuel cell. *Electrochemical and Solid-State Letters*, **3** (9), 407 – 409.

[10] Pichonat, T., Gauthier-Manuel, B., and Hauden, D. (2004) A new protonconducting porous silicon membrane for small fuel cells. *Chemical Engineering Journal*, **101** (1 – 3), 107 – 111.

[11] Gold, S., Chu, K.-L., Lu, C., Shannon, M. A., and Masel, R. I. (2004) Acid loaded porous silicon as a proton exchange membrane for micro-fuel cells. *Journal of Power Sources*, **135** (1 – 2), 198 – 203.

[12] Pennathur, S., Eijkel, J. C. T., and van denBerg, A. (2007) Energy conversion in microsystems: is there a role for micro/nanofluidics? *Lab on a Chip*, **7** (10), 1234 – 1237.

[13] Brody, J. P. and Yager, P. (1997) Diffusionbased extraction in a microfabricated device. *Sensors and Actuators A: Physical*, **58** (1), 13 – 18.

[14] Surmeian, M., Slyadnev, M. N., Hisamoto, H., Hibara, A., Uchiyama, K., and Kitamori, T. (2002) Three-layer flow membrane system on a microchip for investigation of molecular transport. *Analytical Chemistry*, **74** (9), 2014 – 2020.

[15] Weigl, B. H. and Yager, P. (1999) Microfluidic diffusion-based separation and detection. *Science*, **283** (5400), 346 – 347.

[16] Kenis, P. J. A., Ismagilov, R. F., and Whitesides, G. M. (1999) Microfabrication inside capillaries using multiphase laminar flow patterning. *Science*, **285**, 83 – 85.

[17] Song, C., Nguyen, N. T., Tan, S. H., and Asundi, A. K. (2009) Modelling and optimization of micro optofluidic lenses. *Lab on a Chip*, **9**, 1178 – 1184.

[18] Ferrigno, R., Stroock, A. D., Clark, T. D., Mayer, M., and Whitesides, G. M. (2002) Membraneless vanadium redox fuel cell using laminar flow. *Journal of the American Chemical Society*, **124** (44), 12930 – 12931.

[19] Mousavi Shaegh, S. A., Nguyen, N. T., and Chan, S. H. (2011) A review on membraneless laminar flow-based fuel cells. *International Journal of Hydrogen Energy*, **36** (9), 5675 – 5694.

[20] Kjeang, E., Djilali, N., and Sinton, D. (2008) Microfluidic fuel cells: a review. *Journal of Power Sources*, **186** (2), 353 – 369.

[21] Sharp, K. V., Adrian, R. J., Santiago, J. G., and Molho, J. I. (2002) Liquid flows in microchannels,

in *CRC Handbook of MEMS* ( ed. M. Gad-el-Hak ), CRC Press, New York, 6-1 - 6-38.

[22] Nguyen, N. T. and WEreley, S. T. (2006) *Fundamentals and Applications of Microfluidics*, Artech House, Inc.

[23] Atencia, J. and Beebe, D. J. (2005) Controlled microfluidic interfaces. *Nature*, **437** (7059), 648 - 655.

[24] Mench, M. M. (2008) *Fuel Cell Engines*, John Wiley & Sons, Inc., New Jersey.

[25] Tominaka, S., Ohta, S., Obata, H., Momma, T., and Osaka, T. (2008) Onchip fuel cell: micro direct methanol fuel cell of an air-breathing, membraneless, and monolithic design. *Journal of the American Chemical Society*, **130** (32), 10456 - 10457.

[26] Choban, E. R., Spendelow, J. S., Gancs, L., Wieckowski, A., and Kenis, P. J. A. (2005) Membraneless laminar flow-based micro fuel cells operating in alkaline, acidic, and acidic/alkaline media. *Electrochimica Acta*, **50** (27), 5390 - 5398.

[27] Kjeang, E., McKechnie, J., Sinton, D., and Djilali, N. (2007) Planar and threedimensional microfluidic fuel cell architectures based on graphite rod electrodes. *Journal of Power Sources*, **168** (2), 379 - 390.

[28] Cohen, J. L., Westly, D. A., Pechenik, A., and Abruña, H. D. (2005) Fabrication and preliminary testing of a planar membraneless microchannel fuel cell. *Journal of Power Sources*, **139** (1 - 2), 96 - 105.

[29] Kjeang, E., Proctor, B. T. M., Brolo, A. C., Harrington, D. A., Djilali, N., and Sinton, D. (2007) High-performance microfluidic vanadium redox fuel cell. *Electrochimica Acta*, **52** (15), 4942 - 4946.

[30] Kjeang, E., Michel, R., Harrington, D. A., Djilali, N., and Sinton, D. (2008) A microfluidic fuel cell with flow-through porous electrodes. *Journal of the American Chemical Society*, **130** (12), 4000 - 4006.

[31] Morales-Acosta, D., Rodríguez, G., Godinez, H., Luis, A., and Arriaga, L. G. (2010) Performance increase of microfluidic formic acid fuel cell using Pd/MWCNTs as catalyst. *Journal of Power Sources*, **195** (7), 1862 - 1865.

[32] Jayashree, R. S., Gancs, L., Choban, E. R., Primak, A., Natarajan, D., Markoski, L. J., and Kenis, P. J. A. (2005) Air-breathing laminar flow-based microfluidic fuel cell. *Journal of the American Chemical Society*, **127** (48), 16758 - 16759.

[33] Jayashree, R. S., Egas, D., Spendelow, J. S., Natarajan, D., Markoski, L. J., and Kenis, P. J. A. (2006) Air-breathing laminar flowbased direct methanol fuel cell with alkaline electrolyte. *Electrochemical and Solid-State Letters*, **9** (5), A252 - A256.

[34] Brushett, F. R., Jayashree, R. S., Zhou, W. P., and Kenis, P. J. A. (2009) Investigation of fuel and media flexible laminar flow-based fuel cells. *Electrochimica Acta*, **54** (27), 7099 - 7105.

[35] Whipple, D. T., Jayashree, R. S., Egas, D., Alonso-Vante, N., and Kenis, P. J. A. (2009) Ruthenium cluster-like chalcogenide as a methanol tolerant cathode catalyst in airbreathing laminar flow fuel cells. *Electrochimica Acta*, **54** (18), 4384 - 4388.

[36] Hollinger, A. S., Maloney, R. J., Jayashree, R. S., Natarajan, D., Markoski, L. J., and Kenis, P. J. A. (2010) Nanoporous separator and low fuel concentration to minimize crossover in direct methanol laminar flow fuel cells. *Journal of Power Sources*, **195** (11), 3523 - 3528.

[37] Mitrovski, S. M. and Nuzzo, R. G. (2006) A passive microfluidic hydrogen-air fuel cell with exceptional stability and high performance. *Lab on a Chip*, **6** (3), 353 - 361.

[38] Mitrovski, S. M., Elliott, L. C. C., and Nuzzo, R. G. (2004) Microfluidic devices for energy conversion: planar integration and performance of a passive, fully immersed $H_2 - O_2$ fuel cell. *Langmuir*, **20** (17),

6974 – 6976.

[39] Li, A., Chan, S. H., and Nguyen, N. T. (2007) A laser-micromachined polymeric membraneless fuel cell. *Journal of Micromechanics and Microengineering*, **17**, 1107 – 1113.

[40] Mousavi Shaegh, S. A., Nguyen, N. T., and Chan, S. H. (2010) An air-breathing laminar flow-based formic acid fuel cell with porous planar anode: experimental and numerical investigations. *Journal of Micromechanics and Microengineering*, **20**, 12.

[41] Choban, E. R. et al. (2004) Microfluidic fuel cell based on laminar flow. *Journal of Power Sources*, **128** (1), 54 – 60.

[42] Choban, E. R., Waszczuk, P., and Kenis, P. J. A. (2005) Characterization of limiting factors in laminar flow-based membraneless microfuel cells. *Electrochemical and Solid-State Letters*, **8** (7), A348 – A352.

[43] Cohen, J. L., Volpe, D. J., Westly, D. A., Pechenik, A., and Abruña, H. D. (2005) A dual electrolyte $H_2/O_2$ planar membraneless microchannel fuel cell system with open circuit potentials in excess of 1.4V. *Langmuir*, **21** (8), 3544 – 3550.

[44] López-Montesinos, P. O. et al. (2011) Design, fabrication, and characterization of a planar, silicon-based, monolithically integrated micro laminar flow fuel cell with a bridge-shaped microchannel crosssection. *Journal of Power Sources*, **196** (10), 4638 – 4645.

[45] Hasegawa, S., Shimotani, K., Kishi, K., and Watanabe, H. (2005) Electricity generation from decomposition of hydrogen peroxide. *Electrochemical and Solid-State Letters*, **8** (2), A119 – A121.

[46] Sun, M. H., Velve Casquillas, G., Guo, S. S., Shi, J., Ji, H., Ouyang, Q., and Chen, Y. (2007) Characterization of microfluidic fuel cell based on multiple laminar flow. *Microelectronic Engineering*, **84** (5 – 8), 1182 – 1185.

[47] Kjeang, E., Brolo, A. G., Harrington, D. A., Djilali, N., and Sinton, D. (2007) Hydrogen peroxide as an oxidant for microfluidic fuel cells. *Journal of the Electrochemical Society*, **154** (12), B1220 – B1226.

[48] Kjeang, E., Michel, R., Harrington, D. A., Sinton, D., and Djilali, N. (2008) An alkaline microfluidic fuel cell based on formate and hypochlorite bleach. *Electrochimica Acta*, **54** (2), 698 – 705.

[49] Salloum, K. S., Hayes, J. R., Friesen, C. A., and Posner, J. D. (2008) Sequential flow membraneless microfluidic fuel cell with porous electrodes. *Journal of Power Sources*, **180** (1), 243 – 252.

[50] Sung, W. and Choi, J.-W. (2007) A membraneless microscale fuel cell using non-noble catalysts in alkaline solution. *Journal of Power Sources*, **172** (1), 198 – 208.

[51] Xia, Y. N. and Whitesides, G. M. (1998) Soft lithography. *Annual Review of Materials Science*, **28** (1), 153 – 184.

[52] Lee, J. N., Park, C., and Whitesides, G. M. (2003) Solvent compatibility of poly (dimethylsiloxane)-based microfluidic devices. *Analytical Chemistry*, **75** (23), 6544 – 6554.

[53] Merkel, T. C., Bondar, V. I., Nagai, K., Freeman, B. D., and Pinnau, I. (2000) Gas sorption, diffusion, and permeation in poly (dimethylsiloxane). *Journal of Polymer Science, Part B: Polymer Physics*, **38** (3), 415 – 434.

[54] Franssila, S. (2010) *Introduction to Microfabrication*, 2nd edn, John Wiley & Sons, Inc., Hoboken, NJ.

[55] Madou, M. J. (2002) *Fundamental of Microfabrication: The Science of Miniaturization*, 2nd edn, CRC Press, Boca Raton, FL.

[56] Jayashree, R. S., Yoon, S. K., Brushett, F. R., Lopez-Montesinos, P. O., Natarajan, D., Markoski, L. J., and Kenis, P. J. A. (2010) On the performance of membraneless laminar flow-based fuel cells. *Journal of Power Sources*, **195** (11), 3569 – 3578.

[57] Hayes, J. R., Engstrom, A. M., and Friesen, C. (2008) Orthogonal flow membraneless fuel cell. *Jour-*

*nal of Power Sources*, **183** (1), 257 – 259.

[58] Lim, K. G., Palmore, G., and Tayhas, R. (2007) Microfluidic biofuel cells: the influence of electrode diffusion layer on performance. *Biosensors and Bioelectronics*, **22** (6), 941 – 947.

[59] Lee, J., Lim, K. G., Palmore, G. T. R., and Tripathi, A. (2007) Optimization of microfluidic fuel cells using transport principles. *Analytical Chemistry*, **79** (19), 7301 – 7307.

[60] Mitrovski, S. M., Elliott, L. C. C., and Nuzzo, R. G. (2004) Microfluidic devices for energy conversion: Planar integration and performance of a passive, fully immersed $H_2$-$O_2$ fuel cell (Supporting Information). *Langmuir*, **20** (17), 6974 – 6976.

[61] Salloum, K. S. and Posner, J. D. (2010) Counter flow membraneless microfluidic fuel cell. *Journal of Power Sources*, **195** (19), 6941 – 6944.

[62] Choban, E. R., Markoski, L. J., Wieckowski, A., and Kenis, P. J. A. (2004) Microfluidic fuel cell based on laminar flow. *Journal of Power Sources*, **128** (1), 54 – 60.

# 第10章

## 直接醇类燃料电池催化剂的研究进展

Luhua Jiang, Gongquan Sun

## 10.1 引　言

质子交换膜电池(PEMFC)由于其固有的优点,如安静、环保、紧密系统、快速开关机、高功率密度等[1-5],作为电动车、固定和移动终端等的电源受到越来越多的关注。如果PEMFC的正极直接与醇类反应,即称为直接醇类燃料电池(DAFC)。由于其在燃料储存与运输等方面的优点,使得DAFC自20世纪60年代得到广泛的发展[6]。在现有所有研究过的可能的液体燃料中,甲醇与乙醇由于其相对简单的结构和高氧化活性被视为最有潜力[7-16]。直到现在,由于甲醇不需要断开碳碳键,所以现在的研究聚焦在甲醇的运用。优良的甲醇氧化催化剂的特点是低温时既可使甲醇脱氢还可去除甲醇氧化产物如CO类化合物。为了提高Pt的活性,不同的助剂如Ru[17-20]、Sn[21-25]、W[26,27]和Mo[28,29]都被加入以促进Pt在甲醇氧化反应(MOR)过程中的催化活性并减少CO类化合物在Pt活性位点上的沉积。近期,根据双官能团机理的解释,PtRu合金具有最好的催化效果[18]。而对于乙醇来说,由于其氧化必须断开C—C键,所以完全的电化学氧化就相对困难。乙醇氧化催化剂必须要具备多功能,其中包括脱氢、去除CO类化合物、断开C—C键和相对低温下进行反应。这些使得乙醇的电催化方法显得更加重要。因此,阴极电催化剂不仅要考虑其氧气还原电催化活性,还要限制乙醇由阳极向阴极的渗透。Fe、Co、Ni和V修饰的铂[30-33]和一些非贵金属催化剂[34-37]逐渐被用来达到此目的。然而新的挑战出现了,即当非贵金属用作阴极时,这些非贵金属会在酸性的工作环境中腐蚀。

本章综述了过去10年在DICP中关于ORR、MOR和乙醇氧化反应(EOR)催化剂的系统性研究。

## 10.2　高效电化学催化剂制备的发展

颗粒尺度、形态与结构对于催化剂活性都有极大影响。因此,发展一种简单有效的方法来控制颗粒大小甚至是纳米颗粒的微观结构显得非常重要。我们的综述中将提到几种方法,包括浸渍 - 还原法(甲醛作为还原试剂)[38]、过氧化物氧化分解法[39-41]和多元醇还原法[42-44]。

### 10.2.1　碳载铂催化剂

铂是最常用的 ORR 电化学催化剂。为了提高 Pt 的分散性,Pt 纳米颗粒通常分散在碳载体上。本章中,碳载体为 Vulcan XC-72。在实际使用的燃料电池催化剂中,Pt 的含量有时要高达 60% 的质量分数以减少催化层的厚度来最大程度地降低电极的扩散电阻。为了达到如此高的 Pt 含量,较为便利的方法是浸渍 - 还原法,此法使用氢气或是液态还原剂,如 $NaBH_4$ 或 HCHO,但是此法无法制备小而且均一的 Pt 纳米颗粒[45-47]。图 10.1(a)和(b)是以甲醛作为还原剂,通过浸渍法制备的和多元醇还原法制备的质量分数为 20% 的 Pt/C 样品(分别记作 20% Pt/C-HCHO 和 20% Pt/C-EG)的透射电子显微镜图[45,46]。从 20% Pt/C-EG 的 TEM 图中可见,Pt 纳米颗粒高度分散于碳表面,并且颗粒尺寸为 2nm 左右;与其相比,20% Pt/C-HCHO 上的 Pt 纳米颗粒出现轻度聚合,其颗粒尺寸为 5.3nm,远大于 Pt/C-EG 的 Pt 颗粒尺寸。而且,我们采用多元醇还原法合成了一系列 Pt/C 催化剂,Pt 载量为 10%(质量分数)~60%(质量分数),如图 10.1(c)~(g)的 TEM 图所示。所有采用多元醇还原法合成的样品,样品中的 Pt 纳米颗粒均匀分散于碳表面。图 10.1(h)所示为 Pt 颗粒的平均尺寸。由图可见,其颗粒的尺寸也都小于 5nm。

更重要的是,由以上的实验可知颗粒尺寸对于多元醇合成法的溶液中的水分含量十分敏感[48]。通过简单的控制溶液中的水分含量,电催化剂的颗粒大小可以达到纳米级别。图 10.2 是 20% Pt/C 不同颗粒大小的 TEM 图。Pt/C 的颗粒尺寸随着水含量的增加由 2nm 增大到 3nm。

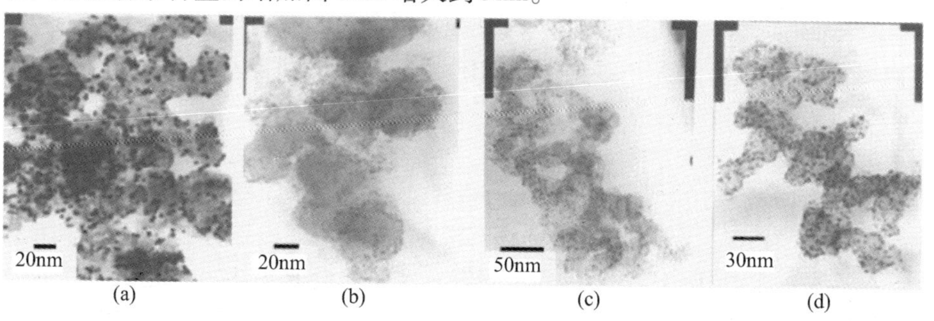

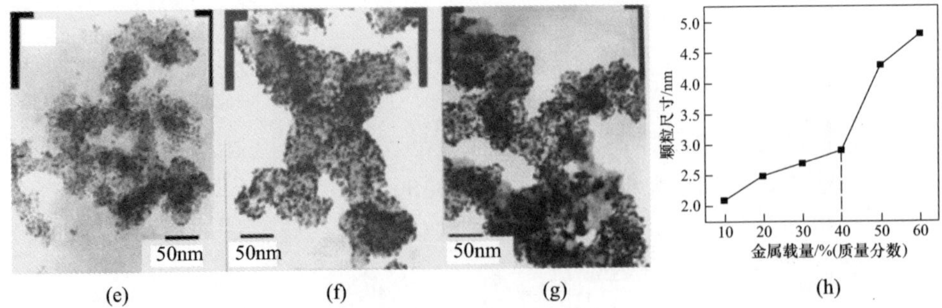


图 10.1 (a) 20%(质量分数)Pt/C-HCHO,(b) 20%(质量分数)Pt/C-EG,(c) 10%(质量分数)Pt/C-EG,(d) 30%(质量分数)Pt/C-EG,(e) 40%(质量分数)Pt/C-EG,(f) 50%(质量分数)Pt/C-EG,(g) 60%(质量分数)Pt/C-EG 的透射电镜图以及(h)金属粒径和金属负载的关系图

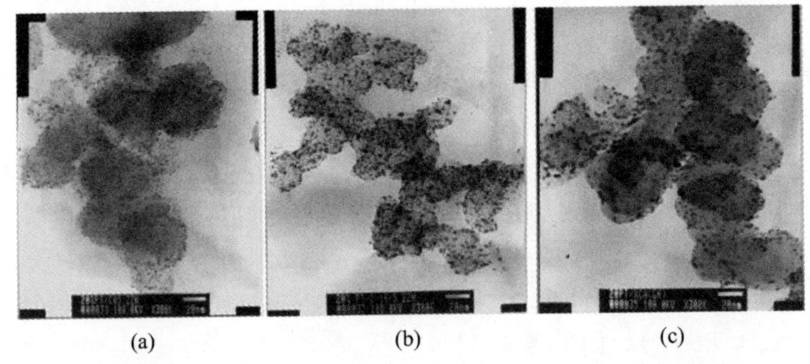


图 10.2 不同粒径 20%(质量分数)Pt/C 催化剂的透射电镜图:(a) $d_{mean}=2.0$nm;(b) $d_{mean}=2.6$nm;(c) $d_{mean}=3.0$nm

### 10.2.2 碳载铂钌合金催化剂

PtRu 是一种高效的甲醇氧化电催化剂,其颗粒分散程度对于其应用与活性十分重要。因此,我们比较了由几种方法制备出的碳载铂钌合金催化剂。与在 10.2.1 节所提到的理由一样,PtRu/C 催化剂同样需要高载量的金属加入。我们以含 20%(质量分数)Pt、10%(质量分数)Ru 的 PtRu/C 为例讲解。

表 10.1 总结了浸渍 - 还原法、过氧化物氧化法和多元醇还原法 3 种方法制备的催化剂的 TEM 和 X 射线衍射(XRD)分析结果。商品化的 PtRu/C 催化剂(Johnson Matthey 公司生产)的数据列于表 10.1 中用以比较。由表可见,由多元醇还原法制备的 PtRu/C 催化剂比起用其他方法制备的催化剂表现出更小的平均颗粒尺寸,并且 PtRu/C 中的 Pt 的晶格参数要比单质 Pt 的要小(3.923Å)。此现象说明 Pt 与 Ru 之间存在强作用力。然而,以浸渍 - 还原法制备的PtRu/C催化剂的晶格参数与 Pt/C 的晶格参数相近。这说明此时 Pt 和 Ru 之间存在着弱相互作用。而且以过氧化物法制备的 PtRu/C 的晶格参数与商品化的 PtRu/C

的晶格参数相近。在表 10.1 中可见，由 XRD 测得的各样品的比表面积相对于 TEM 测出的要略高，这种由于技术不同产生的误差是可以接受的，而且 XRD 技术更适于对块状晶体进行表征。

表 10.1　通过不同方法制备的 PtRu/C 的 XRD 和 TEM 的表征结果

| PtRu/C 样品 | XRD | | | TEM | |
|---|---|---|---|---|---|
| | 晶格参数 /Å① | 平均粒径 /nm | 比表面积 $S_{XRD}$ /($m^2$/g) | 平均粒径 /nm | 比表面积 $S_{TEM}$ /($m^2$/g) |
| 商业催化剂(JM) | 3.890 | 2.4 | 136.1 | 2.7 | 120.9 |
| 浸渍 - 还原法 | 3.906 | 2.7 | 120.9 | 3.5 | 96.1 |
| 过氧化物法 | 3.899 | 2.5 | 130.7 | 2.8 | 116.7 |
| 多元醇还原法 | 3.883 | 1.9 | 171.9 | 2.0 | 163.4 |
| 注：$S_{XRD}$，$S_{TEM}$分别根据 XRD 图表和 TEM 图的平均粒径算得 | | | | | |

多元醇还原法无论用于制备单金属体系还是双金属体系的高金属含量的高分散电化学催化剂都很合适。使用这种方法可以制备包括 PtFe、PtRu、PtSn 和 PtPd 在内的双金属体系催化剂。

## 10.3　ORR 催化剂

### 10.3.1　高反应活性 PtFe 氧还原催化剂

由于要进行氧的还原反应，所以通常采用 Pt 基催化剂作为阴极催化剂用于直接甲醇燃料电池。我们已经研究了 Pt - Fe/C 系统对 ORR 的反应活性[49]。首先通过多元醇还原法制备出 Pt - Fe/C 催化剂，然后将其在氢气/氩气（10%（体积））的环境下于适当的温度（300℃，Pt - Fe/C300）或高温（900℃，Pt - Fe/C900）中进行热处理。作为比较，我们通过两步法制备 Pt - Fe/C 合金催化剂（Pt - Fe/C900B）。如 X 射线衍射和透射电子显微镜图像（图 10.3）所示，该催化剂的粒径随着热处理温度的升高而增加。Pt - Fe/C300 催化剂的平均粒径为 2.8nm（在 XRD 中）和 3.6nm（在 TEM 中）的，并且该样品部分生成了铂铁合金。Pt - Fe/C900B 催化剂具有 6.2nm（XRD）的最大粒径和最佳的铂铁合金形态。循环伏安（CV）（CV 图在本文中没有出示）表明，Pt - Fe/C300 比其他的 Pt - Fe/C 具有较高的电化学表面积，且具有最高利用率 76%。以上述的Pt - Fe/C 作为阴极催化剂组装直接甲醇燃料电池，实验结果表明，当作为阴极催化剂用于直接甲醇燃料电池时，Pt - Fe/C300 比其他 Pt - Fe/C 或 Pt/C 催化剂显示出更高的氧还原活性（图 10.4）。从逻辑上分析，其对电池性能增强的原因主要是由于其较

① 1Å = 0.1nm。

高的氧还原活性，而这种活性则可能源于铁离子刻蚀催化剂之后更多的 $Pt^0$ 原子出现所致。

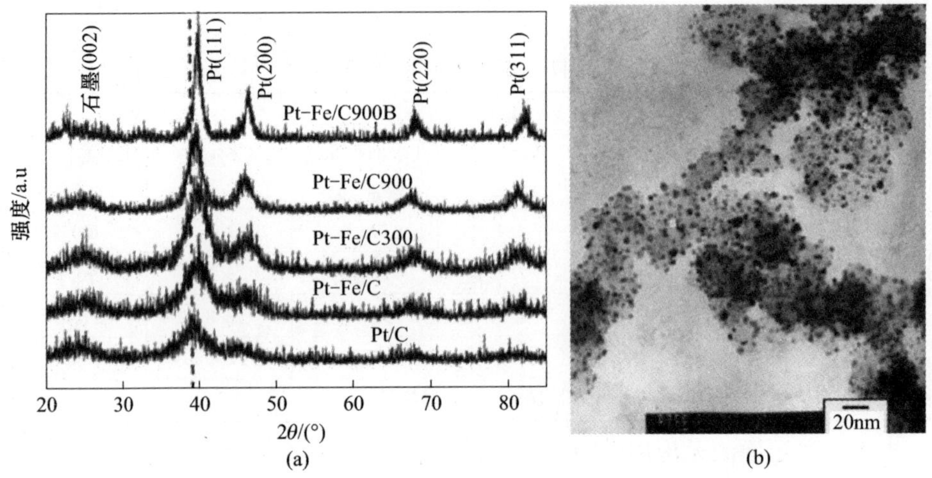


图 10.3　Pt - Fe/C 催化剂的 XRD 图和 Pt - Fe/C300 的透射电镜图[49]（经许可，转自 Elsevier）

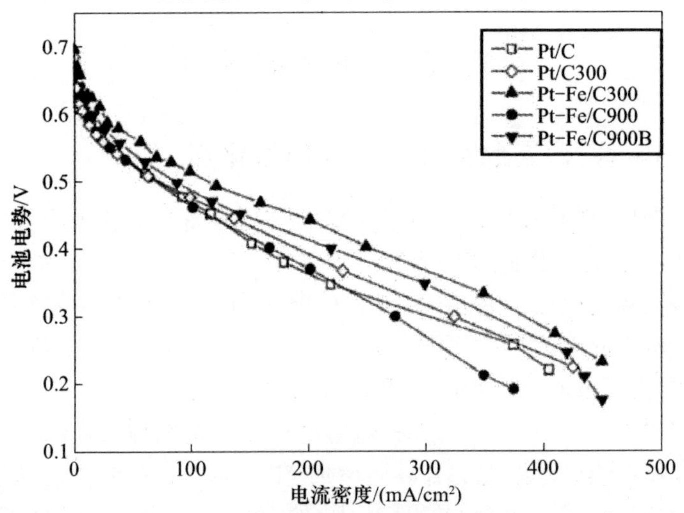


图 10.4　分别以 Pt/C、Pt - Fe/C300、Pt - Fe/C900、Pt - Fe/C900B 为阴极催化剂的单电池 DMFC 的极化曲线（$1mg/cm^2$ Pt）。

阳极：PtRu/C（20%（质量分数）Pt、10%（质量分数）Ru，Johnson Matthey 公司）；

负载催化剂：PtRu，$2mg/cm^2$；电解质膜：Nafion115 膜；工作温度：90℃；

甲醇浓度：1M；流速：1mL/min；氧气压力：0.2MPa[49]（经许可，转自 Elsevier）

## 10.3.2　耐甲醇 PtPd 氧还原催化剂

对于直接甲醇燃料电池来说，甲醇对于电极渗透是其发展的主要障碍之一。

大量的研究与工作致力于避免或减少甲醇的渗透作用对直接甲醇燃料电池的阴极性能的影响[50-55]，其中包括耐甲醇催化剂的研究，如大环化合物或硫化物[34-37]，或通过添加其他金属如铁、钴、镍和钯对铂催化剂进行改性，从而减少从阳极渗透的甲醇对 Pt 阴极催化剂性能的影响。根据之前的研究，尽管 PtFe 是良好的氧还原催化剂并且对甲醇分子表现惰性，但是其在酸性介质中并不稳定，特别是在工作温度为 60～90℃ 范围内更不稳定，这些缺点限制了其在 DMFC 中的应用。正如上面所讨论的，PtFe 似乎是一个具有良好前景的材料，但是其在酸性环境中长期工作会导致铁元素的损失。

近来开发了一种新型碳负载的富钯 $Pd_3Pt_1$ 催化剂，它作为直接甲醇燃料电池催化剂表现出比 Pt/C 更好的催化性能[56]。在有或无甲醇存在的 0.5mol/L 高氯酸溶液中（图 10.5）的氧还原反应极化曲线中，从半波电位来看，$Pd_3Pt_1$/C 的氧还原反应性能与无甲醇的 Pt/C 相近。但在甲醇存在的条件下，Pt/C 的甲醇氧化的峰值很大，以至于它的氧还原活性在 500～900mV 电位范围内性能显著下降；然而，在 $Pd_3Pt_1$/C 的氧还原反应的极化曲线中并没有甲醇氧化峰的出现。这表明，$Pd_3Pt_1$/C 相比 Pt/C 在甲醇存在时可表现出更优异的氧还原选择性。因此，在燃料电池的操作中，当 $Pd_3Pt_1$/C 作为阴极催化剂，与 Pt/C 相比，甲醇渗透对阴极性能的负面影响可以部分地被抑制。我们分别对以 $Pd_3Pt_1$/C 或 Pt/C 作为阴极催化剂的直接甲醇燃料电池进行了测试和比较，图 10.6给出了燃料电池单电池的测试结果。

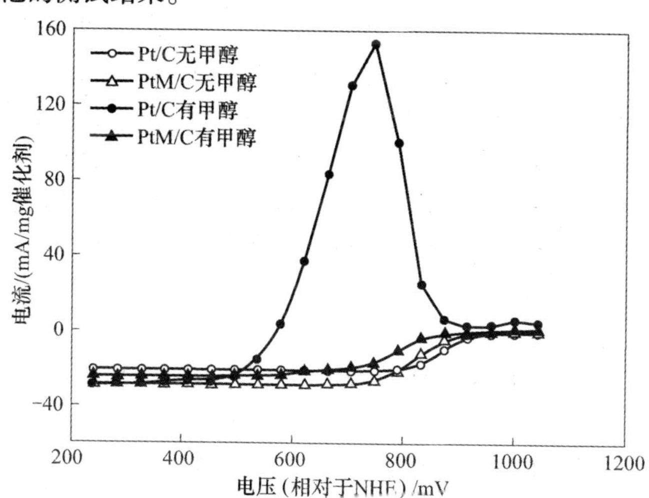


图 10.5　室温下，0.5mol/L 氧饱和的 $HClO_4$ 溶液中，在有无甲醇两种情况下，$Pd_3Pt_1$/C 和 Pt/C 的氧还原极化曲线[56]。
电位扫描：5mV/s，供氧速率：5mL/min；旋转速度：2500r/min
（经许可，转自英国皇家化学学会）

从图 10.6 可见，在所有的电流密度范围内，用 $Pd_3Pt_1$/ C 作阴极催化剂的单

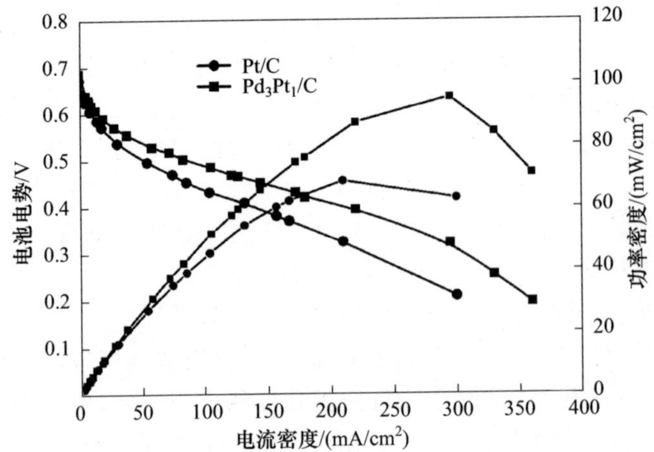


图 10.6 作为 DMFC 阴极催化剂的 $Pd_3Pt_1/C$ 和 Pt/C 的性能测试。

$T_{cell}=75℃$,阳极:PtRu/C(20%(质量分数)Pt、10%(质量分数)Ru,Johnson Matthey 公司),

2.0mg/$cm^2$(Pt+Ru),$C_{methanol}=1.0$mol/L;流速:1.0mL/min;

阴极:Pt/C 或 $Pd_3Pt_1/C$(20%(质量分数)金属),1.0mg/$cm^2$ 金属,$P_{O2}=2.0$atm.;

电解液:Nafion 115 膜[56](经许可,转自皇家化学学会)

个直接甲醇燃料电池表现出比用 Pt/C 作为阴极催化剂更好的性能。单电池测试结果与 RDE 结果一致。$Pd_3Pt_1/C$ 的氧还原活性的提升可能是因为它对于甲醇的氧化反应的惰性,但是其氧还原反应活性与 Pt/C 类似。总之,在甲醇存在的情况下,富钯 $Pd_3Pt_1/C$ 催化剂能提高直接甲醇燃料电池的阴极氧还原反应的选择性,并可作为直接甲醇燃料电池的替代耐甲醇阴极。通过进一步关于 $O_2$ 在 Pt、Pd 族的吸附和解析的理论计算(密度函数理论(DFT))研究发现,除了对甲醇惰性这一优点外,钯原子的存在还有利于氧气在 Pt 活性位点上的离解[57]。

## 10.4 甲醇氧化反应催化剂

### 10.4.1 甲醇氧化反应催化剂的组成筛查

在直接甲醇燃料电池中,与氧的还原相比,甲醇氧化导致的活性降低更为严重,因为在反应条件为单独使用 Pt 作为催化剂并且中间产物(如 CO 类化合物)被完全吸收时,甲醇氧化过程包括了两个部分,即参加反应的每个甲醇分子的 6 电子转移和催化剂的自中毒。从热力学观点来看,甲醇电极氧化是由于在燃料电池中的负吉布斯自由能的变化驱动所致。另一方面,在实际操作条件下,其速率由缓慢的反应动力学控制。为了加快阳极反应速度,有必要开发一种有效的具有高活性的甲醇氧化反应的催化剂。碳载(XC-72C,Cabot 公司)的PtRu、Pt-Pd、PtW 和 PtSn 均由上面所述的改性多元醇还原方法制备[58]。在所有催化剂中 Pt 的含量为20%(质量分数)。

如上面所描述方法制备的 Pt 基催化剂作为阳极催化剂的直接甲醇燃料电池单电池试验如图 10.7 所示。作为比较,Pt/C 作为阳极催化剂也示于图 10.7 中。阴极催化剂为 Pt/C(质量分数 20%,Johnson Matthey 公司)。几何电极面积为 $2.0\times2.0cm^2$。膜电极组件(MEA)制备过程已被详细说明[59]。

从图 10.7 可以看出,在 Pt/C 作为阳极催化剂时,DMFC 电池呈现最差的性能。开路电压(OCV)只有 0.56V,并且该值远离理论值(1.18V),这主要是由于缓慢的反应动力学和甲醇渗透造成的。从图 10.7 也可看出最大放电电流密度约 $165mA/cm^2$,并且在 $120mA/cm^2$ 的情况下,功率密度峰值只有 $17.5mW/cm^2$,从这些实验结果来看,Pt/C 单独使用并不适合作甲醇氧化反应的催化剂。必须通过添加其他物质来提高铂的活性。

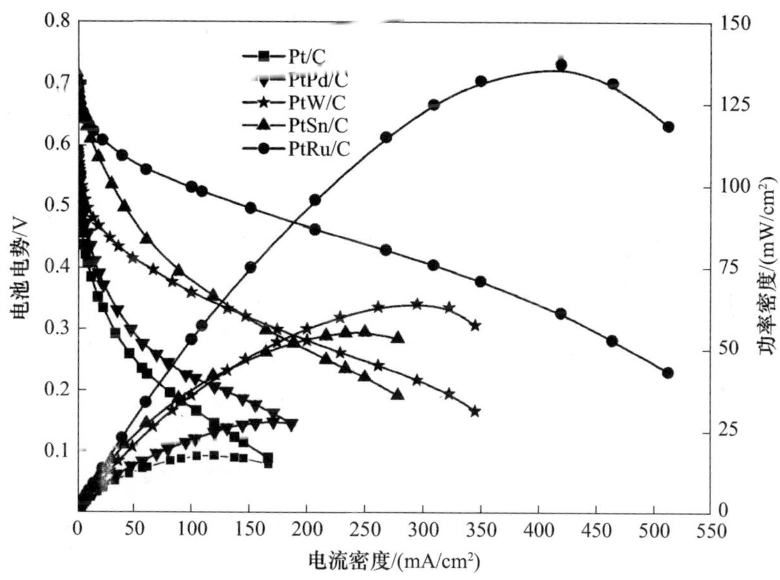


图 10.7 单一 DMFC 的不同阳极催化剂的电池性能。$T_{cell}=90℃$。阳极:含铂催化剂(Pt:M=1:1,20%(质量分数)Pt),$1.33mg/cm^2$ Pt,当 Pt/C 作为阳极时,则为 $2.0mg/cm^2$ Pt。$C_{methanol}=1.0mol/L$,流速:1.0mL/min。阴极:Pt/C(20%(质量分数)Pt,Johnson Matthey 公司),$1.0mg/cm^2$ Pt,$P_{O_2}=2.0atm$。电解液:Nafion 115 膜

对 Pt 来说,Pd、Ru、Sn 和 W 的添加可以整体提高直接甲醇燃料电池的性能。在它们之间,PtPd/C 相比于 Pt/C 仅仅轻微地提高了电池的性能,在 $171mA/cm^2$ 的情况下,开路电压和功率密度峰值分别提高到 0.58V 和$27.9mW/cm^2$,在测试过

的催化剂中,PtSn/C 可以提供最高的开路电压,然而,事实是用 PtSn 作阳极时,开路电压将随电流密度的增加急剧增加。这可能是由于 PtSn/C 催化剂的电子电导率差的缘故,因为锡氧化物在催化剂中存在不同的价态[60]。特别是在更高的电流密度时,锡氧化物欧姆效应变得更为严重,从而导致电池电压下降更快:一方面,W 能提供丰富的含氧物质和其不同价态之间的变化可能加速去除吸附于催化剂上的反应中间体如 $CO_{ads}$;另一方面,从单 DMFC 测试结果来看,PtRu/C 表现出较差的性能。基于这些结果,PtRu 对于直接甲醇燃料电池似乎是效率最高的阳极催化剂,正如 PtRu 是至今为止被广泛接受的甲醇氧化最有效的催化剂的结论是一致的。

### 10.4.2 甲醇氧化反应的碳载铂钌催化剂

为了验证 PtRu/C 催化剂的微观结构和相应的性能影响,对通过浸渍还原法(PtRu-IM)和改性多元醇还原法(PtRu-EG)制备的 PtRu/C 颗粒分别进行了 HRTEM 和 HR-EDS 实验表征。图 10.8 为 PtRu-IM 和 PtRu-EG 的HRTEM图像,从 HRTEM 图中可以看出,PtRu-IM 的金属颗粒大小分布是不均匀的,并且一些小颗粒聚合成体积较大的颗粒,而 PtRu-EG 的颗粒均匀地分散在载体上。为使差异更明显,我们把 HRTEM 和 HR-EDS 的简要数据列在表 10.2 中。PtRu-1 和 PtRu-2 具有相似的平均粒度,然而 PtRu-IM 的粒度分布为1 ~ 6nm,比 PtRu-EG (1.5 ~2.5nm)的粒度要大得多。HR-EDS 颗粒测试的结果清楚地表明 Pt 和 Ru 在 PtRu-IM 样品中的分布不是均匀的。对于较大的颗粒,Pt 与 Ru 比为 8.8:1,而对于较小的颗粒,相应的值是1:8。该结果意味着较大颗粒 Pt 含量高,而较小的颗粒 Ru 含量高。而在 PtRu-EG 的颗粒中,Pt 与 Ru 的含量比较为均匀。对于较大的颗粒,Pt 与 Ru 比为 1.2:1 和对更小的颗粒的相应值是 0.7:1,接近于正常的比率。

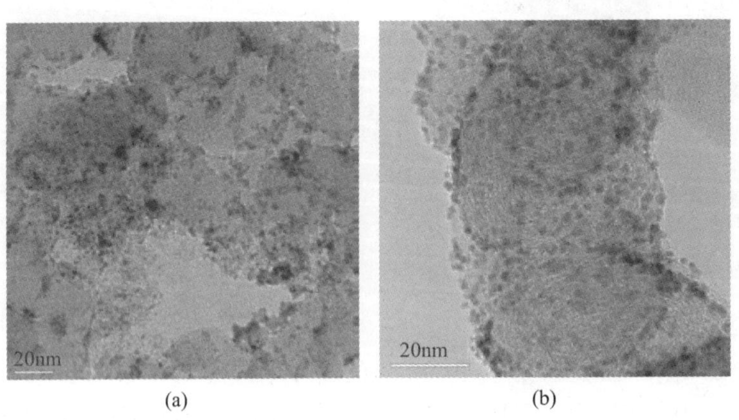


(a) (b)

图 10.8 (a) PtRu/C-IM 和(b) PtRu-EG 的电镜图

表 10.2　HRTEM 和 HR-EDS 结果分析

| 样品 | 平均粒径/nm | HR-EDS 结果(Pt/Ru 原子比) | |
|---|---|---|---|
| | | 小颗粒 | 大颗粒 |
| PtRu/C-IM | 3.5(±2.5) | 1/8.0 | 8.8/1 |
| PtRu/C-EG | 2.5(±0.5) | 0.7/1 | 1.2/1 |

为了提高 PtRu 的合金度,所制备的 PtRu/C-EG 催化剂在氮气中进行 200℃ 热处理。TG 和在线 MS 的结果表明,大量被吸附的物质存在于催化剂表面,通过在氮气气氛下加热 200℃ 处理 2h 可有效地去除这些吸附的物质[61]。TEM 和 X 射线衍射结果(图 10.9)表明,虽然催化剂的粒度在略微热处理后轻微增加,但该颗粒的粒径在碳载体上仍有较好的分散性。PtRu/C-EG 的衍射峰位置在热处理后移向更高角度,Pt 的晶格参数减小(表 10.3),这表明铂和钌之间的相互作用有所增强。电化学和 DMFC 单电池测试结果(图 10.10)表明在热处理后 PtRu/C 的甲醇氧化催化的活性有极大地提高。

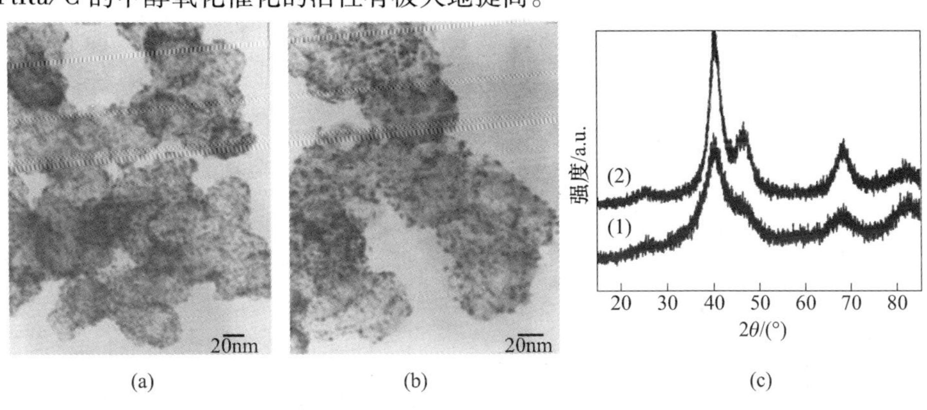


图 10.9　在(a)热处理之前和(b)热处理之后,PtRu/C-EG 催化剂的透射电镜图和(c)XRD 图

表 10.3　PtRu/C-EG 催化剂的 XRD 计算结果表

| | 热处理前 | 热处理后 |
|---|---|---|
| 平均粒径/nm | 2.1 | 2.6 |
| (220)峰位置/(°) | 68.4 | 68.6 |
| 晶格参数/nm | 0.388 | 0.386 |

Watanabe 和 Motoo[62] 提出的双功能机理认为,在 Pt 的活性位点上发生的甲醇脱氢反应可产生类似 CO 的物质。Ru 分解水产生—OH 物质的电位比 Pt 产生—OH 物质的电位更低,Ru 还可以与 Pt 活性位点上产生的 CO 类似化合物反应并使之解毒。在此机理的前提下,该催化剂模型可被看为 PtRu 合金(原子比的 Pt/Ru = 1∶1)的混合是在两个元素的原子尺度之内(图 10.11)。

在该模型的基础上,Pt 和 Ru 应在原子尺度混合以便于顺利除去 CO 类物

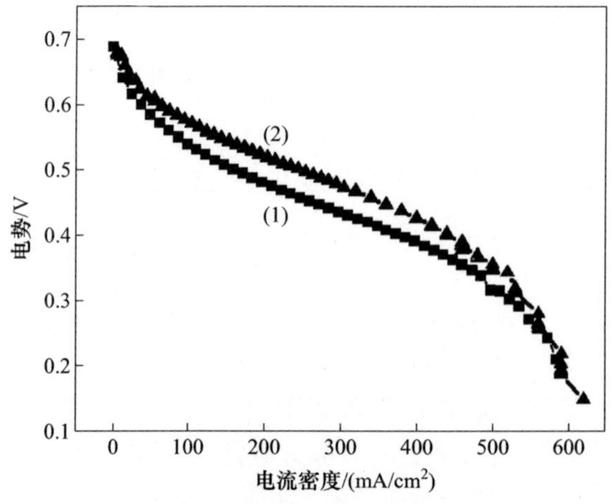


图 10.10 (1)在热处理之前和(2)热处理之后,以 PtRu/C-EG 为阳极催化剂的 DMFC 单电池性能的比较

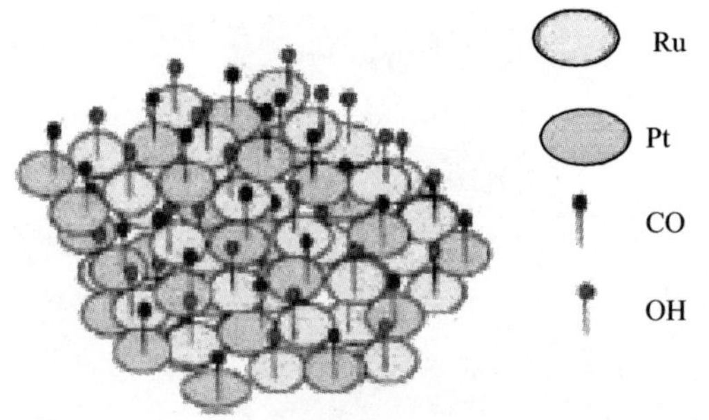


图 10.11 甲醇电化学氧化反应的 PtRu 催化剂模型

质。从之前的 HRTEM 和 HR-EDS 分析结果来看,PtRu-EG 的颗粒尺寸均匀且分布的范围狭窄((2.5 ±0.5)nm)而且 Pt 和 Ru 原子尺寸的分布是均匀的,依据双功能机理,这种结构有利于除去类似 CO 的物质。PtRu-IM 的特征具是较宽的粒度分布((3.5 ±2.5)nm),而 Pt 和 Ru 的分布在原子尺度上是不均匀的,不能顺利消除毒物。这种逐渐积累的毒物会导致 Pt 活性部位逐渐失活。

## 10.5 乙醇氧化的催化剂

为了拓展低温燃料电池的实际应用使其进入运输商业市场,需考虑增加可应用于此类设备上的液体燃料种类。在所有可能的燃料中,乙醇是最有前途的,因为它是一种天然的、可再生燃料,对经济和环境方面有积极的影响[63,64]。此外,乙醇有比甲醇更高的能量密度(分别为 8.01kW · h · $kg^{-1}$ 和 6.09kW · h · $kg^{-1}$)。因此,由于乙醇可以满足大多数的低温燃料电池的要求,使其具有一定吸引力[65,66]。

但是与甲醇电氧化相比，乙醇的电氧化似乎是一个更复杂的过程，因为它涉及每个乙醇分子发生 12 个电子转移和 C—C 单键的裂解。为了促进 DEFC 的发展，开发一种新颖的具有高活性的乙醇氧化催化剂是非常必要的。

### 10.5.1　乙醇氧化反应的催化剂的成分筛选

文献[44,67]报道了碳载 PtPd、PtW、PtRu 和 PtSn 作为乙醇氧化催化剂的性能分析。图 10.12 为不同的碳载 Pt 和 PtM 催化剂的循环伏安法实验结果。当 Pt/C 催化剂作为乙醇的氧化催化剂时，其循环伏安曲线上会出现两个氧化峰。第一个出现在约 0.76V（相对于 SCE），第二个出现在一个更高的电位上。但只有第一个氧化峰在已有的文献上报道过，而且为了防止辅助金属电极的溶解，其所施加的最高电位相对于 SCE 不允许超过 1.0V。在 Pt 催化剂中加入第二种金属元素会导致乙醇的第一个氧化峰向低电位移动。使用 $Pt_1Ru_1/C$ 催化剂会使第一个氧化峰出现在更低的电位上，约 0.53V（相对于 SCE），并且电势的峰值比 Pt 催化剂的值约低 0.23V。而 $Pt_1Pd_1/C$ 的乙醇第一电氧化峰值约 0.65V（相对于 SCE），比 $Pt_1Ru_1/C$ 的峰值高。在第一个乙醇电化学氧化的峰值处，$Pt_1Ru_1/C$ 的电流密度值比 $Pt_1Pd_1/C$ 高，但比 $Pt_1Sn_1/C$、$Pt_1W_1/C$ 和 Pt/C 中任何一个的电流密度都低。当电流值为第一峰值所对应的电流值时，$Pt_1Sn_1/C$ 催化剂对乙醇氧化具有最高的电催化活性，但是同样也具有更高的过电势（0.71V 相对于 SCE）。$Pt_1W_1/C$ 催化剂也具有比 Pt/C、$Pt_1Ru_1/C$ 和 $Pt_1Pd_1/C$ 更高的电流密度，但存在与 Pt/C 一样的高过电位（0.75V 相对于 SCE）。从电流密度大小的角度来看，$Pt_1Sn_1/C$ 似乎是乙醇氧化反应最好的催化剂。在已经测试过的催化剂中，使用 $Pt_1Sn_1/C$ 作为催化剂的乙醇电化学氧化催化过程具有最低的过电势，这表明 $Pt_1Sn_1/C$ 催化剂是一种具有发展前景的乙醇电化学氧化催化剂。

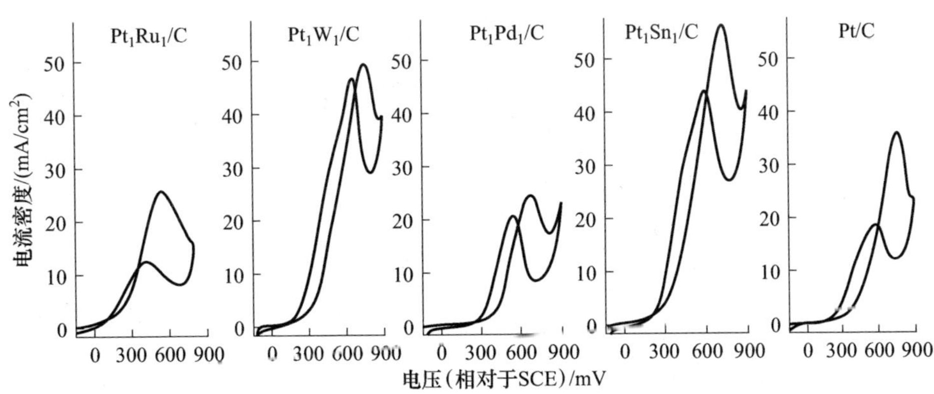


图 10.12　不同阳极催化剂对乙醇电化学氧化反应催化作用的循环伏安曲线。
操作温度：25℃，扫描速度：10mV/s；电解液：1M 乙醇和 0.5M 硫酸[67]
（经许可，转自 Elsevier）

从实际应用方面来说,所有具有潜力的催化剂最终都应该置于燃料电池系统中进行研究。通过一个简单的 DEFC 测试对上面所提到的 5 种催化剂作为乙醇电化学氧化的阳极催化剂的性能进行测评。图 10.13 展示了不同阳极催化剂在单电池中工作温度 90℃时的性能参数。当 Pt/C 作为催化乙醇的阳极催化剂时,单电池的性能很差。OCV 电压大约为 0.55V,远低于标准电动势(1.145V),造成这一现象的原因是 Pt/C 对于电催化甲醇的催化活性较弱并且出现了甲醇从阳极向阴极渗透的情况。其最大输出能量密度仅有 10.8mV/$cm^2$。然而采用 $Pt_1Pd_1/C$ 代替 Pt/C 作为阳极催化剂时,单电池的简单 DEFC 测试性能并没有明显提升,二者的电流密度 - 电压($I-V$)曲线非常相似。当采用 $Pt_1W_1/C$ 作为阳极催化剂时,与 $Pt_1Pd_1/C$ 或 Pt/C 催化剂相比,其性能上有了极大提高,而这种性能的提高主要表现于内电阻控制区和传质区。其在 90℃时的最大输出功率密度可以接近 16.0mW/$cm^2$。当 $Pt_1Ru_1/C$ 或 $Pt_1Sn_1/C$ 用于电池中时,OCV 与最高功率密度都得到了提升。$Pt_1Ru_1/C$ 单电池的 OCV 为 0.67V,高于 Pt/C 单电

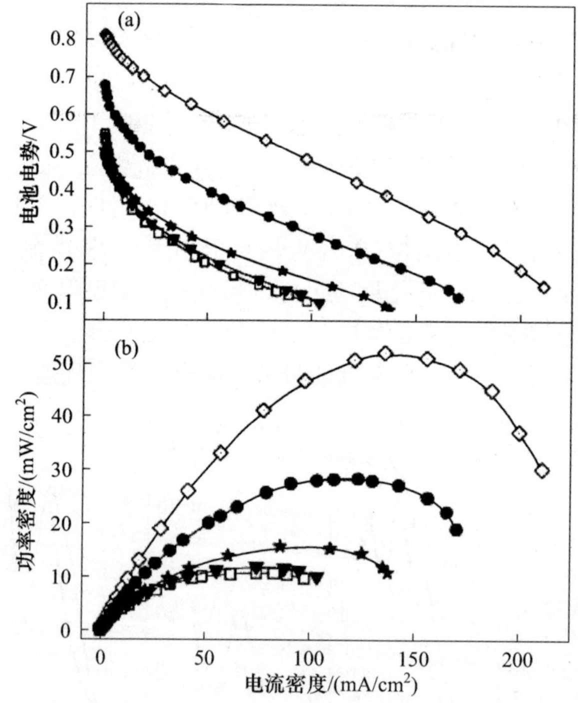


图 10.13　不同阳极催化剂的直接乙醇燃料电池的性能比较,操作温度 90℃。(□)Pt/C, 2mg/$cm^2$Pt;(▼)$Pt_1Pd_1/C$, 1.3mg/$cm^2$Pt;(★)$Pt_1W_1/C$, 2mg/$cm^2$Pt;(●)$Pt_1Ru_1/C$, 1.3mg/$cm^2$Pt;(◇)$Pt_1Sn_1/C$, 1.3mg/$cm^2$Pt;电解液:Nafion 115;乙醇浓度和流速:1M 和 1mL/min;阴极催化剂和金属负载量:1mg/$cm^2$Pt(20%(质量分数)Pt,Johnson Matthey 公司)[67](经许可,转自 Elsevier)

池约 0.12V，并且在 90℃时最大功率密度为 28.6mW/cm$^2$。当 $Pt_1Sn_1/C$ 作为阳极催化剂时，单电池的 OCV 达到了约 0.81V，高于 $Pt_1Ru_1/C$ 阳极催化剂约 0.14mV。而且 $Pt_1Sn_1/C$ 电池的最大功率密度达到 52mW/cm$^2$，约为 $Pt_1Ru_1/C$ 催化剂的两倍。单电池电化学实验结果指出在 90℃的工作温度 $Pt_1Ru_1/C$ 作为 DEFC 催化剂时更稳定。

尽管 $Pt_1Ru_1/C$ 是 DEFC 甲醇电化学氧化的最佳催化剂。但是事实证明其却不是电化学氧化乙醇最好的阳极催化剂。而向 $Pt_1Sn_1/C$ 催化剂中添加 W 和 Mo 可以提升其乙醇电化学氧化活性。即使如此，直接乙醇燃料单电池的 $I-V$ 表征结果仍明确表明 $Pt_1Sn_1/C$ 是比 $Pt_1Ru_1/C$ 以及其他碳载双金属 Pt 基催化剂更好的 DEFC 阳极催化剂。而且从实际应用角度来看，$Pt_1Sn_1/C$ 仍然是这里介绍的阳极催化剂中最好的 DEFC 催化剂。

### 10.5.2　PtSn/C 的乙醇电化学氧化

鉴于 $Pt_1Sn_1/C$ 是 EOR 反应的高活性阳极催化剂，研究者们进行了大量的研究以试图探寻 Pt/Sn 原子比例对催化剂性能的影响[44]。通过 XRD 谱图与 TEM 图所获得的 PtSn/C 催化剂的平均粒径与晶格参数列于表 10.4 中。图 10.14 所示为以不同的 PtSn/C 作阳极催化剂的单电池的性能参数。

表 10.4　碳载铂和铂双金属催化剂的 XRD 和电镜结果分析[44]

| 催化剂 | 平均粒径/nm | | 晶格参数/Å |
|---|---|---|---|
| | TEM | XRD | |
| $Pt_1Sn_1/C$ | 2.3 | 2.1 | 3.987 |
| $Pt_3Sn_2/C$ | 2.2 | 1.9 | 3.973 |
| $Pt_2Sn_1/C$ | 3.0 | 2.6 | 3.956 |
| $Pt_3Sn_1/C$ | 2.2 | 1.9 | 3.953 |
| $Pt_4Sn_1/C$ | 2.3 | 1.9 | 3.938 |
| 注：转自 Elsevier | | | |

在随后的研究中，研究者做了大量的工作，从原子尺度上对 PtSn/C 催化剂的组成进行控制。我们通过调整制备条件获得了均一的 PtSn/C 催化剂分子。图 10.15 所示为 $Pt_3Sn/C$ 纳米颗粒的 HRTEM 图与 HR-EDS 分析。

由图 10.15 可见，金属颗粒是均一的，平均粒径为 3nm。HR-EDS 分析数据是从随机线性区域获得的，并且结果显示 Pt 的分布与 Sn 的分布一致，这表明所有颗粒中 Pt 与 Sn 的分布都是相同的。

我们的初步研究表明氧化锡可以作为乙醇的电化学氧化反应的活性组分[60,68-70]。我们的研究焦点是研究锡的化学态以及 PtSn/C 催化剂的组成对 DEFC 性能的影响。为了做对比，分别制备了两个有氧化锡的 PtSn/C 催化剂和

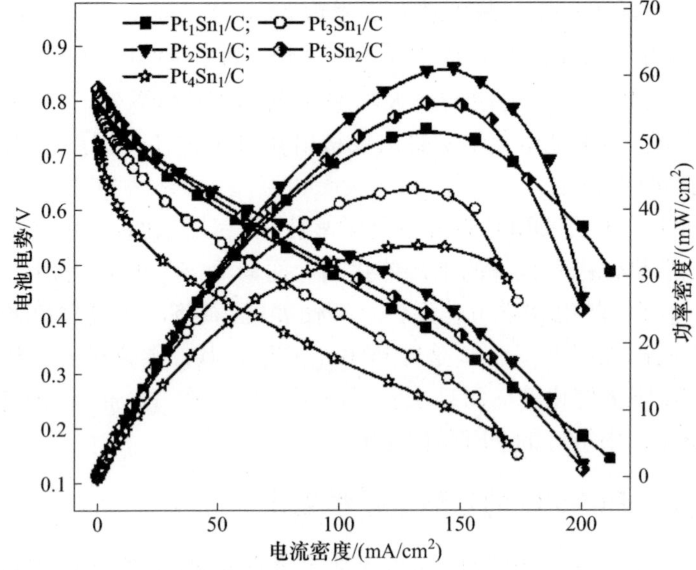


图 10.14 以不同 PtSn/C 作阳极催化剂的直接乙醇燃料电池的性能。

温度为 90℃，阳极是不同 Pt/Sn 原子比（1.33mg/$cm^2$ Pt）的 PtSn/C；固体电解质为 Nafion 115 膜；乙醇水溶液浓度为 1mol/L，流速为 1mL/min；阴极含有 Pt/C（Johnson Matthey 公司），其中铂含量为 1.0mg/$cm^2$ Pt（经许可，转自 Elsevier）

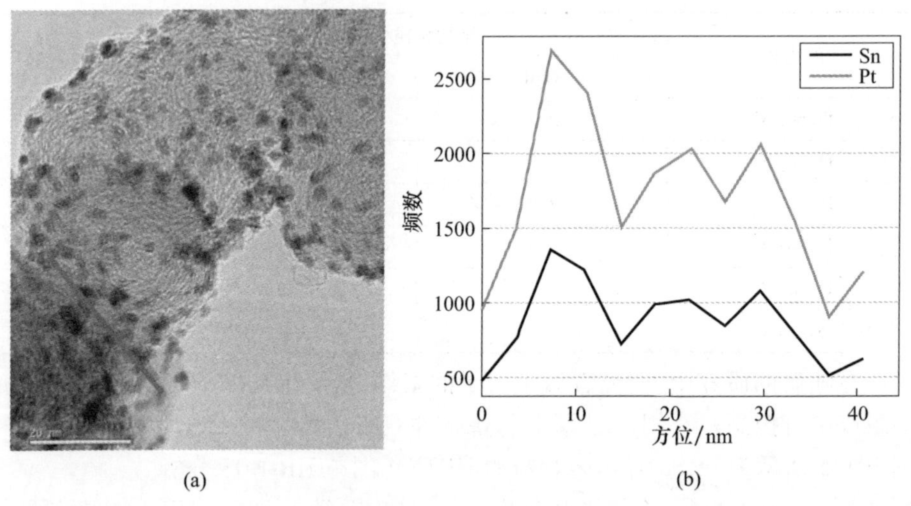


(a) (b)

图 10.15 $Pt_3Sn$/C 纳米颗粒的（a）HRTEM 图和（b）HR-EDS 分析

PtSn 合金。前者，直径为 1nm 的氧化锡预先在乙二醇中制好，然后铂在表面或接近氧化锡的地方被还原（记为 PtSn-1）。后者，首先把铂和锡的前体混合在一起，然后在乙二醇中被还原（记为 PtSn-2）。

PtSn-1 和 PtSn-2 的 X 射线衍射图谱如图 10.16 所示。从图中可以看出，除了 Pt(111)、Pt(200)、Pt(220)和 Pt(311)这些衍射峰外,在 PtSn-1 的图谱中,大约在 34°和 52°(PCPDF#411445)处分别还有 $SnO_2$(101)和 $SnO_2$(211)的衍射峰。通过谢乐公式计算得 PtSn-1 和 PtSn-2 有相同的粒径,大约为 2.3nm。通过 Vegard 定律可得,PtSn-1 和 PtSn-2 的晶格参数分别为 3.928Å 和 3.946Å。与 Pt 的晶格参数 3.923Å 相比,样品 PtSn-2 中 Pt 的晶格有明显增长,而在 PtSn-1 中只有很少的增长。一般情况下,晶格的扩张度反映了两种金属的合金程度。为了验证 PtSn 催化剂的微观结构,PtSn-1 和 PtSn-2 的高分辨率透射电子显微镜图如图 10.17 所示。在 PtSn-1 的电镜图中,可以看到 $SnO_2$ 纳米颗粒在 Pt 微粒的附近。图 10.17(a)中箭头所指之处即为 $SnO_2$(101)晶面标定的 0.264nm 区。图 10.17(a)中亦以箭头标出了面心立方结构的典型粒子的 Pt(111)面和 Pt(200)面的晶面间距,分别为 0.228nm 和 0.198nm。在图 10.17(b)中,只看到了 Pt(111)面的间距为 0.234nm,在 Pt 微粒的附近没有看到单独的 $SnO_2$ 纳米颗粒。基于此,我们可以得出结论:在样品 PtSn-1 中,部分锡存在于氧化锡中,部分锡存在于铂锡合金中;而在样品 PtSn-2 中,绝大部分锡以锡铂合金的形式存在。

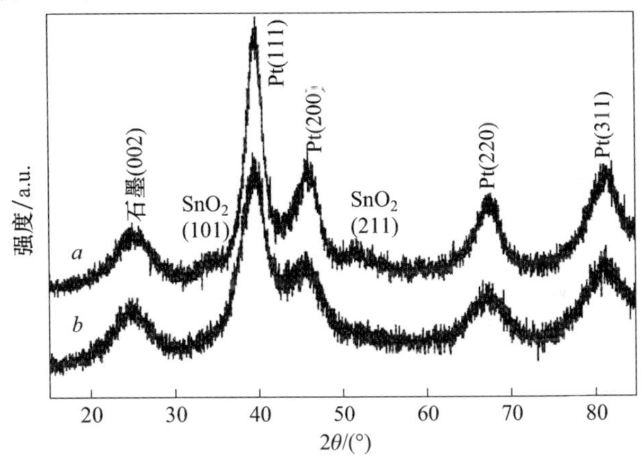


图 10.16　(a)PtSn-1 和(b)PtSn-2 的 X 射线衍射图[70](经许可,转自 Elsevier)

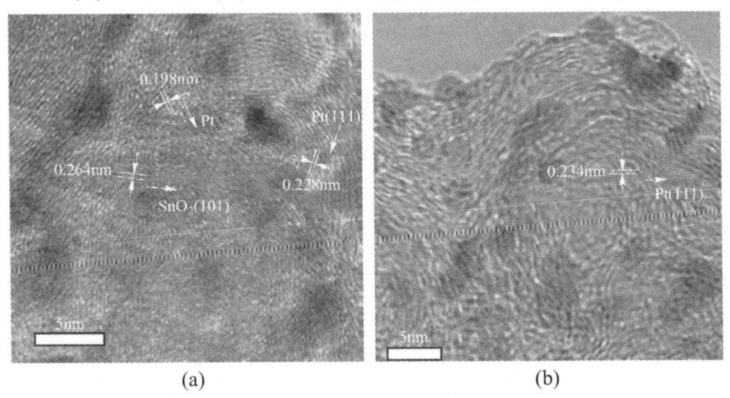


图 10.17　(a)PtSn-1 和(b)PtSn-2 的高分辨透射电镜图[70](经许可,转自 Elsevier)

循环伏安法是一种实用又简单的表征催化剂电催化活性的方法。图 10.18 所示为 PtSn 电极在 0.3M $HClO_4$ 和 1M EtOH 溶液中的 $C-V$ 曲线。PtSn-1 和 PtSn-2 的 EOR 初始电压(在 0.2mA 时)分别为 0.033V 和 0.124V。在全部扫描范围内,PtSn-1 电极的乙醇的电化学氧化电流比 PtSn-2 电极高。

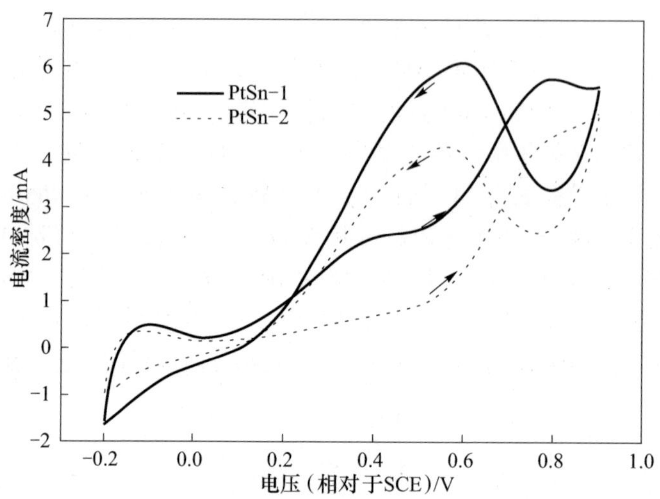


图 10.18　室温下,PtSn 电极在 0.3M $HClO_4$ 和 1M EtOH 溶液中的 $C-V$ 曲线[70](经许可,转自 Elsevier)

图 10.19 所示为分别以 PtSn-1 和 PtSn-2 作阳极催化剂的 DEFC 的 $I-V$ 曲线。由图可知,以 PtSn-1 为阳极催化剂的 DEFC 性能要比以 PtSn-2 为阳极催化剂的高。以 PtSn-1 和 PtSn-2 作阳极催化剂的 DEFC 的最大功率密度和最大交换电流分别为 $81mW/cm^2$($200mA/cm^2$) 和 $47mW/cm^2$($240mA/cm^2$)。在我们之前的研究中[60],我们推断氧化锡在低电压下比铂更易提供含氧物种,该含氧物可以与乙醇的电化学氧化反应生成的有毒 CO 类似物反应。在此,根据之前的研究,PtSn-1 含有 PtSn 合金和氧化锡纳米颗粒,而在 PtSn-2 中,大部分为铂锡合金,这也是它使 Pt 的晶格参数扩大的原因。在 PtSn 催化剂的 MOR 研究中,有人认为 Pt 的晶格参数的扩张限制了 Pt 对甲醇和 C—H 键的吸附能力[71]。同样,由于铂锡的全合金,乙醇的吸附和解离可能也会被限制。我们之前的研究表明,Pt 晶体的晶格参数适当扩张是有利于乙醇吸附的[58]。乙醇分子在 Pt 的活性位点被吸附,脱氢产生 CO 类似物。在样品 PtSn-2 中,乙醇的电化学氧化残留物不能从铂的活性位点除去,因为它的附近没有含氧物。但是,对于 PtSn-1 来说,乙醇的电化学氧化残留物可以和铂微粒附近的氧化锡反应以释放铂的活性位点。基于这个讨论,乙醇电化学氧化反应的理想 PtSn 催化剂应该是适当比例

的铂锡合金和部分锡在氧化物中的组合。

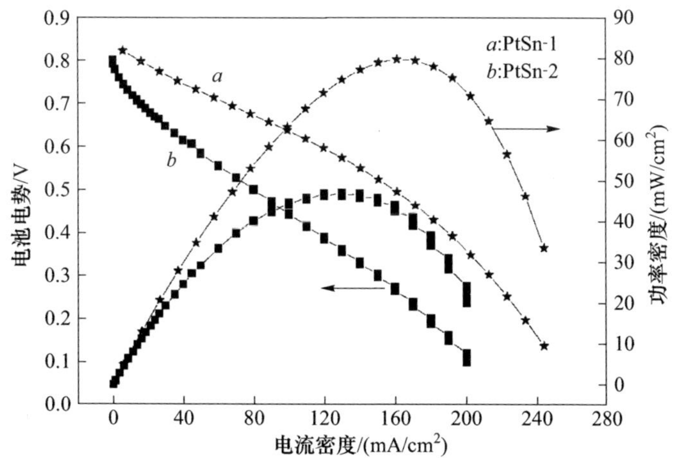


图 10.19　分别以(a)PtSn-1 和(b) PtSn-2 作阳极催化剂的 DEFC 的 $I-V$ 曲线[70](经许可,转自 Elsevier)

通过不同的电化学质谱学分析法(DEMS)[72],在三电极系统中对不同 Pt/Sn 原子比例的 PtSn-1 的 EOR 结果进行了深入研究。结果显示,乙酸和甲醛是两种主要的产物,$CO_2$ 的产量比在 PtSn-1 或 PtSn 合金催化剂上的低 5%,这表明乙酸的生成需要两个活性位点:一个(Pt)脱氢,另一个($SnO_x$)给乙醛提供 OH 物种以生成乙酸。当在实际的燃料电池中测试 PtSn-1 催化剂时结果完全不同。据报道,EOR 的 $CO_2$ 电流效率随着负载的催化剂量和温度的增加而增加。在温度为 90℃,0.1M 乙醇和 5mg/$cm^2$ 铂催化剂负载量的条件下,$CO_2$ 的电流效率已经达到了 75% [73]。$CO_2$ 的产量随着温度和多孔催化剂层厚度的增加而增加是合乎情理的,EOR 的中间体也很有可能被重新吸附,进一步被氧化。

## 10.5.3　IrSn/C 对乙醇电化学氧化的催化

在室温下,以 $NaBH_4$ 为还原剂,乙二醇为保护剂,以碳为载体易制得 $Ir_3Sn$/C 和 Ir/C 催化剂[74]。TEM 和 XRD 数据显示,小粒径的催化剂表现出了典型的铱元素的面心立方结构。它们与 Pt/C 和 $Pt_3Sn$/C 催化剂的电催化活性比较可见线性扫描伏安法数据分析(图 10.20)。结果表明,在低电位区的乙醇氧化催化活性比较中,含铱催化剂相比 Pt/C 和 $Pt_3Sn$/C 催化剂表现出更优异的电催化活性。在 90℃的单电池测试中(图 10.21),与 Pt/C 催化剂相比,作为阳极的铱基催化剂表现出了更好的性能。相比于 $Pt_3Sn$/C,$Ir_3Sn$/C 表现出来的整体性能优异,可为 DEFC 的阳极催化剂提供一个有希望的替代选择。

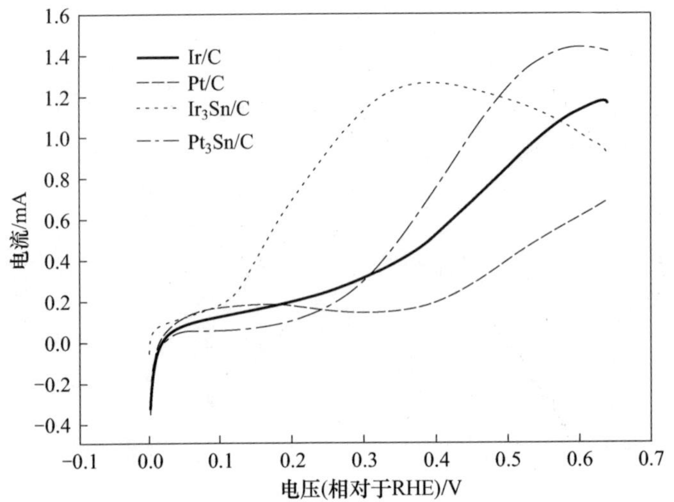


图 10.20 室温下，0.5M $H_2SO_4$ 和 1M 乙醇，分别以 Ir/C，Pt/C，$Ir_3Sn/C$ 和 $Pt_3Sn/C$ 为催化剂的乙醇的电催化氧化反应的线性扫描曲线，扫描速率为 10mV/s[74]（经许可，转自 Elsevier）

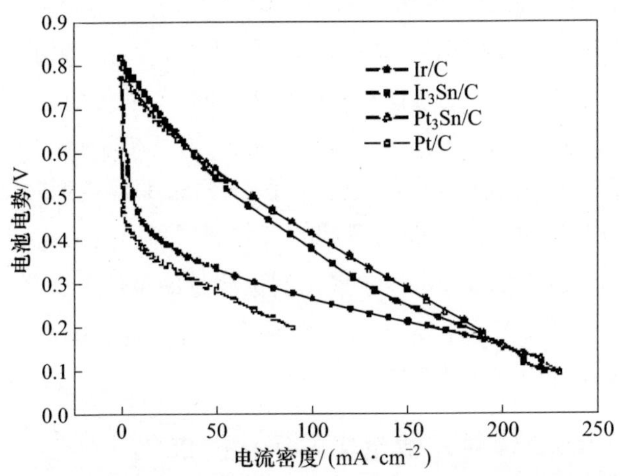


图 10.21 分别以 Ir/C、$Ir_3Sn/C$、$Pt_3Sn/C$ 和 Pt/C 作阳极催化剂的 DEFC 的极化曲线。阳极：1M 乙醇，流速为 1mL/min，电池温度 90℃，阴极：两节负载阳极催化剂的 1.5mg/$cm^2$ 贵金属，1mg/$cm^2$ 铂阴极催化剂（40% 的 Pt/C）[74]（经许可，转自 Elsevier）

## 10.6 本章小结

我们对过去 10 年里 DICP 的直接醇类燃料电池在催化剂领域的研究和发展进行了回顾。总的来说，研究进展包括以下几个方面：①采用一种简便环保的多元醇还原法，可制备具有可控的金属粒径，高分散度和结构均一的催化剂，甚至是高负载量的金属催化剂;②PtFe/C 和 PtPd/C 催化剂对 ORR 表现了一定的活性,并且 PtPd 对 MOR 不反应,使其可以作为有效的耐甲醇 ORR 催化剂;③Pt-Ru/C 被认为是对 MOR 最有效的催化剂之一,经热处理后,不但可以增强 PtRu 合金度,还可提高 MOR 的活性;④PtSn 在乙醇的电化学氧化中比其他含铂催化剂表现出了更高的活性。存在于氧化物中的锡提供的 OH 物种促进了乙酸的生成,同时 PtSn 合金上的乙醇更容易转化生成甲醛。

## 参考文献

[1] Cheng, X., Yi, B., Han, M., Zhang, J., Qiao, Y., and Yu, J. (1999) Investigation of platinum utilization and morphology in catalyst layer of polymer electrolyte fuel cells. *Journal of Power Sources*, **79** (1), 75 – 81.

[2] Wang, X., Hsing, I.-M., and Yue, P. L. (2001) Electrochemical characterization of binary carbon supported electrode in polymer electrolyte fuel cells. *Journal of Power Sources*, **96** (2), 282 – 287.

[3] Schmal, D., Kluiters, C. E., and Barendregt, I. P. (1996) Testing of a De Nora polymer electrolyte fuel cell stack of 1kW for naval applications. *Journal of Power Sources*, **61** (1 – 2), 255 – 257.

[4] Fournier, J., Faubert, G., Tilquin, J. Y., Côte, R., Guay, D., and Dodelet, J. P. (1997) High-performance, low Pt content catalysts for the electroreduction of oxygen in polymer-electrolyte fuel cells. *Journal of the Electrochemistry Society*, **144**, 145 – 154.

[5] Susai, T., Kawakami, A., Hamada, A., Miyake, Y., and Azegami, Y. (2001) Development of a 1kW PEM fuel cell power source. *Fuel Cells Bulletin*, **3**, 7 – 11.

[6] Glaebrook, W. (1982) Efficiencies of heat engines and fuel cells: the methanol fuel cell as a competitor to otto and diesel engines. *Journal of Power Sources*, **7** (3), 215 – 256.

[7] Ren, X., Zelenay, P., Thomas, S. C., Davey, J., and Gottesfeld, S. (2000) Recent advances in direct methanol fuel cells at Los Alamos National Laboratory. *Journal of Power Sources*, **86**, 111 – 116.

[8] Scott, K., Argyropoulos, P., and Sundmacher, K. (1999) A model for the liquid feed direct methanol fuel cell. *Journal of Electroanalytical Chemistry*, **477**, 97 – 110.

[9] Scott, K., Taama, W. M., Argyropoulos, P., and Sundmacher, K. (1999) The impact of mass transport and methanol crossover on the direct methanol fuel cell. *Journal of Power Sources*, **83**, 204 – 216.

[10] Baldauf, M. and Preidel, W. (1999) Status of the development of a direct methanol fuel cell. *Journal of Power Sources*, **84**, 161 – 166.

[11] Gao, P., Chang, S., Zhou, Z., and Weaver, M. J. (1989) Electrooxidation pathways of simple alcohols at platinum in pure nonaqueous and concentrated aqueous environments as studied by real-time FTIR spectroscopy. *Journal of Electroanalytical Chemistry*, **272**, 161 – 178.

[12] Wang, J., Wasmus, S., and Savinell, R. F. (1995) Evaluation of ethanol, 1-propanol, and 2-propanol in a direct oxidation polymer-electrolyte fuel cell. *Journal of the Electrochemistry Society*, **142**, 4218 – 4224.

[13] Qi, Z., Hollett, M., Atita, A., and Kaufman, A. (2002) Low temperature direct 2-propanol fuel cells. *Electrochemical Solid-State Letters*, **5**, A129 – A130.

[14] Peled, E., Duvdevani, T., Aharon, A., and Melman, A. (2001) New fuels as alternatives to methanol for direct oxidation fuel cells. *Electrochemical Solid-State Letters*, **4**, A38 – A41.

[15] Narayanan, S. R., Vamos, E., Surampudi, S., Frank, H., Halpert, G., Prakash, G. K. S., Smart,

M. C., Knieler, R., Olah, G. A., Kosek, J., and Cropley, C. (1997) Direct electro-oxidation of dimethoxymethane, trimethoxymethane, and trioxane and their application in fuel cells. *Journal of the Electrochemistry Society*, **144**, 4195 – 4201.

[16] Rice, C., Ha, S., Masel, R. I., Waszczuk, P., Wieckowski, A., and Barnard, T. (2002) Direct formic acid fuel cells. *Journal of Power Sources*, **111**, 83 – 89.

[17] Iwasita, T., Hoster, H., John-Anacker, A., Lin, W. F., and Vielstich, W. (2000) Methanol oxidation on PtRu electrodes. Influence of surface structure and Pt – Ru atom distribution. *Langmuir*, **16**, 522 – 529.

[18] Watanabe, M., Uchida, M., and Motoo, S. (1987) Preparation of highly dispersed Pt + Ru alloy clusters and the activity for the electrooxidation of methanol. *Journal of Electroanalytical Chemistry*, **229**, 395 – 406.

[19] Iwasita, T., Nart, F. C., and Vielstich, W. (1990) An FTIR study of the catalytic activity of a 85:15 Pt:Ru alloy for methanol oxidation. *Physics and Chemistry*, **94**, 1030 – 1034.

[20] Rolison, D. R., Hagans, P. L., Swider, K. E., and Long, J. W. (1999) Role of hydrous ruthenium oxide in Pt – Ru direct methanol fuel cell anode electrocatalysts: the importance of mixed electron/proton conductivity. *Langmuir*, **15**, 774 – 779.

[21] Frelink, T. and Visscher, W. (1994) The effect of Sn on Pt/C catalysts for the methanol electro-oxidation. *Electrochimica Acta*, **39**, 1871 – 1875.

[22] Aricò, A. S., Antonucci, V., and Giordano, N. (1994) Methanol oxidation on carbonsupported platinum-tin electrodes in sulfuric acid. *Journal of Power Sources*, **50**, 295 – 309.

[23] Janssen, M. M. P. and Moolhuysen, J. (1976) Platinum-tin catalysts for methanol fuel cells prepared by a novel immersion technique, by electrocodeposition and by alloying. *Electrochimica Acta*, **21**, 861 – 868.

[24] Janssen, M. M. P. and Moolhuysen, J. (1976) Binary systems of platinum and a second metal as oxidation catalysts for methanol fuel cells. *Electrochimica Acta*, **21**, 869 – 878.

[25] Watanabe, M., Furuuchi, Y., and Motoo, S. (1985) Electrocatalysis by AD-atoms: Part XIII. Preparation of ad-electrodes with tin ad-atoms for methanol, formaldehyde and formic acid fuel cells. *Journal of Electroanalytical Chemistry*, **191**, 367 – 375.

[26] Shukla, A. K., Ravikumar, M. K., and Aricò, A. S. (1995) Methanol electrooxidation on carbon-supported Pt-$WO_{3-x}$ electrodes in sulphuric acid electrolyte. *Journal of Applied Electrochemistry*, **25** (6), 528 – 532.

[27] Shen, P. K. and Tseung, A. C. C. (1994) Anodic oxidation of methanol on Pt/WO3 in acidic media. *Journal of the Electrochemistry Society*, **141**, 3082 – 3090.

[28] Wang, J., Nakajima, H., and Kita, H. (1990) Metal electrodes bonded on solid polymer electrolyte membrane (SPE): VI. Methanol oxidation on molybdenum modified Pt-SPE electrode. *Electrochimica Acta*, **35**, 323 – 328.

[29] Grgur, B. N., Zhuang, G., Marvovich, M. M., Jr., and Rose, P. N. (1997) Electrooxidation of $H_2$/CO mixtures on a well-characterized Pt75Mo25 alloy surface. *Journal of the Physical Chemistry B*, **101**, 3910 – 3913.

[30] Toda, T., Igarashi, H., and Watanabe, M. (1999) Enhancement of the electroreduction of oxygen on Pt alloys with Fe, Ni, and Co. *Journal of the Electrochemistry Society*, **146**, 3750 – 3756.

[31] Toda, T., Igarashi, H., and Watanabe, M. (1999) Enhancement of the electrocatalytic $O_2$ reduction on Pt-Fe alloys. *Journal of Electroanalytical Chemistry*, **460**, 258 – 262.

[32] Wan, L., Moriyama, T., Ito, M., Uchida, H., and Watanabe, M. (2002) In situ STM imaging of sur-

face dissolution and rearrangement of a Pt-Fe alloy electrocatalyst in electrolyte solution. *Chemical Communications*, 58.

[33] Uchiida, H., Ozuka, H., and Watanabe, M. (2002) Electrochemical quartz crystal microbalance analysis of CO-tolerance at Pt-Fe alloy electrodes. *Electrochimica Acta*, **47**, 3629 – 3636.

[34] Jasinski, R. (1964) A new fuel cell cathode catalyst. *Nature*, **201**, 1212 – 1213.

[35] Putten, V. D., Elzing, A., and Visscher, W. (1986) Oxygen reduction on pyrolysed carbon-supported transition metal chelates. *Journal of Electroanalytical Chemistry*, **205**, 233 – 244.

[36] Durand, R. R., Bencosme, C. S., and Collman, J. P. (1983) Mechanistic aspects of the catalytic reduction of dioxygen by cofacial metalloporphyrins. *Journal of the American Chemical Society*, **105**, 2710.

[37] Behret, H., Binder, H., and Sandstede, G. (1981) On the mechanism of electrocatalytic oxygen reduction at metal chelates: Part III. Metal phthalocyanines. *Journal of Electroanalytical Chemistry*, **117**, 29 – 42.

[38] Tang, S., Xiong, G. X., Cai, H. L., and Wang, H. L. (1987) The metal-support interaction on the model catalysts Pt-$TiO_2$, Pt-$Ti_2O_3$ and Pt-TiO CO-sputtering film: I. TEM and HEED studies. *Chinese Journal of Catalysis*, **8**, 225 – 233.

[39] Aricò, A. S., Créti, P., Giordano, N., Antonucci, V., Antonucci, P. L., and Chuvilin, A. (1996) Chemical and morphological characterization of a direct methanol fuel cell based on a quaternary Pt-Ru-Sn-W/C anode. *Journal of Applied Electrochemistry*, **26** (9), 959 – 967.

[40] Chen, K. Y., Shen, P. K., and Tseung, A. C. C. (1995) Anodic oxidation of formic acid on electrodeposited Pt/$WO_3$ electrode at room temperature. *Journal of the Electrochemistry Society*, **142**, L54 – L56.

[41] Xin, Q., Zhou, W. J., Zhou, Z. H., and Li, W. Z. (2004) A preparation method for bi-/multi-metallic noble metal electrocatalysts with high loadings. Chinese Patent ZL 02106201. 3.

[42] Xin, Q., Zhou, W. J., Zhou, Z. H., and Wei, Z. B. (2005) A preparation method of electrocatalyst for proton-exchangemembrane fuel cell. Chinese Patent ZL 01138909. 5.

[43] Xin, Q., Zhou, Z. H., Jiang, L. H., Zhou, W. J., and Sun, G. Q. (2010) A highly active Pt-based electrocatalyst for fuel cell and its preparation method. Chinese Patent ZL 03143681. 1.

[44] Zhou, W., Zhou, Z., Song, S., Li, W., Sun, G., Tsiakaras, P., and Xin, Q. (2003) Pt based anode catalysts for direct ethanol fuel cells. *Applied Catalysis B*, **46**, 273 – 285.

[45] Zhou, Z. H., Wang, S. L., Zhou, W. J., Wang, G. X., Jiang, L. H., Li, W. Z., Song, S. Q., Liu, J. G., Sun, G. Q., and Xin, Q. (2003) Novel synthesis of highly active Pt/C cathode electrocatalyst for direct methanol fuel cell. *Chemical Communications*, 394 – 395.

[46] Zhou, Z. H., Zhou, W. J., Wang, S. L., Wang, G. X., Jiang, L. H., Li, H. Q., Sun, G. Q., and Xin, Q. (2004) Preparation of highly active 40wt. % Pt/C cathode electrocatalysts for DMFC via different routes. *Catalyst Today*, **93 – 95**, 523 – 528.

[47] Zhou, Z. H. (2003) Direct methanol fuel cells: studies on the preparation of highly dispersed Pt based electrocatalysts with high loading. PhD thesis. Dalian Institute of Physical Chemistry.

[48] Li, W. Z. (2003) Carbon supported Pt based cathode catalysts for direct methanol fuel cells (DMFCs). Ph. D. thesis. Dalian Institute of Physical Chemistry.

[49] Li, W. Z., Zhou, W., Li, H., Zhou, Z., Zhou, B., Sun, G., and Xin, Q. (2004) Nano-structured Pt-Fe/C as cathode catalyst in direct methanol fuel cell. *Electrochimica Acta*, **49**, 1045 – 1055.

[50] Hobson, L. J., Ozu, H., Yamaguchi, M., and Hayase, S. (2001) Modified Nafion 117 as an improved polymer electrolyte membrane for direct methanol fuel cells. *Journal of the Electrochemistry Society*, **148**, A1185 – A1190.

[51] Uchida, H., Mizuno, Y., and Watanabe, M. (2000) Suppression of methanol crossover in Pt-dispersed polymer electrolyte membrane for direct methanol fuel cells. *Chemical Letters*, 1268 – 1269.

[52] Ryszard, W. and Peter, N. (1996) Sulfonated polyphosphazene ion-exchange membranes. *Journal of Membrane Science*, **119**, 155 – 160.

[53] Nolte, R., Ledjeff, K., Bauer, M., and Mülhaupt, R. (1993) Partially sulfonated poly(arylene ether sulfone): a versatile proton conducting membrane material for modern energy conversion technologies. *Journal of Membrane Science*, **83**, 211 – 220.

[54] Peled, E., Duvdevani, T., and Melman, A. (1998) A novel proton-conducting membrane. *Electrochemical Solid-State Letters*, **1**, 210 – 211.

[55] Argyropoulos, P., Scott, K., and Taama, W. M. (2000) Dynamic response of the direct methanol fuel cell under variable load conditions. *Journal of Power Sources*, **87**, 153 – 161.

[56] Li, H., Xin, Q., Li, W., Zhou, Z., Jiang, L., Yang, S., and Sun, G. (2004) An improved palladium-based DMFCs cathode catalyst. *Chemical Communications*, 2776 – 2777.

[57] Li, H., Sun, G. Q., Li, N., Sun, S. G., Su, D. S., and Xin, Q. (2007) Design and preparation of highly active Pt-Pd/C catalyst for the oxygen reduction reaction. *Journal of the Physical Chemistry C*, **111**, 5605 – 5617.

[58] Zhou, W. J. (2003) Research on the anode catalysts for low-temperature direct alcohol fuel cells. Ph. D. thesis. Dalian Institute of Physical Chemistry.

[59] Wei, Z. B., Liu, J. G., Qiao, Y. G., Zhou, W. J., Li, W. Z., Chen, L. K., Xin, Q., and Yi, B. L. (2001) Performance of a direct methanol fuel cell. *Chinese Electrochemistry*, **7** (2), 228 – 233.

[60] Jiang, L., Zhou, Z., Li, W., Zhou, W., Song, S., Li, H., Sun, G., and Xin, Q. (2004) Effects of treatment in different atmosphere on Pt3Sn/C electrocatalysts for ethanol electro-oxidation. *Energy & Fuel*, **18**, 866 – 871.

[61] Yan, S. Y., Sun, G. Q., Qi, J., Gao, Y., and Xin, Q. (2009) Effect of heat-treatment on polyol-synthesized PtRu/C electrocatalyst. *Chinese Journal of Catalysis*, **30** (11), 1109 – 1113.

[62] Watanabe, M. and Motoo, S. (1975) Electrocatalysis by ad-atoms: Part III. Enhancement of the oxidation of carbon monoxide on platinum by ruthenium ad- atoms. *Journal of Electroanalytical Chemistry*, **60**, 275 – 283.

[63] Douvartzides, S. L., Coutelieris, F. A., Demin, A. K., and Tsiakaras, P. E. (2004) Electricity from ethanol fed SOFCs: the expectations for sustainable development and technological benefits. *International Journal of Hydrogen Energy*, **29**, 375 – 379.

[64] Goula, M. A., Kontou, S. K., and Tsiakaras, P. E. (2004) Hydrogen production by ethanol steam reforming over a commercial Pd/$\gamma$-$Al_2O_3$ catalyst. *Applied Catalysis B*, **49**, 135 – 144.

[65] Ogden, J. M., Steinbugler, M. M., and Kreutz, T. G. (1999) A comparison of hydrogen, methanol and gasoline as fuels for fuel cell vehicles: implications for vehicle design and infrastructure development. *Journal of Power Sources*, **79**, 143 – 168.

[66] Aricò, A. S., Cretì, P., Antonucci, P. L., and Antonucci, V. (1998) Comparison of ethanol and methanol oxidation in a liquid-feed solid polymer electrolyte fuel cell at high temperature. *Electrochemical Solid-State Letters*, **1**, 66 – 68.

[67] Zhou, W. J., Li, W. Z., Song, S. Q., Zhou, Z. H., Jiang, L. H., Sun, G. Q., Xin, Q., Poulianitis, K., Kontou, S., and Tsiakaras, P. (2004) Bi- and tri-metallic Pt-based anode catalysts for direct ethanol fuel cells. *Journal of Power Sources*, **131**, 217 – 223.

[68] Jiang, L. H., Sun, G. Q., Zhou, Z. H., Zhou, W. J., and Xin, Q. (2004) Preparation and character-

ization of PtSn/C anode electrocatalysts for direct ethanol fuel cell. *Catalysis Today*, **93 – 95**, 665 – 670.

[69] Jiang, L. H., Sun, G. Q., Zhou, Z. H., Sun, S. G., Wang, Q., Yan, S. Y., Li, H. Q., Tian, J., Guo, J. S., Zhou, B., and Xin, Q. (2005) Size-controllable synthesis of monodispersed $SnO_2$ nanoparticles and application in electrocatalysts. *Journal of the Physical Chemistry B*, **109**, 8774 – 8778.

[70] Jiang, L. H., Sun, G. Q., Sun, S. G., Liu, J. G., Tang, S. H., Li, H. Q., Zhou, B., and Xin, Q. (2005) Structure and chemical composition of supported Pt-Sn electrocatalysts for ethanol oxidation. *Electrochimica Acta*, **50**, 5384 – 5389.

[71] Mukerjee, S. and McBreen, J. (1999) An in situ X-ray absorption spectroscopy investigation of the effect of Sn additions to carbon-supported Pt electrocatalysts. *Journal of the Electrochemistry Society*, **146**, 600 – 606.

[72] Jiang, L. H., Colmenares, L., Jusys, Z., Sun, G. Q., and Behm, R. J. (2007) Ethanol electrooxidation on novel carbon supported Pt/SnOx/C catalysts with varied Pt:Sn ratio. *Electrochimica Acta*, **53**, 377 – 389.

[73] Rao, V., Cremers, C., Stimming, U., Cao, L., Sun, S., Yan, S. Y., Sun, G. Q., and Xin, Q. (2007) Electro-oxidation of ethanol at gas diffusion electrodes A DEMS study. *Journal of the Electrochemical Society*, **154** (11), B1138 – B1147.

[74] Cao, L., Sun, G. Q., Li, H., and Xin, Q. (2007) Carbon-supported IrSn catalysts for a direct ethanol fuel cell. *Electrochemistry Communications*, **9**, 2541 – 2546.

# 内 容 简 介

本书为澳大利亚莫纳什大学的 Bradley Ladewig 副教授、科廷大学蒋三平教授(澳籍)和特拉华大学严玉山教授(美籍)的最新著作。本书首先介绍了在低温燃料电池中常见的关键性材料及其发展的重要意义。其次,介绍了碱性阴离子交换膜燃料电池技术中的电池原理和 ORR、HOR 两种反应机理,以及水性电解液、碱性膜等材料技术在碱性阴离子交换膜燃料电池的实例应用。然后,详细分析了催化剂载体材料及碳载体的研究现状,主要介绍了新型的碳材料作为燃料电池中电催化载体的实例分析,包括介孔碳载体、石墨纤维载体、碳纳米管载体、石墨烯载体、氮掺杂碳载体等。同时,本书还对燃料电池中阴极材料、阳极材料和电池膜材料的发展现状及关键技术的关键科学问题进行了详细的阐述,涵盖的主题包括质子交换膜燃料电池、直接甲醇和乙醇燃料电池、微生物燃料电池、微流体燃料电池、碱性膜燃料电池等。

这是一本聚焦于燃料电池关键材料要求的科研用书。本书对于从事低温燃料电池领域研究的专业技术人员及相关材料研究的工程技术人员、各类高校科研工作者及相关专业学生具有重要的参考价值。本书亦可作为高等院校研究生教材。

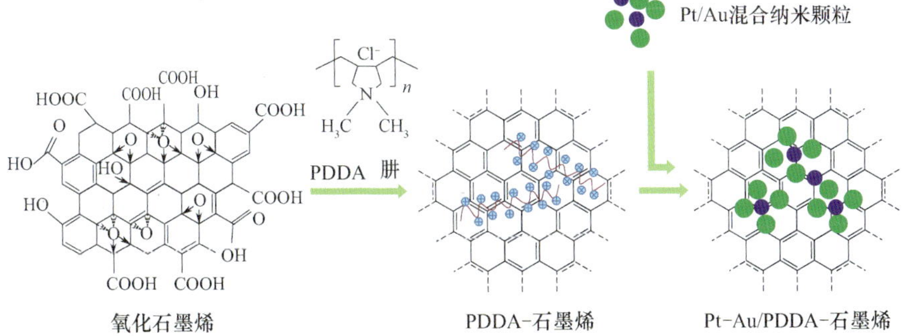


图 3.13 PDDA 官能化石墨烯上的 Pt 和 Au 纳米颗粒自组装[45]

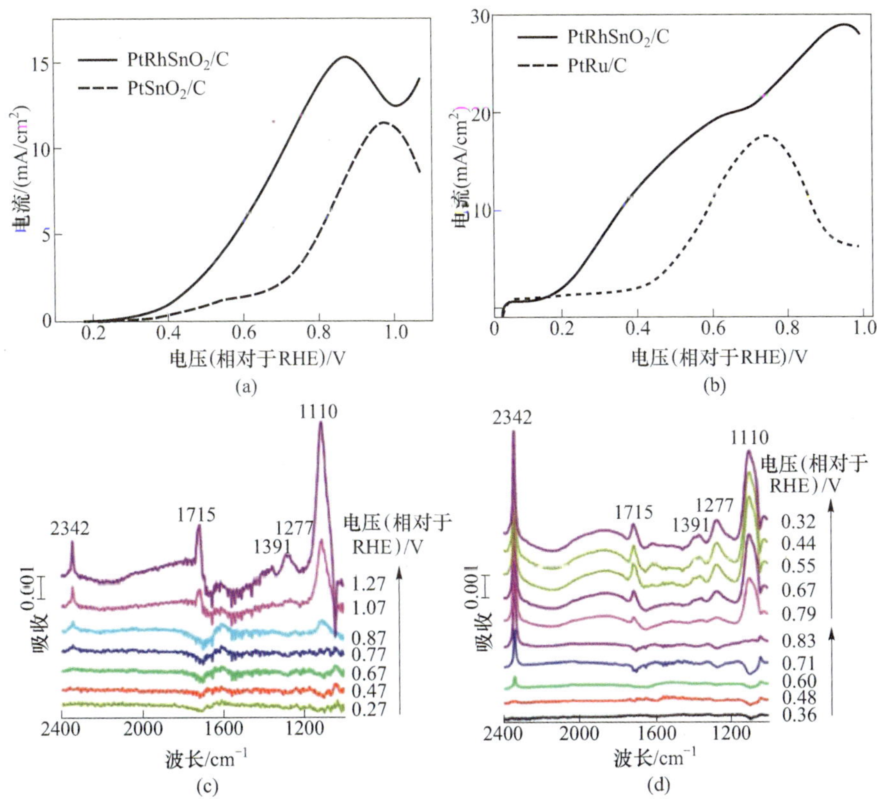


图 4.2 $PtRhSnO_2/C$ 和其他几种催化剂的用于比较乙醇氧化活性的电流-电压曲线；测定条件为 0.1M $HClO_4$ +0.2M 乙醇，$50mV \cdot s^{-1}$；催化剂组成：(a) $PtRhSnO_2/C$ 为 30nmol Pt、8nmol Rh 和 60nmol $SnO_2$，$PtSnO_2/C$ 为 30nmol Pt 和 60nmol $SnO_2$；(b) $PtRhSnO_2/C$ 为 25nmol Pt、5nmol Rh 和 20 nmol $SnO_2$，PtRu/C 为 25nmol Pt 和 25nmol Ru。(c) Pt(111) 电极和 (d) $PtRhSnO_2/C$ 在 0.1M $HClO_4$ +0.2M 乙醇溶液中的电化学氧化原位红外反射吸收光谱

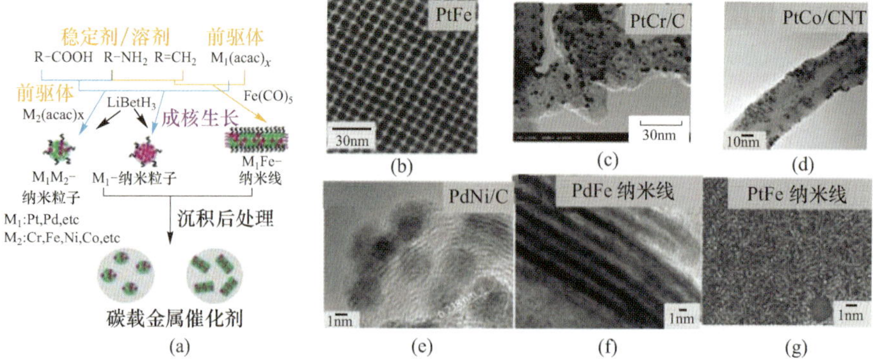


图 4.7 (a)有机相合成法的示意图,通过此种方法制备催化剂;(b)PtFe 纳米颗粒[129];(c) PtCr/C (28%(质量分数), $D=2.3$nm)[187];(d) PtCo/CNT (20%(质量分数)Pt, $D=2.0$nm)[180];(e) PdNi/C (20%(质量分数)Pt, $D=2.4$nm)[187];(f) PdFe 纳米叶($D=1.8$nm, $L=100$nm)[190];(g) PtFe 纳米线($D=2.7$nm)[188]

图 4.9 (a) PtRu/DWNT(50%(质量分数))的 TEM 图;(b) PtRu/DWNT 薄膜的 SEM 图;(c)在 0.5M $H_2SO_4$ +0.5M 甲醇中 PtRu/CNT 催化剂的甲醇氧化曲线;(d) PtRu/CNT 为阳极催化剂的单电池极化曲线[177]

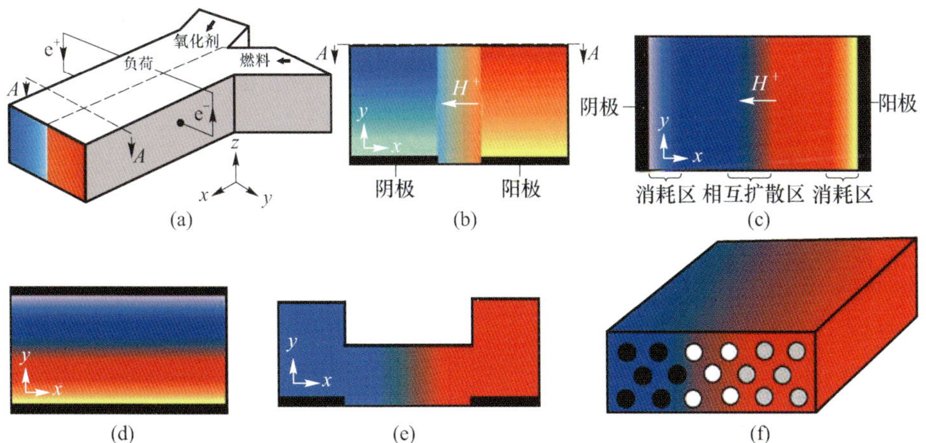


图 9.1　溢出设计图。(a)具有两种物质流的 LFFC 通过 Y 形孔道的原理图;(b)两电极都在底部的两物流平行穿过通道的截面原理图;(c)垂直于侧壁电极的液-液界面区,阳极、阴极损耗边界及相互扩散区域的截面图;(d)上下电极组通道的截面图;(e)两电极都在通道凹槽底部的截面图;(f)石墨棒为电极的通道截面图[19](版权 2011,Elsevier)

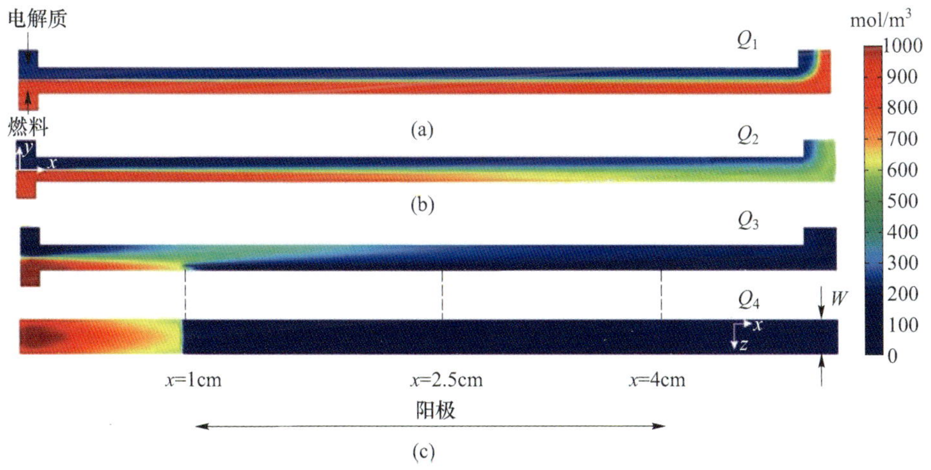


图 9.3　无膜 LFFC 中甲酸浓度的数值模拟,阳极接入口浓度为 1M。(a) $Q_1$ = 500μL/min,最大电流密度为 68mA/cm;(b) $Q_2$ = 100μL/min,最大电流密度为 58mA/cm;(c) $Q_3$ = 10μL/min,最大电流密度为 40mA/cm[40](版权 2010,英国物理学会)

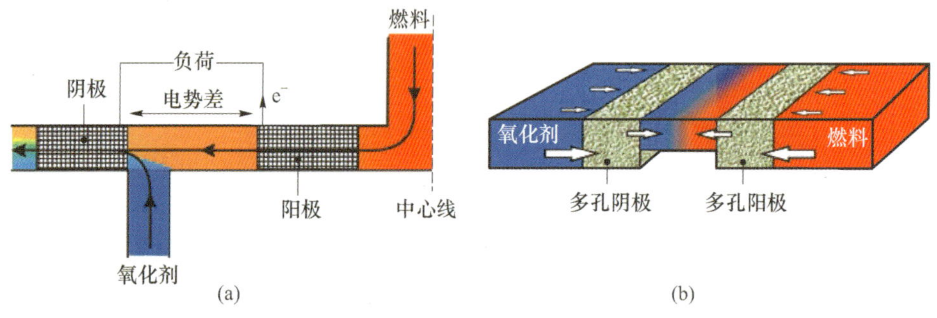


图 9.4 溢流设计。(a)无膜 LFFC 的放射状流体构造;(b)具有“多重入口”概念的三维多孔电极[19](版权 2011,Elsevier)

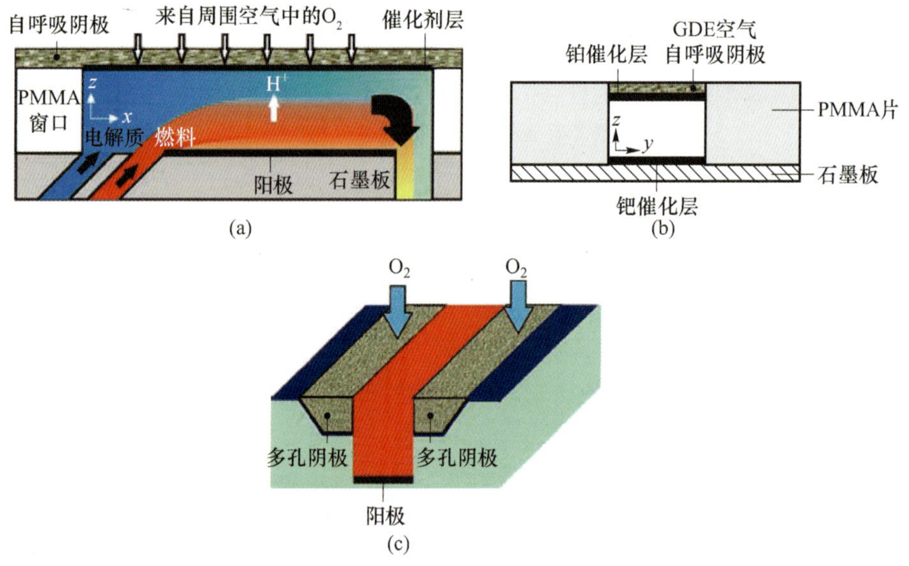


图 9.5 吸气式阴极无膜 LFFC。(a) 通道中燃料和电解液流的安排示意图[32];(b)对图(a)中通道的截面和载体基质的描述;(c)具有吸气式阴极的整体设计的电池[19](版权 2011,Elsevier)